AF885848

To Dr. Basim Uthman

with best wishes

Panos M. Pardalos

Nov. 15, 2007

DATA MINING, SYSTEMS ANALYSIS, AND OPTIMIZATION IN BIOMEDICINE

To learn more about the AIP Conference Proceedings, including the Conference Proceedings Series, please visit the webpage
http://proceedings.aip.org/proceedings

DATA MINING, SYSTEMS ANALYSIS, AND OPTIMIZATION IN BIOMEDICINE

Gainesville, Florida, U. S. A. 28 – 30 March 2007

EDITORS

Onur Seref
O. Erhun Kundakcioglu
Panos M. Pardalos
University of Florida
Gainesville, Florida, U. S. A.

SPONSORING ORGANIZATIONS
Center for Applied Optimization, University of Florida
College of Engineering, University of Florida
National Science Foundation

AMERICAN INSTITUTE OF PHYSICS

Melville, New York, 2007
AIP CONFERENCE PROCEEDINGS ■ 953

Editors

Onur Seref
O. Erhun Kundakcioglu
Panos M. Pardalos

Department of Industrial and Systems Engineering
University of Florida
303 Weil Hall
P. O. Box 116595
Gainesville, FL 32611-6595
U. S. A.
E-mail: seref@ufl.edu

L.C. Catalog Card No. 2007938607
ISBN 978-0-7354-0467-0
ISSN 0094-243X

Printed in the United States of America

This book is dedicated to the memory of our colleague and friend

Ioannis (Yanni) Constantinidis,

who died suddenly and untimely on April 16, 2007.

CONTENTS

PREFACE

Recent developments in biomedicine have put this relatively new field among the leading research fields for the past two decades. Today, challenges faced in medicine and biology can only find solutions through interdisciplinary collaboration. The size and diversity of the problems arising from such challenges require a well coordinated systems approach. Such interdisciplinary systems are formed all around the world integrating medical and biological sciences with engineering and computer sciences.

The speed and detail achieved in medical recording devices as a result of recent breakthroughs across these disciplines have given rise to medical databases exponentially increasing in size. The majority of these databases is comprised of neural signals, biomedical images, microarrays, protein sequences, and genomic data. Extracting and interpreting the information content from such ever-growing massive databases has caused a proliferation in new high-speed data mining methods. A wide spectrum of general and specialized techniques has been developed in data mining research to attack the growing set of biomedical problems. Optimization plays a crucial role in developing efficient, consistent, and robust methods that can perform well across different databases. The future of biomedical research heavily depends on the advancement of an array of harmonically integrated and optimized data mining techniques.

This volume contains a comprehensive selection of original biomedical research. The chapters of the book consist of contributions from the participants of the "Data Mining, Systems Analysis and Optimization in Biomedicine, 2007" Conference, and from other invited authors. The chapters are grouped as survey studies, biomedical applications, and neuroscience applications. Biomedical applications traverse a diverse range of problems from protein folding to genomics, drug design, and cellular processes. A number of chapters describe new techniques for analyzing magnetic resonance images, including applications on automatic detection of brain tumors and other malignancies and disorders in the human body. Applications in neuroscience involve chapters on quantification, classification, and analysis of electroencephalogram (EEG) recordings. Chapters on EEG recordings are mainly focused on epilepsy monitoring and treatment. Novel theoretical and computational methods are introduced in a large selection of chapters. These methods include mathematical programming and optimization methods in classification, statistical analysis, approximation algorithms, and new metaheuristics.

This book is intended to collect a sample of the latest research methods in data mining across diverse fields of biomedicine, neuroscience, engineering, and computer science. The problems and methods discussed in this book are of great interest to new and established theoreticians and practitioners in these fields, providing them with new research directions. The interdisciplinary nature of the material included in the book may help spark new ideas to bridge these disciplines in new ways.

We would like to take the opportunity to thank the authors of the chapters for their valuable contributions, the National Science Foundation and the US Air Force for their financial support, the Center for Applied Optimization and the University of Florida for supporting this effort. Finally, we would like to thank the American Institute of Physics for their assistance in producing this book.

Onur Seref, O. Erhun Kundakcioglu, Panos M. Pardalos
September, 2007

Classification and disease prediction via mathematical programming

Eva K. Lee and Tsung-Lin Wu

Center for Operations Research in Medicine and HealthCare
School of Industrial and Systems Engineering
Georgia Institute of Technology, Atlanta, Georgia 30332-0205

Abstract. In this chapter, we present classification models based on mathematical programming approaches. We first provide an overview on various mathematical programming approaches, including linear programming, mixed integer programming, nonlinear programming and support vector machines. Next, we present our effort of novel optimization-based classification models that are general purpose and suitable for developing predictive rules for large heterogeneous biological and medical data sets. Our predictive model simultaneously incorporates (1) the ability to classify any number of distinct groups; (2) the ability to incorporate heterogeneous types of attributes as input; (3) a high-dimensional data transformation that eliminates noise and errors in biological data; (4) the ability to incorporate constraints to limit the rate of misclassification, and a reserved-judgment region that provides a safeguard against over-training (which tends to lead to high misclassification rates from the resulting predictive rule); and (5) successive multi-stage classification capability to handle data points placed in the reserved judgment region. To illustrate the power and flexibility of the classification model and solution engine, and its multigroup prediction capability, application of the predictive model to a broad class of biological and medical problems is described. Applications include: the differential diagnosis of the type of erythemato-squamous diseases; predicting presence/absence of heart disease; genomic analysis and prediction of aberrant CpG island meythlation in human cancer; discriminant analysis of motility and morphology data in human lung carcinoma; prediction of ultrasonic cell disruption for drug delivery; identification of tumor shape and volume in treatment of sarcoma; multistage discriminant analysis of biomarkers for prediction of early atherosclerois; fingerprinting of native and angiogenic microvascular networks for early diagnosis of diabetes, aging, macular degeneracy and tumor metastasis; prediction of protein localization sites; and pattern recognition of satellite images in classification of soil types. In all these applications, the predictive model yields correct classification rates ranging from 80% to 100%. This provides motivation for pursuing its use as a medical diagnostic, monitoring and decision-making tool.

Keywords: Classification, Support Vector Machines, data mining, Integer Programming
PACS: 87.55.de, 87.55.kd, 87.55.kh, 87.85.-d

CP953, *Data Mining, Systems Analysis, and Optimization in Biomedicine*
edited by O. Seref, O. E. Kundakcioglu, P. M. Pardalos

1. INTRODUCTION

Classification is a fundamental machine learning task whereby rules are developed for the allocation of independent observations to groups. Classic examples of applications include medical diagnosis - the allocation of patients to disease classes based on symptoms and lab tests, and credit screening - the acceptance or rejection of credit applications based on applicant data. Data are collected concerning observations with known group membership. This *training data* is used to develop rules for the classification of future observations with unknown group membership.

In the introduction section, we briefly describe some terminologies related to classification, and provide a brief organization of the materials written in this chapter.

1.1. Pattern recognition, discriminant analysis, and statistical pattern classification

Cognitive science is the science of learning, knowing, and reasoning. *Pattern recognition* is a broad field within *cognitive science* that is concerned with the process of recognizing, identifying, and categorizing input information. These areas intersect with computer science, particularly in the closely related areas of *artificial intelligence*, *machine learning*, and *statistical pattern recognition*. Artificial intelligence is associated with constructing machines and systems that reflect human abilities in cognition. Machine learning refers to how these machines and systems replicate the learning process, which is often achieved by seeking and discovering patterns in data, or statistical pattern recognition.

Discriminant analysis is the process of discriminating between categories or populations. Associated with discriminant analysis as a statistical tool are the tasks of determining the features that best discriminate between populations, and the process of classifying new objects based on these features. The former is often called *feature selection* and the latter is referred to as *statistical pattern classification*. This work will be largely concerned with the development of a viable statistical pattern classifier.

As with many computationally intensive tasks, recent advances in computing power have led to a sharp increase in the interest and application of discriminant analysis techniques. The reader is referred to Duda et al. [1] for an introduction to various techniques for pattern classification, and to Zopounidis et al. [2] for examples of applications of pattern classification.

1.2. Supervised learning, training, and cross validation

An *entity* or *observation* is essentially a data point as commonly understood in statistics. In the framework of statistical pattern classification, an entity is a set of quantitative measurements (or qualitative measurements expressed quantitatively) of *attributes* for a particular object. As an example, in medical diagnosis an entity could be the various blood chemistry levels of a patient. With each entity is associated one or more *groups* (or *populations, classes, categories*) to which it belongs. Continuing with the medical diagnosis example, the groups could be the various classes of heart disease. Statistical classification seeks to determine rules for associating entities with the groups to which they belong. Ideally, these associations align with the associations that human reasoning would produce based on information gathered on objects and their apparent categories.

Supervised learning is the process of developing classification rules based on entities for which the classification is already known. Note that the process implies that the populations are already well-defined. *Unsupervised learning* is the process of discovering patterns from unlabeled entities and thereby discovering and describing the underlying populations. Models derived using supervised learning can be used for both functions of discriminant analysis - feature selection and classification. The model that we consider is a method for supervised learning, so we assume that populations are previously defined.

The set of entities with known classification that is used to develop classification rules is the *training set*. The training set may be partitioned so that some entities are withheld during the model-development process, also known as the *training* of the model. The withheld entities form a *test set* that is used to determine the validity of the model, a process known as *cross validation*. Entities from the test set are subjected to the rules of classification to measure the performance of the rules on entities with unknown group membership.

Validation of classification models is often performed using m-fold cross validation where the data with known classification is partitioned into m *folds* (subsets) of approximately equal size. The classification model is trained m times, with the mth fold withheld during each run for testing. The performance of the model is evaluated by the classification accuracy on the m test folds, and can be represented using a *classification matrix* or *confusion matrix*.

The classification matrix is a square matrix with the number of rows and columns equal to the number of groups. The ij^{th} entry of the classification matrix contains the number or proportion of test entities from group i that were classified by the model as belonging to group j. Therefore, the number or proportion of correctly classified entities are contained in the diagonal elements of the classification matrix, and the number or proportion of misclassified entities are in the off-diagonal entries.

1.3. Bayesian inference and classification

The popularity of *Bayesian inference* has risen drastically over the past several decades, perhaps in part due to its suitability for statistical learning. The reader is referred to O'Hagan's volume [3] for a thorough treatment of Bayesian inference. Bayesian inference is usually contrasted against *classical inference*, though in practice they often imply the same methodology.

The Bayesian method relies on a *subjective* view of probability, as opposed to the *frequentist* view upon which classical inference is based [3]. A subjective probability describes a degree of belief in a *proposition* held by the investigator based on some information. A frequency probability describes the likelihood of an *event* given an infinite number of trials.

In Bayesian statistics, inferences are based on the *posterior distribution*. The posterior distribution is the product of the *prior probability* and the *likelihood function*. The prior probability distribution represents the initial degree of belief in a proposition, often before empirical data is considered. The likelihood function describes the likelihood that the behavior is exhibited, given that the proposition is true. The posterior distribution describes the likelihood that the proposition is true, given the observed behavior.

Suppose we have a proposition or random variable θ about which we would like to make inferences, and data x. Application of Bayes' Theorem gives

$$dF(\theta|x) = \frac{dF(\theta)dF(x|\theta)}{dF(x)}$$

Here, F denotes the (cumulative) distribution fuction. For ease of conceptualization, assume that F is differentiable, then $dF = f$, and the above equality can be rewritten as

$$f(\theta|x) = \frac{f(\theta)f(x|\theta)}{f(x)}.$$

For classification, a prior probability function $\pi(g)$ describes the likelihood that an entity is allocated to group g regardless of its exhibited feature values x. A group density function $f(x|g)$ describes the likelihood that an entity exhibits certain measurable attribute values, given that it belongs to population g. The posterior distribution for a group $P(g|x)$ is given by the product of the prior probability and group density function, normalized over the groups to obtain a unit probability over all groups. The observation x is allocated to group h if $h = \arg\max_{g \in \mathcal{G}} P(g|x) = \arg\max_{g \in \mathcal{G}} \frac{\pi(g)f(x|g)}{\sum_{j \in \mathcal{G}} \pi(j)f(x|j)}$, where $\mathcal{G}$ denotes the set of groups.

1.4. Discriminant functions

Most classification methods can be described in terms of *discriminant functions*. A discriminant function takes as input an observation and returns information about the classification of the observation. For data from a set of groups $\mathcal{G}$, an observation x is assigned to group h if $h = \arg\max_{g \in \mathcal{G}} l_g(x)$ where the functions l_g are the discriminant functions. Classification methods restrict the form of the discriminant functions, and training data is used to determine the values of parameters that define the functions.

The optimal classifier in the Bayesian framework can be described in terms of discriminant functions. Let $\pi_g = \pi(g)$ be the prior probability that an observation is allocated to group g and let $f_g(x) = f(x|g)$ be the likelihood that data x is drawn from population g. If we wish to minimize the probability of misclassification given x, then the optimal allocation for an entity is to the group $h = \arg\max_{g \in \mathcal{G}} P(g|x) = \arg\max_{g \in \mathcal{G}} \frac{\pi_g f_g(x)}{\sum_{j \in \mathcal{G}} \pi_j f_j(x)}$. Under the Bayesian framework,

$$P(g|x) = \frac{\pi_g f(x|g)}{f(x)} = \frac{\pi_g f(x|g)}{\sum\limits_{j \in \mathcal{G}} \pi_j f(x|j)}$$

The discriminant functions can be $l_g(x) = P(g|x)$ for $g \in \mathcal{G}$. The same classification rule is given by $l_g(x) = \pi_g f(x|g)$ and $l_g(x) = \log f(x|g) + \log \pi_g$. The problem then becomes finding the form of the prior functions and likelihood functions that match the data.

If the data are multivariate normal with equal covariance matrices ($f(x|g) \sim N(\mu_g, \Sigma)$), then a linear discriminant function is optimal:

$$\begin{aligned} l_g(x) &= \log f(x|g) + \log \pi_g \\ &= -1/2(x-\mu_g)^T \Sigma^{-1}(x-\mu_g) - 1/2 \log |\Sigma_g| - d/2 \log 2\pi + \log \pi_g \\ &= w_g^T x + w_{g0} \end{aligned}$$

where d is the number of attributes, $w_g = \Sigma^{-1}\mu_g$, and $w_{g0} = -1/2\mu_g^T \Sigma^{-1}\mu_g + \log \pi_g + x^T \Sigma^{-1} x - d/2 \log 2\pi$. Note that the last two terms of w_{g0} are constant for all g and need not be calculated. When there are 2 groups ($\mathcal{G} = \{1, 2\}$) and the priors are equal ($\pi_1 = \pi_2$), the discriminant rule is equivalent to Fisher's linear discriminant rule [4]. Fisher's rule can also be derived, as it was by Fisher, by choosing w so that $\frac{(w^T\mu_1 - w^T\mu_2)^2}{w^T \Sigma w}$ is maximized.

These linear and quadratic discriminant functions are often applied to data sets that are not multivariate normal or continuous (see [5], pages 234-235) by using approximations for the means and covariances. Regardless, these models are *parametric* in that they incorporate assumptions about the distribution of the data. Fisher's linear discriminant is *non-parametric* because no assumptions are made about the underlying distribution of the data. Thus, for a special case, a parametric and non-parametric model coincide to produce the same discriminant rule. The linear discriminant function derived above is also called the *homoscedastic model*, and the quadratic discriminant function is called the *heteroscedastic model*. The exact form of discriminant functions in the Bayesian framework can be derived for other distributions [1].

Some classification methods are essentially methods for finding coefficients for linear discriminant functions. In other words, they seek coefficients w_g and constants w_{g0} such that $l_g(x) = w_g x + w_{g0}$, $g \in \mathcal{G}$, is an optimal set of discriminant functions. The criteria for optimality is different for different methods. Linear discriminant functions project the data onto a linear subspace and then discriminate between entities in that subspace. For example, Fisher's linear discriminant projects two-group data on an optimal line, and discriminates on that line. A good linear subspace may not exist for data with overlapping distributions between groups and therefore the data will not be classified accurately using these methods. The hyperplanes defined by the discriminant functions form boundaries between the group regions. A large portion of the literature concerning the use of mathematical programming models for classification describe methods for finding coefficients of linear discriminant functions [2].

Other classification methods seek to determine parameters to establish quadratic discriminant functions. The general form of a quadratic discriminant function is $l_g(x) =$

$x^T W_g x + w_g^T x + w_{g0}$. The boundaries defining the group regions can assume any hyperquadric form, as can the Bayes decision rules for arbitrary multivariate normal distributions [1].

In this chapter, we survey the development and advances of classification models via the mathematical programming techniques, and summarize our experience in classification models applied to prediction in biological and medical applications. The rest of this chapter is organized as follows. Section 2 first provides a detailed overview of the development and advances of mathematical programming-based classification models, including linear programming, mixed integer programming, nonlinear programming, and support vector machine approaches. In Section 3, we describe our effort in developing optimization-based multigroup multi-stage discriminant analysis predictive models for classification. The use of the predictive models on various biological and medical problems are presented. Section 4 provides several tables to summarize the progress of mathematical-programming based classification models and their characteristics. This is followed by a brief description of other classification methods in Section 5, and summary and concluding remarks in Section 6.

2. MATHEMATICAL PROGRAMMING APPROACHES

Mathematical programming methods for statistical pattern classification emerged in the 1960's, gained popularity in the 1980's, and have grown drastically since. Most of the mathematical programming approaches are non-parametric which has been cited as an advantage when analyzing contaminated data sets over methods that require assumptions about the distribution of the data [6]. Most of the literature about mathematical programming methods is concerned with either using mathematical programming to determine the coefficients of linear discriminant functions or with *support vector machines*.

The following notation will be used. The subscripts i, j, and k are used for the observation, attribute, and group, respectively. Let x_{ij} be the value of attribute j of observation i. Let m be the number of attributes, K be the number of groups, G_k represent the set of data from group k, M be a big positive number, and ε be a small positive number. The abbreviation "urs" is used in reference to a variable to denote "unrestricted in sign".

2.1. Linear programming classification models

The use of linear programs to determine the coefficients of linear discriminant functions has been widely studied [7, 8, 9, 10]. The methods determine the coefficients for different objectives, including minimizing the sum of the distances to the separating hyperplane, minimizing the maximum distance of an observation to the hyperplane, and minimizing other measures of badness of fit or maximizing measures of goodness of fit.

2.1.1. Two-group classification

One of the earliest linear programming (LP) classification models was proposed by Mangasarian [10] to construct a hyperplane to separate two groups of data. Separation by a nonlinear surface using LP was also proposed when the surface parameters appear linearly. Two sets of points may be inseparable by one hyperplane or surface through a single-step LP approach, but they can be strictly separated by more planes or surfaces via a multi-step LP approach (Mangasarian [11]). In [11] real problems with up to 117 data points, 10 attributes, and 3 groups were solved. The 3-group separation was achieved by separating group 1 from groups 2 and 3, and then group 2 from group 3.

Studies of LP models for the discriminant problem in the early 1980's was carried out by Hand [12], Freed and Glover [7, 13], and Bajgier and Hill [14]. Three LP models for the two-group classification problem, including minimizing the sum of deviations (MSD), minimizing the maximum deviation (MMD), and minimizing the sum of interior distances (MSID) were proposed. Freed and Glover [15] provided computational studies of these models where the test conditions involved normal and nonnormal populations.

MSD (Minimize the sum of deviations)

$$\begin{aligned}
\text{Min} \quad & \textstyle\sum_i d_i \\
\text{s.t.} \quad & w_0 + \textstyle\sum_j x_{ij} w_j - d_i \le 0 \quad \forall i \in G_1 \\
& w_0 + \textstyle\sum_j x_{ij} w_j + d_i \ge 0 \quad \forall i \in G_2 \\
& w_j \text{ urs} \quad \forall j \\
& d_i \ge 0 \quad \forall i
\end{aligned}$$

MMD (Minimize the maximum deviation)

$$\begin{aligned}
\text{Min} \quad & d \\
\text{s.t.} \quad & w_0 + \textstyle\sum_j x_{ij} w_j - d \le 0 \quad \forall i \in G_1 \\
& w_0 + \textstyle\sum_j x_{ij} w_j + d \ge 0 \quad \forall i \in G_2 \\
& w_j \text{ urs} \quad \forall j \\
& d \ge 0
\end{aligned}$$

MSID (Minimize the sum of interior distances)

$$\begin{aligned}
\text{Min} \quad & pd - \textstyle\sum_i e_i \\
\text{s.t.} \quad & w_0 + \textstyle\sum_j x_{ij} w_j - d + e_i \le 0 \quad \forall i \in G_1 \\
& w_0 + \textstyle\sum_j x_{ij} w_j + d - e_i \ge 0 \quad \forall i \in G_2 \\
& w_j \text{ urs} \quad \forall j \\
& d \ge 0 \\
& e_i \ge 0 \quad \forall i
\end{aligned}$$

where p is a weight constant.

The objective function of the MSD model is the L_1-norm distance while the objective function of MMD is the L_∞-norm distance. They are special cases of L_p-norm classification [9, 16].

In some models the constant term of the hyperplane is a fixed number instead of a decision variable. The model MSD^0 shown below is an example where the cut-off score b replaces w_0 in the formulation. The same replacement could be used in other formulations.

MSD^0 (Minimize the sum of deviations with constant cut-off score)

$$\begin{aligned} \text{Min} \quad & \textstyle\sum_i d_i \\ \text{s.t.} \quad & \textstyle\sum_j x_{ij} w_j - d_i \leq b \quad \forall i \in G_1 \\ & \textstyle\sum_j x_{ij} w_j + d_i \geq b \quad \forall i \in G_2 \\ & w_j \text{ urs} \quad \forall j \\ & d_i \geq 0 \quad \forall i \end{aligned}$$

A gap can be introduced between the two regions determined by the separating hyperplane to prevent degenerate solutions. Take MSD as an example, the separation constraints become

$$w_0 + \textstyle\sum_j x_{ij} w_j - d_i \leq -\varepsilon \quad \forall i \in G_1$$
$$w_0 + \textstyle\sum_j x_{ij} w_j + d_i \geq \varepsilon \quad \forall i \in G_2$$

The small number ε can be normalized to 1.

Besides introducing a gap, another normalization approach is to include constraints such as $\sum_{j=0}^{m} w_j = 1$ or $\sum_{j=1}^{m} w_j = 1$ in the LP models to avoid unbounded or trivial solutions.

Specifically, Glover et al. [17] gave the hybrid model, as follows.

Hybrid model

$$\begin{aligned} \text{Min} \quad & pd + \textstyle\sum_i p_i d_i - qe - \sum_i q_i e_i \\ \text{s.t.} \quad & w_0 + \textstyle\sum_j x_{ij} w_j - d - d_i + e + e_i = 0 \quad \forall i \in G_1 \\ & w_0 + \textstyle\sum_j x_{ij} w_j + d + d_i - e - e_i = 0 \quad \forall i \in G_2 \\ & w_j \text{ urs} \quad \forall j \\ & d, e \geq 0 \\ & d_i, e_i \geq 0 \quad \forall i \end{aligned}$$

where p, p_i, q, q_i are the cost for different deviations. Including different combinations of deviation terms in the objective function then leads to variant models.

Joachimsthaler and Stam [9] review and summarize LP formulations applied to two-group classification problems in discriminant analysis, including MSD, MMD, MSID, MIP models, and the hybrid model. They summarize the performance of the LP methods together with the traditional classification methods such as Fisher's linear discriminant function (LDF) [4], Smith's quadratic discriminant function (QDF) [18], and a logistic discriminant method. In their review, MSD sometimes but not uniformly improves classification accuracy, compared with traditional methods. On the other hand, MMD is found to be inferior to MSD. Erenguc and Koehler [19] present a unified survey of LP models and their experimental results, in which the LP models include several versions of MSD, MMD, MSID, and hybrid models. Rubin [20] provides experimental results of comparing these LP models with Fisher's LDF and Smith's QDF. He concludes that QDF performs best when the data follow normal distributions and that QDF could be the benchmark when seeking situations for advantageous LP methods. In summary, the above review papers [19, 9, 20] describe previous work on LP classification models and their comparison with traditional methods. However, it is difficult to make definitive statements about conditions under which one LP model is superior to others, as stated in [6].

Stam and Ungar [21] introduce a software package RAGNU, a utility program in conjunction with the LINDO optimization software, for solving two-group classification problems using LP-based methods. LP formulations such as MSD, MMD, MSID, hybrid models and their variants are contained in the package.

There are some difficulties in LP-based formulations, in that some models could result in unbounded, trivial, or unacceptable solutions [22, 23], but possible remedies are proposed. Koehler [24, 25, 26] and Xiao [27, 28] characterize the conditions of unacceptable solutions in two-group LP discriminant models, including MSD, MMD, MISD, the hybrid model, and their variants. Glover [29] proposes the normalization constraint, $\sum_{j=1}^{m}(-|G_2|\sum_{i\in G_1} x_{ij} + |G_1|\sum_{i\in G_2} x_{ij})w_j = 1$, which is more effective and reliable. Rubin [30] examines the separation failure for two-group models and suggests to apply the models twice, reversing the group designations the second time. Xiao and Feng [31] propose a regularization method to avoid multiple solutions in LP discriminant analysis by adding the term $\varepsilon \sum_{j=1}^{m} w_j^2$ in the objective functions.

Bennett and Mangasarian [32] propose the following model which minimizes the average of the deviations, which is called robust linear programming.

RLP (Robust linear programming)

$$
\begin{array}{lll}
\text{Min} & \frac{1}{|G_1|}\sum_{i\in G_1} d_i + \frac{1}{|G_2|}\sum_{i\in G_2} d_i & \\
\text{s.t.} & w_0 + \sum_j x_{ij} w_j - d_i \le -1 & \forall i \in G_1 \\
 & w_0 + \sum_j x_{ij} w_j + d_i \ge 1 & \forall i \in G_2 \\
 & w_j \text{ urs } \ \forall j & \\
 & d_i \ge 0 \ \ \forall i &
\end{array}
$$

It is shown that this model gives the null solution $w_1 = \cdots = w_m = 0$ if and only if $\frac{1}{|G_1|}\sum_{i\in G_1} x_{ij} = \frac{1}{|G_2|}\sum_{i\in G_2} x_{ij}$ for all j, in which case the solution $w_1 = \cdots = w_m = 0$ is guaranteed to be not unique. Data of different diseases are tested by the proposed classification methods, as in most of Mangasarian's papers.

Mangasarian et al. [33] describe two applications of LP models in the field of breast cancer research, one in diagnosis and the other in prognosis. The first application is to discriminate benign from malignant breast lumps, while the second one is to predict when breast cancer is likely to recur. Both of them work successfully in clinical practice. The RLP model [32] together with the multisurface method tree algorithm (MSMT) [34] is used in the diagnostic system.

Duarte Silva and Stam [35] include the second-order (i.e., quadratic and cross-product) terms of the attribute values in the LP-based models such as MSD and hybrid models and compare them with linear models, Fisher's LDF, and Smith's QDF. The results of the simulation experiments show that the methods which include second-order terms perform much better than first-order methods, given that the data substantially violate the multivariate normality assumption. Wanarat and Pavur [36] investigate the effect of the inclusion of the second-order terms in the MSD, MIP, and hybrid models when sample size is small to moderate. However, the simulation study shows that second-order terms may not always improve the performance of a first-order LP model even with data configurations that are more appropriately classified by Smith's QDF. Another result of the simulation study is that inclusion of the cross-product terms may hurt the model's accuracy, while omission of these terms causes the model to be not invariant with respect to a nonsingular transformation of the data.

Pavur [37] studies the effect of the position of the contaminated normal data in the two-group classification problem. The methods for comparison in their study include MSD, MM (described in the mixed integer programming part), Fisher's LDF, Smith's QDF, and nearest neighbor models. The nontraditional methods such as LP models have potential for outperforming the standard parametric procedures when nonnormality is present, but this study shows that no one model is consistently superior in all cases.

Asparoukhov and Stam [38] propose LP and MIP models to solve the two-group classification problem where the attributes are binary. In this case the training data can be partitioned into multinomial cells, allowing for a substantial reduction in the number of variables and constraints. The proposed models not only have the usual geometric interpretation, but also possess a strong probabilistic foundation. Let s be the index of the cells, n_{1s}, n_{2s} be the number of data points in cell s from groups 1 and 2, respectively, and $(b_{s1}, \ldots, b_{sm})$ be the binary digits representing cell s. The model shown below is the LP model of minimizing the sum of deviations for two-group classification with binary attributes.

Cell convensional MSD

$$\begin{aligned}
\text{Min} \quad & \textstyle\sum_{s:\, n_{1s}+n_{2s}>0} (n_{1s}d_{1s} + n_{2s}d_{2s}) \\
\text{s.t.} \quad & w_0 + \textstyle\sum_j b_{sj}w_j - d_{1s} \leq 0 \quad \forall s : n_{1s} > 0 \\
& w_0 + \textstyle\sum_j b_{sj}w_j + d_{2s} > 0 \quad \forall s : n_{2s} > 0 \\
& w_j \text{ urs} \quad \forall j \\
& d_{1s}, d_{2s} \geq 0 \quad \forall s
\end{aligned}$$

Binary attributes are usually found in medical diagnoses data. In this study three real data sets about disease discrimination are tested: developing postoperative pulmonary embolism or not, having dissecting aneurysm or other diseases, and suffering from post-traumatic epilepsy or not. In these data sets the MIP model for binary attributes (BMIP), which will be described later, performs better than other LP models or traditional methods.

2.1.2. Multigroup classification

Freed and Glover [13] extend the LP classification models from two-group to multi-group problems. One formulation which uses a single discriminant function is given below.

$$\begin{aligned}
\text{Min} \quad & \textstyle\sum_{k=1}^{K-1} c_k\alpha_k \\
\text{s.t.} \quad & \textstyle\sum_j x_{ij}w_j \leq U_k \quad \forall i \in G_k \; \forall k \\
& \textstyle\sum_j x_{ij}w_j \geq L_k \quad \forall i \in G_k \; \forall k \\
& U_k + \varepsilon \leq L_{k+1} + \alpha_k \quad \forall k = 1, \ldots, K-1 \\
& w_j \text{ urs} \quad \forall j \\
& U_k, L_k \text{ urs} \quad \forall k \\
& \alpha_k \text{ urs} \quad \forall k = 1, \ldots, K-1
\end{aligned}$$

where the number ε could be normalized to be 1, and c_k is the misclassification cost. However, single function classification is not as flexible and general as multiple function classification. Another extension from the two-group case to multigroup in [13] is to solve two-group LP models for all pairs of groups and determine classification rules based on these solutions. However, in some cases the group assignment is not clear and the resulting classification scheme may be sub-optimal [6].

For the multigroup discrimination problem, Bennett and Mangasarian [39] define the piecewise-linear separability of data from K groups as the following: The data from K groups are piecewise-linear separable if and only if there exist $(w_0^k, w_1^k, \ldots, w_m^k) \in R^{m+1}$, $k = 1, \ldots, K$, such that $w_0^h + \sum_j x_{ij} w_j^h \geq w_0^k + \sum_j x_{ij} w_j^k + 1, \ \forall i \in G_h \, \forall h, k \neq h$. The following LP will generate a piecewise-linear separation for the K groups if one exists, otherwise it will generate an error-minimizing separation.

$$\begin{aligned} \text{Min} \quad & \textstyle\sum_h \sum_{k \neq h} \frac{1}{|G_h|} \sum_{i \in G_h} d_i^{hk} \\ \text{s.t.} \quad & d_i^{hk} \geq -(w_0^h + \textstyle\sum_j x_{ij} w_j^h) + (w_0^k + \sum_j x_{ij} w_j^k) + 1 \quad \forall i \in G_h \, \forall h, k \neq h \\ & w_j^k \text{ urs} \quad \forall j, k \\ & d_i^{hk} \geq 0 \quad \forall i \in G_h \, \forall h, k \neq h \end{aligned}$$

The method is tested in three data sets. It performs pretty well in two of the data sets which are totally (or almost totally) piecewise-linear separable. The classification result is not good in the third data set, which is inherently more difficult. However, by combining the multisurface method tree algorithm (MSMT) [34], the performance improves.

Gochet et al. [8] introduce an LP model for the general multigroup classification problem. The method separates the data with several hyperplanes by sequentially solving LPs. The vectors w^k, $k = 1, \ldots, K$, are estimated for the classification decision rule. The rule is to classify an observation i into group s where $s = \arg\max_k \{w_0^k + \sum_j x_{ij} w_j^k\}$.

Suppose observation i is from group h. Denote the goodness of fit for observation i with respect to group k as

$$G_{hk}^i(w^h, w^k) = [(w_0^h + \textstyle\sum_j x_{ij} w_j^h) - (w_0^k + \sum_j x_{ij} w_j^k)]^+, \text{ where } [a]^+ = \max\{0, a\}.$$

Likewise, denote the badness of fit for observation i with respect to group k as

$$B_{hk}^i(w^h, w^k) = [(w_0^h + \textstyle\sum_j x_{ij} w_h^h) - (w_0^k + \sum_j x_{ij} w_j^k)]^-, \text{ where } [a]^- = -\min\{0, a\}.$$

The total goodness of fit and total badness of fit are then defined as

$$G(w) = G(w^1, \ldots, w^K) = \textstyle\sum_h \sum_{k \neq h} \sum_{i \in G_h} G_{hk}^i(w^h, w^k)$$

$$B(w) = B(w^1, \ldots, w^K) = \textstyle\sum_h \sum_{k \neq h} \sum_{i \in G_h} B_{hk}^i(w^h, w^k)$$

The LP is to minimize the total badness of fit, subject to a normalization equation, in which $q > 0$.

$$\begin{aligned} \text{Min} \quad & B(w) \\ \text{s.t.} \quad & G(w) - B(w) = q \\ & w \text{ urs} \end{aligned}$$

Expanding $G(w)$ and $B(w)$ and substituting $G_{hk}^i(w^h, w^k)$ and $B_{hk}^i(w^h, w^k)$ by γ_{hk}^i and β_{hk}^i respectively, the LP becomes

$$\text{Min} \quad \textstyle\sum_h \sum_{k \neq h} \sum_{i \in G_h} \beta_{hk}^i$$

s.t. $(w_0^h + \sum_j x_{ij} w_j^h) - (w_0^k + \sum_j x_{ij} w_j^k) = \gamma_{hk}^i - \beta_{hk}^i \quad \forall i \in G_h \, \forall h, k \neq h$
$\sum_h \sum_{k \neq h} \sum_{i \in G_h} (\gamma_{hk}^i - \beta_{hk}^i) = q$
w_j^k urs $\forall j, k$
$\gamma_{hk}^i, \beta_{hk}^i \geq 0 \quad \forall i \in G_h \, \forall h, k \neq h$

The classification results for two real data sets show that this model can compete with Fisher's LDF and the nonparametric *k*-nearest neighbor method.

The LP-based models for classification problems highlighted above are all nonparametric models. In Section 3, we describe LP-based and MIP-based classification models that utilize a parametric multigroup discriminant analysis approach [40, 41, 42, 43]. These latter models have been employed successfully in various multigroup disease diagnosis and biological/medical prediction problems [44, 45, 46, 47, 48, 49, 43, 50, 51].

2.2. Mixed integer programming classification models

While linear programming offers a polynomial-time computational guarantee, mixed integer programming (MIP) allows more flexibility in (among other things) modeling misclassified observations and/or misclassification costs.

2.2.1. Two-group classification

In the two-group classification problem, binary variables can be used in the formulation to track and minimize the exact number of misclassifications. Such an objective function is also considered as the L_0-norm criterion [6].

MM (Minimizing the number of misclassifications)

Min $\sum_i z_i$
s.t. $w_0 + \sum_j x_{ij} w_j \leq M z_i \quad \forall i \in G_1$
$w_0 + \sum_j x_{ij} w_j \geq -M z_i \quad \forall i \in G_2$
w_j urs $\forall j$
$z_i \in \{0,1\} \quad \forall i$

The vector w is required to be a nonzero vector to prevent the trivial solution.

In the MIP formulation the objective function could include the deviation terms, such as those in the hybrid models, as well as the number of misclassifications [14]; or it could represent expected cost of misclassification [52, 53, 54, 55]. In particular, there are some variant versions of the basic model.

Stam and Joachimsthaler [56] study the classification performance of MM and compare it with MSD, Fisher's LDF, and Smith's QDF. In some cases the MM model performs better, but in some cases it does not. MIP formulations are in the review studies of Joachimsthaler and Stam [9] and Erenguc and Koehler [19], and contained in the software developed by Stam and Ungar [21]. Computational experiments show that the MIP model performs better when the group overlap is higher [9, 56], although it is still not easy to reach general conclusions [6].

Since the MIP model is $\mathcal{NP}$-hard, exact algorithms and heuristics are proposed to solve it efficiently. Koehler and Erenguc [57] develop a procedure to solve MM in which the condition of nonzero w is replaced by the requirement of at least one violation of the constraints $w_0 + \sum_j x_{ij} w_j \leq 0$ for $i \in G_1$ or $w_0 + \sum_j x_{ij} w_j \geq 0$ for $i \in G_2$. Banks and Abad [52] solve the MIP of minimizing the expected cost of misclassification by an LP-based algorithm. Abad and Banks [53] develop three heuristic procedures to the problem of minimizing the expected cost of misclassification. They also include the interaction terms of the attributes in the data and apply the heuristics [58]. Duarte Silva and Stam [54] introduce the Divide and Conquer algorithm for the classification problem of minimizing the misclassification cost by solving MIP and LP subproblems. Rubin [55] solves the same problem by using a decomposition approach, and tests this procedure on some data sets, including two breast cancer data sets. Yanev and Balev [59] propose exact and heuristic algorithms for solving MM, which are based on some specific properties of the vertices of a polyhedral set neatly connected with the model.

For the two-group classification problem where the attributes are binary, Asparoukhov and Stam [38] propose LP and MIP models which partition the data into multinomial cells and result in fewer number of variables and constraints. Let s be the index of the cells, n_{1s}, n_{2s} be the number of data points in cell s from groups 1 and 2, respectively, and $(b_{s1}, \ldots, b_{sm})$ be the binary digits representing cell s. Below is the MIP model for binary attributes (BMIP), which performs best in three real data sets in [38].

BMIP

$$\begin{aligned}
\text{Min} \quad & \textstyle\sum_{s:\, n_{1s}+n_{2s}>0} \{|n_{1s} - n_{2s}| z_s + \min(n_{1s}, n_{2s})\} \\
\text{s.t.} \quad & w_0 + \textstyle\sum_j b_{sj} w_j \leq M z_s \quad \forall s: n_{1s} \geq n_{2s};\ n_{1s} > 0 \\
& w_0 + \textstyle\sum_j b_{sj} w_j > -M z_s \quad \forall s: n_{1s} < n_{2s} \\
& w_j \text{ urs} \quad \forall j \\
& z_s \in \{0,1\} \quad \forall s: n_{1s} + n_{2s} > 0
\end{aligned}$$

Pavur et al. [60] include different secondary goals in the model MM and compare their misclassification rates. A new secondary goal is proposed, which maximizes the difference between the means of the discriminant scores of the two groups. In this model the term $-\delta$ is added to the minimization objective function as a secondary goal with a constant multiplier while the constraint $\sum_j \bar{x}_j^{(2)} w_j - \sum_j \bar{x}_j^{(1)} w_j \geq \delta$ is included, where $\bar{x}_j^{(k)} = \frac{1}{|G_k|} \sum_{i \in G_k} x_{ij}\ \forall j$, for $k = 1, 2$. The results of simulation study show that an MIP model with the proposed secondary goal has better performance than other studied models.

Glen [61] proposes IP techniques for normalization in the two-group discriminant analysis models. One technique is to add the constraint $\sum_{j=1}^{m} |w_j| = 1$. In the proposed model, w_j for $j = 1, \ldots, m$ is represented by $w_j = w_j^+ - w_j^-$, where $w_j^+, w_j^- \geq 0$, and binary variables δ_j and γ_j are defined such that $\delta_j = 1 \Leftrightarrow w_j^+ \geq \varepsilon$ and $\gamma_j = 1 \Leftrightarrow w_j^- \geq \varepsilon$. The IP normalization technique is applied to MSD and MMD, and the MSD version is presented below.

MSD – with IP normalization

$$\text{Min} \quad \textstyle\sum_i d_i$$

s.t. $w_0 + \sum_{j=1}^{m} x_{ij}(w_j^+ - w_j^-) - d_i \leq 0 \;\; \forall i \in G_1$
$w_0 + \sum_{j=1}^{m} x_{ij}(w_j^+ - w_j^-) + d_i \geq 0 \;\; \forall i \in G_2$
$\sum_{j=1}^{m}(w_j^+ + w_j^-) = 1$
$w_j^+ - \varepsilon\delta_j \geq 0 \;\; \forall j = 1,\ldots,m$
$w_j^+ - \delta_j \leq 0 \;\; \forall j = 1,\ldots,m$
$w_j^- - \varepsilon\gamma_j \geq 0 \;\; \forall j = 1,\ldots,m$
$w_j^- - \gamma_j \leq 0 \;\; \forall j = 1,\ldots,m$
$\delta_j + \gamma_j \leq 1 \;\; \forall j = 1,\ldots,m$
w_0 urs
$w_j^+, w_j^- \geq 0 \;\; \forall j = 1,\ldots,m$
$d_i \geq 0 \;\; \forall i$
$\delta_j, \gamma_j \in \{0,1\} \;\; \forall j = 1,\ldots,m$

The variable coefficients of the discriminant function generated by the models are invariant under origin shifts. The proposed models are validated using two data sets from [17, 22]. The models are also extended for attribute selection by adding the constraint $\sum_{j=1}^{m}(\delta_j + \gamma_j) = p$, which allows only a constant number, p, of attributes to be used for classification.

Glen [62] develops MIP models which determine the thresholds for forming dichotomous variables as well as the discriminant function coefficients, w_j. For each continuous attribute to be formed as a dichotomous attribute, the model finds the threshold among possible thresholds while determining the separating hyperplane and optimizing the objective function such as minimizing the sum of deviations or minimizing the number of misclassifications. Computational results of a real data set and some simulated data sets show that the MSD model with dichotomous categorical variable formation can improve classification performance. The reason for the potential of this technique is that the generated linear discriminant function is a nonlinear function of the original variables.

2.2.2. *Multigroup classification*

Gehrlein [63] proposes MIP formulations of minimizing the total number of misclassifications in the multigroup classification problem. He gives both a single function classification scheme and a multiple function classification scheme, as follows.
GSFC (General single function classification – minimizing the number of misclassifications)
Min $\sum_i z_i$

$$
\begin{aligned}
\text{s.t.} \quad & w_0 + \sum_j x_{ij} w_j - M z_i \leq U_k \quad \forall i \in G_k \\
& w_0 + \sum_j x_{ij} w_j + M z_i \geq L_k \quad \forall i \in G_k \\
& U_k - L_k \geq \delta' \quad \forall k \\
& \left.\begin{array}{l} L_g - U_k + M y_{gk} \geq \delta \\ L_k - U_g + M y_{kg} \geq \delta \\ y_{gk} + y_{kg} = 1 \end{array}\right\} \quad \forall g, k,\ g \neq k \\
& w_j \text{ urs} \quad \forall j \\
& U_k, L_k \text{ urs} \quad \forall k \\
& z_i \in \{0,1\} \quad \forall i \\
& y_{gk} \in \{0,1\} \quad \forall g, k,\ g \neq k
\end{aligned}
$$

where U_k, L_k denote the upper and lower endpoints of the interval assigned to group k, and $y_{gk} = 1$ if the interval associated with group g precedes that with group k and $y_{gk} = 0$ otherwise. The constant δ' is the minimum width of an interval of a group and the constant δ is the minimum gap between adjacent intervals.

GMFC (General multiple function classification – minimizing the number of misclassifications)

$$
\begin{aligned}
\text{Min} \quad & \sum_i z_i \\
\text{s.t.} \quad & w_0^h + \sum_j x_{ij} w_j^h - w_0^k - \sum_j x_{ij} w_j^k + M z_i \geq \varepsilon \quad \forall i \in G_h,\ \forall h, k \neq h \\
& w_j^k \text{ urs} \quad \forall j, k \\
& z_i \in \{0,1\} \quad \forall i
\end{aligned}
$$

Both models work successfully on the iris data set provided by Fisher [4].

Pavur [64] solves the multigroup classification problem by sequentially solving GSFC in one dimension each time. Linear discriminant functions are generated by successively solving GSFC with the added constraints that all linear discriminants are uncorrelated to each other for the total data set. This procedure could be repeated for the number of dimensions that is believed to be enough. According to simulation results, this procedure substantially improves the GSFC model and sometimes outperforms GMFC, Fisher's LDF, or Smith's QDF.

To solve the three-group classification problem more efficiently, Loucopoulos and Pavur [65] make a slight modification on GSFC and propose the model MIP3G, which also minimizes the number of misclassifications. Compared to GSFC, MIP3G is also a single function classification model, but it reduces the possible group orderings from six to three in the formulation and thus becomes more efficient. Loucopoulos and Pavur [66] report the results of a simulation experiment on the performance of GMFC, MIG3G, Fisher's LDF and Smith's QDF for three-group classification problem with small training samples. Second-order terms are also considered in the experiment. Simulation results show that GMFC and MIP3G can outperform the parametric procedures in some nonnormal data sets and that the inclusion of second-order terms can improve the performance of MIP3G in some data sets. Pavur and Loucopoulos [67] investigate the effect of the gap size in the MIP3G model for the three-group classification problem. A simulation study illustrates that for fairly separable data, or data with small sample sizes, a nonzero-gap model can improve the performance. A possible reason for this result is that the zero-gap model may be over-fitting the data.

Gallagher et al. [40, 41], Lee et al. [42], and Lee [49, 43] propose MIP models, both heuristic and exact, as a computational approach to solving the constrained discriminant method described by Anderson [68]. These models are described in detail in Section 3.

2.3. Nonlinear programming classification models

Nonlinear programming approaches are natural extensions for some of the LP-based models. Thus far, nonlinear programming approaches have been developed for 2-group classification.

Stam and Joachimsthaler [16] propose a class of nonlinear programming methods to solve the two-group classification problem under the L_p-norm objective criterion. This is an extension of MSD and MMD, for which the objectives are the L_1-norm and L_∞-norm, respectively.

Minimize the general L_p-norm distance

$$\begin{aligned} \text{Min} \quad & (\textstyle\sum_i d_i^p)^{1/p} \\ \text{s.t.} \quad & \textstyle\sum_j x_{ij} w_j - d_i \le b \quad \forall i \in G_1 \\ & \textstyle\sum_j x_{ij} w_j + d_i \ge b \quad \forall i \in G_2 \\ & w_j \text{ urs} \quad \forall j \\ & d_i \ge 0 \quad \forall i \end{aligned}$$

The simulation results show that, in addition to the L_1-norm and L_∞-norm, it is worth the effort to compute other L_p-norm objectives. Restricting the analysis to $1 \le p \le 3$, plus $p = \infty$, is recommended. This method is reviewed by Joachimsthaler and Stam [9] and Erenguc and Koehler [19].

Mangasarian et al. [69] propose a nonconvex model for the two-group classification problem:

$$\begin{aligned} \text{Min} \quad & d^1 + d^2 \\ \text{s.t.} \quad & \textstyle\sum_j x_{ij} w_j - d^1 \le 0 \qquad \forall i \in G_1 \\ & \textstyle\sum_j x_{ij} w_j + d^2 \ge 0 \qquad \forall i \in G_2 \\ & \max_{j=1,\ldots,m} |w_j| = 1 \\ & w_j \text{ urs} \quad \forall j \\ & d^1, d^2 \text{ urs} \end{aligned}$$

This model can be solved in polynomial-time by solving $2m$ linear programs, which generate a sequence of parallel planes, resulting in a piecewise-linear nonconvex discriminant function. The model works successfully in clinical practice for the diagnosis of breast cancer.

Further, Mangasarian [70] also formulates the problem of minimizing the number of misclassifications as a linear program with equilibrium constraints (LPEC) instead of the MIP model MM described previously.

MM-LPEC (Minimizing the number of misclassifications - Linear program with equilibrium constraints)

$$\text{Min} \quad \textstyle\sum_{i \in G_1 \cup G_2} z_i$$

$$
\begin{array}{lll}
\text{s.t.} & w_0+\sum_j x_{ij}w_j-d_i\le -1 & \forall i\in G_1\\
& z_i(w_0+\sum_j x_{ij}w_j-d_i+1)=0 & \forall i\in G_1\\
& w_0+\sum_j x_{ij}w_j+d_i\ge 1 & \forall i\in G_2\\
& z_i(w_0+\sum_j x_{ij}w_j+d_i-1)=0 & \forall i\in G_2\\
& d_i(1-z_i)=0 & \forall i\in G_1\cup G_2\\
& 0\le z_i\le 1 & \forall i\in G_1\cup G_2\\
& d_i\ge 0 & \forall i\in G_1\cup G_2\\
& w_j \text{ urs } \forall j &
\end{array}
$$

The general LPEC can be converted to an exact penalty problem with a quadratic objective and linear constraints. A stepless Frank-Wolfe-type algorithm is proposed for the penalty problem, terminating at a stationary point or a global solution. This method is called the parametric misclassification minimization (PMM) procedure, and numerical testing is included in [71].

To illustrate the next model, we first define the step function $s : R \rightarrow \{0,1\}$ as

$$
s(u)=\begin{cases} 1 & \text{if } u>0\\ 0 & \text{if } u\le 0\end{cases}
$$

The problem of minimizing the number of misclassifications is equivalent to

$$
\begin{array}{lll}
\text{Min} & \sum_{i\in G_1\cup G_2} s(d_i) & \\
\text{s.t.} & w_0+\sum_j x_{ij}w_j-d_i\le -1 & \forall i\in G_1\\
& w_0+\sum_j x_{ij}w_j+d_i\ge 1 & \forall i\in G_2\\
& d_i\ge 0 & \forall i\in G_1\cup G_2\\
& w_j \text{ urs } \forall j &
\end{array}
$$

Mangasarian [71] proposes a simple concave approximation of the step function for nonnegative variables: $t(u,\alpha)=1-e^{-\alpha u}$, where $\alpha>0, u\ge 0$. Let $\alpha>0$ and approximate $s(d_i)$ by $t(d_i,\alpha)$. The problem then reduces to minimizing a smooth concave function bounded below on a nonempty polyhedron, which has a minimum at a vertex of the feasible region. A finite successive linearization algorithm (SLA) is proposed, terminating at a stationary point or a global solution. Numerical tests of SLA are done and compared with the PMM procedure described above. The results show that the much simpler SLA obtains a separation that is almost as good as PMM in considerably less computing time.

Chen and Mangasarian [72] propose an algorithm on a defined hybrid misclassification minimization problem, which is more computationally tractable than the $\mathcal{NP}$-hard misclassification minimization problem. The basic idea of the hybrid approach is to obtain iteratively w_0 and $(w_1,\ldots,w_m)$ of the separating hyperplane: (1) For a fixed w_0, solve RLP (Bennett and Mangasarian [32]) to determine $(w_1,\ldots,w_m)$, and (2) for this $(w_1,\ldots,w_m)$, solve the one-dimensional misclassification minimization problem to determine w_0. Comparison of the hybrid method is made with the RLP method and the PMM procedure. The hybrid method performs better in the testing sets of the tenfold cross-validation and is much faster than PMM.

Mangasarian [73] proposes the model of minimizing the sum of arbitrary-norm distances of misclassified points to the separating hyperplane. For a general norm $||\cdot||$ on R^m, the dual norm $||\cdot||'$ on R^m is defined as $||x||'=\max_{||y||=1}x^Ty$. Define $[a]^+=\max\{0,a\}$ and let $w=(w_1,\ldots,w_m)$. The formulation can then be written as:

$$\begin{array}{ll} \text{Min} & \sum_{i \in G_1} [w_0 + \sum_j x_{ij} w_j]^+ + \sum_{i \in G_2} [-w_0 - \sum_j x_{ij} w_j]^+ \\ \text{s.t.} & ||w||' = 1 \\ & w_0, w \text{ urs} \end{array}$$

The problem is to minimize a convex function on a unit sphere. A related decision problem to this minimization problem is shown to be $\mathcal{NP}$-complete, except for $p = 1$. For a general p-norm, the minimization problem can be transformed via an exact penalty formulation to minimizing the sum of a convex function and a bilinear function on a convex set.

2.4. Support vector machines

A support vector machine is a type of mathematical programming approach (Vapnik [74]). It has been widely studied, and has become popular in many application fields in recent years. The introductory description of support vector machines (SVM) given here is summarized from the tutorial by Burges [75]. In order to maintain consistency with SVM studies in published literature, the notation used below is slightly different than the notation used to describe the mathematical programming methods in earlier sections.

In the two-group separable case, the objective function is to maximize the margin of a separating hyperplane, $2/||w||$, which is equivalent to minimizing $||w||^2$.

$$\begin{array}{lll} \text{Min} & w^T w & \\ \text{s.t.} & x_i^T w + b \geq +1 & \text{for } y_i = +1 \\ & x_i^T w + b \leq -1 & \text{for } y_i = -1 \\ & w, b \text{ urs} & \end{array}$$

where $x_i \in R^m$ represents the values of attributes of observation i and $y_i \in \{-1, 1\}$ represents the group of observation i.

This problem can be solved by solving its Wolfe dual problem.

$$\begin{array}{ll} \text{Max} & \sum_i \alpha_i - \frac{1}{2} \sum_{i,j} \alpha_i \alpha_j y_i y_j x_i^T x_j \\ \text{s.t.} & \sum_i \alpha_i y_i = 0 \\ & \alpha_i \geq 0 \ \ \forall i \end{array}$$

Here, α_i is the Lagrange multiplier for the training point i, and the points with $\alpha_i > 0$ are called the support vectors (analogous to the support of a hyperplane, and thus the introduction of the name "support vector"). The primal solution w is given by $w = \sum_i \alpha_i y_i x_i$. b can be computed by solving $y_i(w^T x_i + b) - 1 = 0$ for any i with $\alpha_i > 0$.

For the non-separable case, slack variables ξ_i are introduced to handle the errors. Let C be the penalty for the errors. The problem becomes

$$\begin{array}{lll} \text{Min} & \frac{1}{2} w^T w + C(\sum_i \xi_i)^k & \\ \text{s.t.} & x_i^T w + b \geq +1 - \xi_i & \text{for } y_i = +1 \\ & x_i^T w + b \leq -1 + \xi_i & \text{for } y_i = -1 \\ & w, b \text{ urs} & \\ & \xi_i \geq 0 \ \ \forall i & \end{array}$$

When k is chosen to be 1, neither the ξ_i's nor their Lagrange multipliers appear in the Wolfe dual problem:

$$\text{Max} \quad \sum_i \alpha_i - \frac{1}{2} \sum_{i,j} \alpha_i \alpha_j y_i y_j x_i^T x_j$$

s.t. $\sum_i \alpha_i y_i = 0$
$0 \leq \alpha_i \leq C \quad \forall i$

The data points can be separated nonlinearly by mapping the data into some higher dimensional space and applying linear SVM to the mapped data. Instead of knowing explicitly the mapping Φ, SVM needs only the dot products of two transformed data points $\Phi(x_i) \cdot \Phi(x_j)$. The kernel function K is introduced such that $K(x_i, x_j) = \Phi(x_i) \cdot \Phi(x_j)$. Replacing $x_i^T x_j$ by $K(x_i, x_j)$ in the above problem, the separation becomes nonlinear while the problem to be solved remains a quadratic program. In testing a new data point x after training, the sign of the function $f(x)$ is computed to determine the group of x:

$$f(x) = \sum_{i=1}^{N_s} \alpha_i y_i \Phi(s_i) \cdot \Phi(x) + b = \sum_{i=1}^{N_s} \alpha_i y_i K(s_i, x) + b$$

where s_i's are the support vectors and N_s is the number of support vectors. Again the explicit form of $\Phi(x)$ is avoided.

Mangasarian provides a general mathematical programming framework for SVM, called generalized support vector machine or GSVM [76, 77]. Special cases can be derived from GSVM, including the standard SVM.

Many SVM-type methods have been developed by Mangasarian and other authors to solve huge-sized classification problems more efficiently. These methods include: successive overrelaxation for SVM [78], proximal SVM [79, 80], smooth SVM [81], reduced SVM [82], Lagrangian SVM [83], incremental SVMs [84], and other methods [85, 86]. Mangasarian summarizes some of the developments in [87]. Examples of applications of SVM include breast cancer studies [88, 89] and genome research [90].

Hsu and Lin [91] compare different methods for multigroup classification using support vector machines. Three methods studied are based on several binary classifiers: one-against-one, one-against-all, and directed acyclic graph (DAG) SVM. The other two methods studied are altogether methods with decomposition implementation. The experiment results show that the one-against-one and DAG methods are more suitable for practical use than the other methods. Lee et al. [92] propose a generic approach to multigroup problems with some theoretical properties, and the proposed method is well applied to microarray data for cancer classification and satellite radiance profiles for cloud classification.

Gallagher et al 1996, 1997 and Lee et al 2003 [40, 41, 42] offer the first discrete support vector machine for multigroup classification with reserved judgement. The approach has been successfully applied to a diverse variety of biological and medical applications (see Section 3).

3. MIP-BASED MULTIGROUP CLASSIFICATION MODELS AND APPLICATIONS TO MEDICINE AND BIOLOGY

Commonly-used methods for classification, such as linear discriminant functions, decision trees, mathematical programming approaches, support vector machines, and artificial neural networks (ANN), can be viewed as attempts at approximating a *Bayes optimal rule* for classification; that is, a rule that maximizes (minimizes) the total probability of correct classification (misclassification). Even if a Bayes optimal rule is known, inter-group misclassification rates may be higher than desired. For example, in a population that is mostly healthy, a Bayes optimal rule for medical diagnosis might misdiagnose sick patients as healthy in order to maximize total probability of correct diagnosis. As a remedy, a constrained discriminant rule that limits the misclassification rate is appealing.

Assuming that the group density functions and prior probabilities are known, Anderson [68] showed that an optimal rule for the problem of maximizing the probability of correct classification subject to constraints on the misclassification probabilities must be of a specific form when discriminating among multiple groups with a simplified model. The formulae in Anderson's result depend on a set of parameters satisfying a complex relationship between the density functions, the prior probabilities, and the bounds on the misclassification probabilities. Establishing a viable mathematical model to describe Anderson's result, and finding values for these parameters that yield an optimal rule are challenging tasks. The first computational models utilizing Anderson's formulae were proposed in [40, 41].

3.1. Discrete support vector machine predictive models

As part of the work carried out at Georgia Institute of Technology's Center for Operations Research in Medicine, we have developed a general-purpose discriminant analysis modeling framework and computational engine that are applicable to a wide variety of applications, including biological, biomedical and logistics problems. Utilizing the technology of large-scale discrete optimization and support-vector machines, we have developed novel classification models that simultaneously include the following features: 1) the ability to classify any number of distinct groups; 2) the ability to incorporate heterogeneous types of attributes as input; 3) a high-dimensional data transformation that eliminates noise and errors in biological data; 4) constraints to limit the rate of misclassification, and a reserved-judgment region that provides a safeguard against over-training (which tends to lead to high misclassification rates from the resulting predictive rule); and 5) successive multi-stage classification capability to handle data points placed in the reserved judgment region. Studies involving tumor volume identification, ultrasonic cell disruption in drug delivery, lung tumor cell motility analysis, CpG island aberrant methylation in human cancer, predicting early atherosclerosis using biomarkers, and fingerprinting native and angiogenic microvascular networks using functional perfusion data indicate that our approach is adaptable and can produce effective and reliable predictive rules for various biomedical and bio-behavior phenomena [44, 45, 46, 47, 48, 50, 51, 49, 43].

Based on the description in [40, 41, 42, 49, 43], we summarize below some of the classification models we have developed.

3.1.1. Modeling of reserved judgment region for general groups

When the population densities and prior probabilities are known, the constrained rules with a reject option (reserved judgment), based on Anderson's results, calls for finding a partition $\{R_0, ..., R_G\}$ of $\mathbb{R}^k$ that maximizes the probability of correct allocation subject to constraints on the misclassification probabilities; i.e.,

$$\text{maximize} \quad \sum_{g=1}^{G} \pi_g \int_{R_g} f_g(w)\, dw \tag{1}$$

$$\text{subject to} \quad \int_{R_g} f_h(w) dw \leq \alpha_{hg},\ h,\ g = 1, ..., G,\ h \neq g, \tag{2}$$

where f_h, $h \in \{1, ..., G\}$, are the group conditional density functions, π_g denotes the prior probability that a randomly selected entity is from group g, $g \in \{1, ..., G\}$, and α_{hg}, $h \neq g$, are constants between zero and one. Under quite general assumptions, it was shown that there exist unique (up to a set of measure zero) nonnegative constants λ_{ih}, i, $h \in \{1, ..., G\}$, $i \neq h$, such that the optimal rule is given by

$$R_g = \{x \in \mathbb{R}^k : L_g(x) = \max_{h \in = \{0,1,...,G\}} L_h(x)\},\ g = 0, ..., G \tag{3}$$

where

$$L_0(x) = 0 \tag{4}$$

$$L_h(x) = \pi_h f_h(x) - \sum_{i=1, i \neq h}^{G} \lambda_{ih} f_i(x),\ h = 1, ..., G \tag{5}$$

For $G = 2$ the optimal solution can be modeled rather straightforward. However, finding optimal λ_{ih}'s for the general case, $G \geq 3$, is a difficult problem, with the difficulty increasing as G increases. Our model offers an avenue for modeling and finding the optimal solution in the general case. It is the first such model to be computationally viable [40, 41].

Before proceeding, we note that R_g can be written as $R_g = \{x \in \mathbb{R}^k : L_g(x) \geq L_h(x)$ for all $h \in \{0, \ldots, G\}$. So, since $L_g(x) \geq L_h(x)$ if, and only if, $(1/\sum_{t=1}^{G} f_t(x))L_g(x) \geq (1/\sum_{t=1}^{G} f_t(x))L_h(x)$, the functions L_h, $h = 1, ..., G$, can be redefined as

$$L_h(x) = \pi_h p_h(x) - \sum_{i=1, i \neq h}^{G} \lambda_{ih} p_i(x),\ h = 1, ..., G \tag{6}$$

where $p_i(x) = f_i(x)/\sum_{t=1}^{G} f_t(x)$. We assume that L_h is defined as in equation (6) in our model.

3.1.2. *Mixed integer programming formulations*

Assume that we are given a training sample of N entities whose group classifications are known; say n_g entities are in group g, where $\sum_{g=1}^{G} n_g = N$. Let the k dimensional vectors x^{gj}, $g = 1,...,G$, $j = 1,...,n_g$, contain the measurements on k available characteristics of the entities. Our procedure for deriving a discriminant rule proceeds in two stages. The first stage is to use the training sample to compute estimates, $\hat{f}_h$, either parametrically or nonparametrically, of the density functions f_h (e.g., see [93]) and estimates, $\hat{\pi}_h$, of the prior probabilities π_h, $h = 1,...,G$. The second stage is to determine the optimal λ_{ih}'s given these estimates. This stage requires being able to estimate the probabilities of correct classification and misclassification for any candidate set of λ_{ih}'s. One could, in theory, substitute the estimated densities and prior probabilities into equations (5), and directly use the resulting regions R_g in the integral expressions given in (1) and (2). This would involve, even in simple cases such as normally distributed groups, the numerical evaluation of k-dimensional integrals at each step of a search for the optimal λ_{ih}'s. Therefore, we have designed an alternative approach. After substituting the $\hat{f}_h$'s and $\hat{\pi}_h$'s into equation (5), we simply calculate the proportion of training sample points which fall in each of the regions $R_1,...,R_G$. The mixed integer programming (MIP) models discussed below attempt to maximize the proportion of training sample points correctly classified while satisfying constraints on the proportions of training sample points misclassified. This approach has two advantages. First, it avoids having to evaluate the potentially difficult integrals in Equations (1) and (2). Second, it is nonparametric in controlling the training sample misclassification probabilities. That is, even if the densities are poorly estimated (by assuming, for example, normal densities for non-normal data), the constraints are still satisfied for the training sample. Better estimates of the densities may allow a higher correct classification rate to be achieved, but the constraints will be satisfied even if poor estimates are used. Unlike most support vector machine models that minimize the sum of errors, our objective is driven by the number of correct classifications, and will not be biased by the distance of the entities from the supporting hyperplane.

A word of caution is in order. In traditional unconstrained discriminant analysis, the true probability of correct classification of a given discriminant rule tends to be smaller than the rate of correct classification for the training sample from which it was derived. One would expect to observe such an effect for the method described herein as well. In addition, one would expect to observe an analogous effect with regard to constraints on misclassification probabilities – the true probabilities are likely to be greater than any limits imposed on the proportions of training sample misclassifications. Hence, the α_{hg} parameters should be carefully chosen for the application in hand.

Our first model is a nonlinear 0/1 MIP model with the nonlinearity appearing in the constraints. Model 1 maximizes the number of correct classifications of the given N training entities. Similarly, the constraints on the misclassification probabilities are modeled by ensuring that the number of group g training entities in region R_h is less than or equal to a pre-specified percentage, $\alpha_{hg}(0 < \alpha_{hg} < 1)$, of the total number, n_g, of group g entities, $h,\ g \in \{1,...,G\}$, $h \neq g$.

For notational convenience, let $\mathbf{G} = \{1,...,G\}$ and $\mathbf{N}_g = \{1,...,n_g\}$, for $g \in \mathbf{G}$. Also, analogous to the definition of p_i, define $\hat{p}_i$ by $\hat{p}_i = \hat{f}_i(x)/\sum_{t=1}^{G} \hat{f}_t(x)$. In our model, we use binary indicator variables to denote the group classification of entities. Mathematically, let u_{hgj} be a binary variable indicating whether or not x^{gj} lies in region R_h; i.e., whether or not the jth entity from group g is allocated to group h. Then Model

1 can be written as follows:
DAMIP

$$\text{Maximize} \quad \sum_{g \in G} \sum_{j \in N_g} u_{ggj}$$

Subject to

$$L_{hgj} = \hat{\pi}_h \hat{p}_h(x^{gj}) - \sum_{i \in G \setminus h} \lambda_{ih} \hat{p}_i(x^{gj}), \quad h, g \in \mathbf{G},\ j \in \mathbf{N}_g \tag{7}$$

$$y_{gj} = \max\{0, L_{hgj} : h = 1, ..., G\}, \quad g \in \mathbf{G},\ j \in \mathbf{N}_g \tag{8}$$

$$y_{gj} - L_{ggj} \leq M(1 - u_{ggj}), \quad g \in \mathbf{G},\ j \in \mathbf{N}_g \tag{9}$$

$$y_{gj} - L_{hgj} \geq \varepsilon(1 - u_{hgj}), \quad h, g \in \mathbf{G},\ j \in \mathbf{N}_g,\ h \neq g \tag{10}$$

$$\sum_{j \in N_g} u_{hgj} \leq \lfloor \alpha_{hg} n_g \rfloor, \quad h, g \in \mathbf{G},\ h \neq g \tag{11}$$

$$-\infty < L_{hgj} < \infty,\ y_{gj} \geq 0,\ \lambda_{ih} \geq 0, u_{hgj} \in \{0, 1\}$$

Constraint (7) defines the variable L_{hgj} as the value of the function L_h evaluated at x^{gj}. Therefore, the continuous variable y_{gj}, defined in constraint (8), represents $\max\{L_h(x^{gj}) : h = 0, ..., G\}$; and consequently, x^{gj} lies in region R_h if, and only if, $y_{gj} = L_{hgj}$. The binary variable u_{hgj} is used to indicate whether or not x^{gj} lies in region R_h; i.e., whether or not the jth entity from group g is allocated to group h. In particular, constraint (9), together with the objective, force u_{ggj} to be 1 if, and only if, the jth entity from group g is correctly allocated to group g; and constraints (10) and (11) ensure that at most $\lfloor \alpha_{hg} n_g \rfloor$ (i.e., the greatest integer less than or equal to $\alpha_{hg} n_g$) group g entities are allocated to group h, $h \neq g$. One caveat regarding the indicator variables u_{hgj} is that although the condition $u_{hgj} = 0$, $h \neq g$, implies (by constraint (10)) that $x^{gj} \notin R_h$, the converse need not hold. As a consequence, the number of misclassifications may be overcounted. However, in our preliminary numerical study we found that the actual amount of overcounting is minimal. One could force the converse (thus, $u_{hgj} = 1$ if and only if $x^{gj} \in R_h$) by adding constraints $y_{gj} - L_{hgj} \leq M(1 - u_{hgj})$, for example. Finally, we note that the parameters M and ε are extraneous to the discriminant analysis problem itself, but are needed in the model to control the indicator variables u_{hgj}. The intention is for M and ε to be, respectively, large and small positive constants.

3.1.3. Model variations

We explore different variations in the model to grasp the quality of the solution and the associated computational effort.

A first variation involves transforming Model 1 to an equivalent linear mixed integer model. In particular, Model 2 replaces the N constraints defined in (8) with the following system of $3GN + 2N$ constraints:

$$
\begin{aligned}
y_{gj} &\geq L_{hgj}, & h,g \in \mathbf{G},\ j \in \mathbf{N}_g \quad (12)\\
\tilde{y}_{hgj} - L_{hgj} &\leq M(1 - v_{hgj}), & h,g \in \mathbf{G},\ j \in \mathbf{N}_g \quad (13)\\
\tilde{y}_{hgj} &\leq \hat{\pi}_h \hat{p}_h(x^{gj}) v_{hgj}, & h,g \in \mathbf{G},\ j \in \mathbf{N}_g \quad (14)\\
\sum_{h \in G} v_{hgj} &\leq 1, & g \in \mathbf{G},\ j \in \mathbf{N}_g \quad (15)\\
\sum_{h \in G} \tilde{y}_{hgj} &= y_{gi}, & g \in \mathbf{G},\ j \in \mathbf{N}_g \quad (16)
\end{aligned}
$$

where $\tilde{y}_{hgj} \geq 0$ and $v_{hgj} \in \{0,1\}$, $h,g \in \mathbf{G}, j \in \mathbf{N}_g$. These constraints, together with the non-negativity of y_{gj} force $y_{gj} = \max\{0, L_{hgj} : h = 1,...,G\}$.

The second variation involves transforming Model 1 to a heuristic linear MIP model. This is done by replacing the nonlinear constraint (8) with $y_{gj} \geq L_{hgj}$, $h,g \in \mathbf{G}, j \in \mathbf{N}_g$, and including penalty terms in the objective function. In particular, Model 3 has the objective

$$\text{Maximize} \sum_{g \in G} \sum_{j \in N_g} \beta u_{ggj} - \sum_{g \in G} \sum_{j \in N_g} \gamma y_{gj},$$

where β and γ are positive constants. This model is heuristic in that there is nothing to force $y_{gj} = \max\{0, L_{hgj} : h = 1,...,G\}$. However, since in addition to trying to force as many u_{ggj}'s to one as possible, the objective in Model 3 also tries to make the y_{gj}'s as small as possible, and the optimizer tends to drive y_{gj} towards $\max\{0, L_{hgj} : h = 1,...,G\}$. We remark that β and γ could be stratified by group (i.e., introduce possibly distinct $\beta_g, \gamma_g, g \in \mathbf{G}$) to model the relative importance of certain groups to be correctly classified.

A reasonable modification to Models 1, 2 and 3 involves relaxing the constraints specified by (11). Rather than placing restrictions on the number of type g training entities classified into group h, for all $h,g \in \mathbf{G}, h \neq g$, one could simply place an upper bound on the *total* number of misclassified training entities. In this case, the $G(G-1)$ constraints specified by (11) would be replaced by the single constraint

$$\sum_{g \in G} \sum_{h \in G \setminus \{g\}} \sum_{j \in N_g} u_{hgj} \leq \lfloor \alpha N \rfloor \tag{17}$$

where α is a constant between 0 and 1. We will refer to Models 1, 2 and 3, modified in this way, as Models 1T, 2T and 3T, respectively. Of course, other modifications are also possible. For instance, one could place restrictions on the total number of type g points misclassified for each $g \in \mathbf{G}$. Thus, in place of the constraints specified in (17), one would include the constraints $\sum_{h \in G \setminus \{g\}} \sum_{j \in N_g} u_{hgj} \leq \lfloor \alpha_g N \rfloor$, $g \in \mathbf{G}$, where $0 < \alpha_g < 1$.

We also explore a heuristic linear model of Model 1. In particular, consider the linear program (DALP):

$$
\begin{aligned}
\text{Maximize} \quad & \sum_{g \in G} \sum_{j \in N_g} (c_1 w_{gj} + c_2 y_{gj}) && (18)\\
\text{Subject to} \quad & && \\
L_{hgj} \quad & = \pi_h \hat{p}_h(x^{gj}) - \sum_{i \in G \setminus h} \lambda_{ih} \hat{p}_i(x^{gj}), \quad h,g \in \mathbf{G},\ j \in \mathbf{N}_g && (19)\\
L_{ggj} - L_{hgj} + w_{gj} \quad & \geq 0, \quad h,g \in \mathbf{G},\ h \neq g,\ j \in \mathbf{N}_g && (20)\\
L_{ggj} \quad + w_{gj} \quad & \geq 0, \quad g \in \mathbf{G},\ j \in \mathbf{N}_g, && (21)\\
-L_{hgj} + y_{gj} \quad & \geq 0, \quad h,g \in \mathbf{G},\ j \in \mathbf{N}_g && (22)\\
& -\infty < L_{hgj} < \infty,\ w_{gj},\ y_{gj},\ \lambda_{ih} \geq 0 &&
\end{aligned}
$$

Constraint (19) defines the variable L_{hgj} as the value of the function L_h evaluated at x^{gj}. As the optimization solver searches through the set of feasible solutions, the λ_{ih} variables will vary, causing the L_{hgj} variables to assume different values. Constraints (20), (21) and (22) link the objective-function variables with the L_{hgj} variables in such a way that correct classification of training entities, and allocation of training entities into the reserved-judgment region, are captured by the objective-function variables. In particular, if the optimization solver drives w_{gj} to zero for some g, j pair, then constraints (20) and (21) imply that $L_{ggj} = \max\{0, L_{hgj} : h \in \mathbf{G}\}$. Hence, the jth entity from group g is correctly classified. If, on the other hand, the optimal solution yields $y_{gj} = 0$ for some g, j pair, then constraint (22) implies that $\max\{0, L_{hgj} : h \in \mathbf{G}\} = 0$. Thus, the jth entity from group g is placed in the reserved-judgment region. (Of course, it is possible for both w_{gj} and y_{gj} to be zero. One should decide prior to solving the linear program how to interpret the classification in such cases.) If both w_{gj} and y_{gj} are positive, the jth entity from group g is misclassified.

The optimal solution yields a set of λ_{ih}'s that best allocates the training entities (i.e., "best" in terms of minimizing the penalty objective function). The optimal λ_{ih}'s can then be used to define the functions L_h, $h \in G$, which in turn can be used to classify a new entity with feature vector $x \in \mathbb{R}^k$ by simply computing the index at which $\max\{L_h(x) : h \in \{0, 1, \ldots, G\}\}$ is achieved.

Note that Model DALP places no a priori bound on the number of misclassified training entities. However, since the objective is to minimize a weighted combination of the variables w_{gj} and y_{gj}, the optimizer will attempt to drive these variables to zero. Thus, the optimizer is, in essence, attempting either to correctly classify training entities ($w_{gj} = 0$), or to place them in the reserved-judgment region ($y_{gj} = 0$). By varying the weights c_1 and c_2, one has a means of controlling the optimizer's emphasis for correctly classifying training entities versus placing them in the reserved-judgment region. If $c_2/c_1 < 1$, the optimizer will tend to place a greater emphasis on driving the w_{gj} variables to zero than driving the y_{gj} variables to zero (conversely, if $c_2/c_1 > 1$). Hence, when $c_2/c_1 < 1$, one should expect to get relatively more entities correctly classified, fewer placed in the reserved-judgment region, and more misclassified, than when $c_2/c_1 > 1$. An extreme case is when $c_2 = 0$. In this case, there is no emphasis on driving y_{gj} to zero (the reserved-judgment region is thus ignored), and the full emphasis of the optimizer is to drive w_{gj} to zero.

Table 1 summarizes the number of constraints, the total number of variables, and the number of 0/1 variables in each of the discrete support vector machine models, and in the heuristic LP model (DALP). Clearly, even for moderately-sized discriminant analysis problems, the MIP instances are relatively large. Also, note that Model 2 is larger than Model 3, both in terms of the number of constraints and the number of variables. However, it is important to keep in mind that the difficulty of solving an MIP problem cannot, in general, be predicted solely by its size; problem structure has a direct and substantial bearing on the effort required to find optimal solutions. The LP relaxation of these MIP models pose computational challenges as commercial LP solvers return (optimal) LP solutions that are infeasible, due to the equality constraints, and the use of big M and small ε in the formulation.

It is interesting to note that the set of feasible solutions for Model 2 is "tighter" than that for Model 3. In particular, if F_i denotes the set of feasible solutions of Model i, then

$$F_1 = \{(L, \lambda, u, y) : \text{ there exists } \tilde{y}, v \text{ such that } (L, \lambda, u, y, \tilde{y}, v) \in F_2\} \subsetneq F_3. \tag{23}$$

Table 1: Model size

Model	Type	Constraints	Total Variables	0/1 Variables
1	nonlinear MIP	$2GN+N+G(G-1)$	$2GN+N+G(G-1)$	GN
2	linear MIP	$5GN+2N+G(G-1)$	$4GN+N+G(G-1)$	$2GN$
3	linear MIP	$3GN+G(G-1)$	$2GN+N+G(G-1)$	GN
1T	nonlinear MIP	$2GN+N+1$	$2GN+N+G(G-1)$	GN
2T	linear MIP	$5GN+2N+1$	$4GN+N+G(G-1)$	$2GN$
3T	linear MIP	$3GN+1$	$2GN+N+G(G-1)$	GN
DALP	linear program	$3GN$	$NG+N+G(G-1)$	0

The novelties of the classification models developed herein include: 1) they are suitable for discriminant analysis given any number of groups, 2) they accept heterogeneous types of attributes as input, 3) they use a parametric approach to reduce high-dimensional attribute spaces, and 4) they allow constraints on the number of misclassifications, and utilize a reserved judgment to facilitate the reduction of misclassifications. The latter point opens the possibility of performing multistage analysis.

Clearly, the advantage of an LP model over an MIP model is that the associated problem instances are computationally much easier to solve. However, the most important criterion in judging a method for obtaining discriminant rules is how the rules perform in correctly classifying new unseen entities. Once the rule is developed, applying it to a new entity to determine its group is trivial. Extensive computational experiments have been performed to gauge the qualities of solutions of different models [41, 42, 49, 43, 94, 95].

3.1.4. Validation of model and computational effort

We performed ten-fold cross validation, and designed simulation and comparison studies on our models. Results reported in [41, 42] demonstrate that our approach works well when applied to both simulated data and datasets from the machine learning database repository [96]. In particular, our methods compare favorably to other mathematical programming methods, including the general single function classification model (GSFC) by Gehrlein [63], and the LP model by Gochet et al. [8], as well as Fisher's LDF, artificial neural networks, quadratic discriminant analysis, tree classification, and other support vector machines, on real biological and medical data.

3.2. Classification results on real-world biological and medical applications

The main objective in discriminant analysis is to derive rules that can be used to classify entities into groups. Computationally, the challenge lies in the effort expended to develop such a rule. Once the rule is developed, applying it to a new entity to determine its group is trivial. Feasible solutions obtained from our classification models correspond to predictive rules. Empirical results [41, 42] indicate that the resulting classification model instances are computationally very challenging, and even intractable by competitive commercial MIP solvers. However, the resulting predictive rules prove to be very promising, offering correct classification rates on new unknown data ranging from 80% to 100% on various types of biological/medical problems. Our results indicate that the general-purpose classification framework that we have designed has the potential to be a very powerful predictive method for clinical settings.

The choice of mixed integer programming (MIP) as the underlying modeling and optimization technology for our support vector machine classification model is guided by the desire to simultaneously incorporate a variety of important and desirable properties of predictive models within a general framework. MIP itself allows for incorporation of continuous and discrete variables, and linear and nonlinear constraints, providing a flexible and powerful modeling environment.

Our mathematical modeling and computational algorithm design shows great promise as the resulting predictive rules are able to produce higher rates of correct classification on new biological data (with unknown group status) compared to existing classification methods. This is partly due to the transformation of raw data via the set of constraints in (7). While most mathematical programming approaches directly determine the hyperplanes of separation using raw data, our approach transforms the raw data via a probabilistic model, before the determination of the supporting hyperplanes. Further, the separation is driven by maximizing the sum of binary variables (representing correct classification or not of entities), instead of minimizing a sum of errors (representing distances of entities from hyperplanes), as in other support vector machines. The combination of these two strategies offers better classification capability. Noise in the transformed data is not as profound as in raw data. And the magnitudes of the errors do not skew the determination of the separating hyperplanes, as all entities have *equal* importance when correct classification is being counted.

To highlight the broad applicability of our approach, below, we briefly summarize the application of our predictive models and solution algorithms to ten different biological problems. Each of the projects was carried out in close partnership with experimental biologists and/or clinicians. Applications to finance and other industry applications are described elsewhere [41, 42, 94].

Determining the Type of Erythemato-Squamous Disease The differential diagnosis of erythemato-squamous diseases is an important problem in dermatology [43]. They all share the clinical features of erythema and scaling, with very little differences. The 6 groups are psoriasis, seboreic dermatitis, lichen planus, pityriasis rosea, cronic dermatitis, and pityriasis rubra pilaris. Usually a biopsy is necessary for the diagnosis but unfortunately these diseases share many histopathological features as well. Another difficulty for the differential diagnosis is that a disease may show the features of another disease at the beginning stage and may have the characteristic features at the following stages [96].

The 6 groups consist of 366 subjects (112,61,72,49,52,20 respectively) with 34 clinical attributes. Patients were first evaluated clinically with 12 features. Afterwards, skin samples were taken for the evaluation of 22 histopathological features. The values of the histopathological features are determined by an analysis of the samples under a microscope. The 34 attributes include 1) clinical attributes: erythema, scaling, definite borders, itching, koebner phenomenon, polygonal papules, follicular papules, oral mucosal involvement, knee and elbow involvement, scalp involvement, family history, age; and 2) histopathological attributes: melanin incontinence, eosinophils in the infiltrate, PNL infiltrate, fibrosis of the papillary dermis, exocytosis, acanthosis, hyperkeratosis, parakeratosis, clubbing of the rete ridges, elongation of the rete ridges, thinning of the suprapapillary epidermis, spongiform pustule, munro microabcess, focal hypergranulosis, disappearance of the granular layer, vacuolisation and damage of basal layer, spongiosis, saw-tooth appearance of retes, follicular horn plug, perifollicular parakeratosis, inflammatory monoluclear infiltrate, band-like infiltrate.

Our multigroup classification model selected 27 discriminatory attributes, and successfully classified the patients into 6 groups, each with an unbiased correct classification of greater than 93% (with 100% correct rate for groups 1, 3, 5, 6) with an average overall accuracy of 98%. Using 250 subjects to develop the rule, and testing the remaining 116 patients, we obtain a prediction accuracy of 91%.

Predicting Presence/Absence of Heart Disease The four databases concerning heart disease diagnosis were collected by Dr. Janosi of Hungarian Institute of Cardiology, Budapest; Dr. Steinbrunn of University Hospital, Zurich; Dr. Pfisterer of University Hospital, Basel, Switzerland; and Dr. Detrano of V.A. Medical Center, Long Beach and Cleveland Clinic Foundation [43]. Each database contains the same 76 attributes. The "goal" field refers to the presence of heart disease in the patient. The classification attempts to distinguish presence (values 1,2,3,4, involving a total of 509 subjects) from absence (value 0, involving 411 subjects) [96]. The attributes include demographics, physio-cardiovascular conditions, traditional risk factors, family history, personal lifestyle, and cardiovascular exercise measurements. This data set has posed some challenges to past analysis via various classification approaches, resulting in less than 80% correct classification. Applying our classification model without reserved judgment, we obtain 79% and 85% correct classification for each group respectively. To gauze the usefulness of multi-stage analysis, we apply 2-stage classification. In the first stage,

14 attributes were selected as discriminatory. 135 Group absence subjects were placed into the reserved judgment region, with 85% of the remaining were classified as Group absence correctly; while 286 Group presence subjects were placed into the reserved judgment region, and 91% of the remaining classified correctly into the Group presence. In the second stage, 11 attributes were selected with 100 and 229 classified into Group absence and presence respectively. Combining the two stages, we obtained a correct classification of 82% and 85% respectively for diagnosis of absence or presence of heart disease. Fig. 1 illustrates the 2-stage classification.

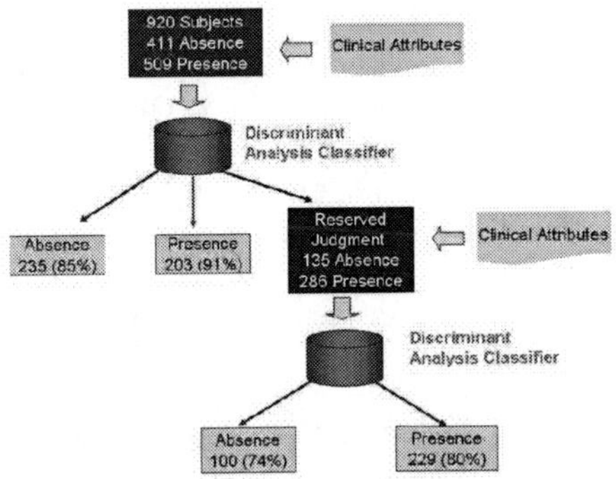

FIGURE 1. A tree diagram for 2-stage classification and prediction of heart disease

Predicting Aberrant CpG Island Methylation in Human Cancer More details of this work can be found in [45, 46]. Epigenetic silencing associated with aberrant methylation of promoter region CpG islands is one mechanism leading to loss of tumor suppressor function in human cancer. Profiling of CpG island methylation indicates that some genes are more frequently methylated than others, and that each tumor type is associated with a unique set of methylated genes. However, little is known about why certain genes succumb to this aberrant event. To address this question, we used Restriction Landmark Genome Scanning (RLGS) to analyze the susceptibility of 1749 unselected CpG islands to de novo methylation driven by overexpression of DNMT1. We found that, whereas the overall incidence of CpG island methylation was increased in cells overexpressing DNMT1, not all loci were equally affected. The majority of CpG islands (69.9%) were resistant to de novo methylation, regardless of DNMT1 overexpression. In contrast, we identified a subset of methylation-prone CpG islands (3.8%) that were consistently hypermethylated in multiple DNMT1 overexpressing clones. Methylation-prone and methylation-resistant CpG islands were not significantly different with respect to size, C+G content, CpG frequency, chromosomal location, or gene- or promoter-association. To discriminate methylation-prone from methylation-resistant CpG islands, we developed a novel DNA pattern recognition model and algorithm [97], and coupled our predictive model described herein with the patterns found. We were able to derive a classification function based on the frequency of seven novel sequence patterns that was capable of discriminating methylation-prone from methylation-resistant CpG islands with 90% correctness upon cross-validation, and 85% accuracy when tested against blind CpG islands unknown to us on the methylation status. The data indicate that CpG islands differ in their intrinsic susceptibility to de novo methylation, and suggest that the propensity for a CpG island to become aberrantly methylated can be predicted based on its sequence context.

The significance of this research is two-fold. First, the identification of sequence patterns/attributes that distinguish methylation-prone CpG islands will lead to a better understanding of the basic mechanisms underlying aberrant CpG island methylation. Because genes that are silenced by methylation are otherwise structurally sound, the potential for reactivating these genes by blocking or reversing the methylation process represents an exciting new molecular target for chemotherapeutic intervention. A better understanding of the factors that contribute to aberrant methylation, including the identification of sequence elements that may act to target aberrant methylation, will be an important step in achieving this long-term goal. Secondly, the classification of the more than 29,000 known (but as yet unclassified) CpG islands in human chromosomes will provide an important resource for the identification of novel gene targets for further study as potential molecular markers that could impact on both cancer prevention and treatment. Extensive RLGS fingerprint information (and thus potential training sets of methylated CpG islands) already exists for a number of human tumor types, including breast, brain, lung, leukemias, hepatocellular carcinomas, and PNET [98, 99, 100, 101]. Thus, the methods and tools developed are directly applicable to CpG island methylation data derived from human tumors. Moreover, new microarray-based techniques capable of 'profiling' more than 7000 CpG islands have been developed and applied to human breast cancers [102, 103, 104]. We are uniquely poised to take advantage of the tumor CpG island methylation profile information that will likely be generated using these techniques over the next several years. Thus, our general-predictive modeling framework has the potential to lead to improved diagnosis and prognosis and treatment planning for cancer patients.

Discriminant Analysis of Cell Motility and Morphology Data in Human Lung Carcinoma Refer to [44] for more detail of this work. This study focuses on the differential effects of extracellular matrix proteins on the motility and morphology of human lung epidermoid carcinoma cells. The behavior of carcinoma cells is contrasted with that of normal L-132 cells, resulting in a method for the prediction of metastatic potential. Data collected from time-lapsed videomicroscopy were used to simultaneously produce quantitative measures of motility and morphology. The data were subsequently analyzed using our discriminant analysis model and algorithm to discover relationships between motility, morphology, and substratum. Our discriminant analysis tools enabled the consideration of many more cell attributes than is customary in cell motility studies. The observations correlate with behaviors seen in vivo and suggest specific roles for the extracellular matrix proteins and their integrin receptors in metastasis. Cell translocation in vitro has been associated with malignancy, as has an elongated phenotype [105] and a rounded phenotype [106]. Our study suggests that extracellular matrix proteins contribute in different ways to the malignancy of cancer cells, and that multiple malignant phenotypes exist.

Ultrasonic Assisted Cell Disruption for Drug Delivery Reference [48] discusses this in detail. Although biological effects of ultrasound must be avoided for safe diagnostic applications, ultrasound's ability to disrupt cell membranes has attracted interest as a method to facilitate drug and gene delivery. This preliminary study seeks to develop rules for predicting the degree of cell membrane disruption based on specified ultrasound parameters and measured acoustic signals. Too much ultrasound destroys cells, while cell membranes will not open up for absorption of macromolecules when too little ultrasound is applied. The key is to increase cell permeability to allow absorption of macromolecules, and to apply ultrasound transiently to disrupt viable cells so as to enable exogenous material to enter without cell damage. Thus our task is to uncover a "predictive rule" of ultrasound-mediated disruption of red blood cells using acoustic spectrums and measurements of cell permeability recorded in experiments.

Our predictive model and solver for generating prediction rules are applied to data obtained from a sequence of experiments on bovine red blood cells. For each experiment, the attributes consist of 4 ultrasound parameters, acoustic measurements at 400 frequencies, and a measure of cell membrane disruption. To avoid over-training, various feature combinations of the 404 predictor variables are selected when developing the classification rule. The results indicate that the variable combination consisting of ultrasound exposure time and acoustic signals measured at the driving frequency and its higher harmonics yields the best rule, and our method compares favorably with classification tree and other ad hoc approaches, with correct classification rate of 80% upon cross-validation and 85% when classifying new unknown entities. Our methods used for deriving the prediction rules are broadly applicable, and could be used to develop prediction rules in other scenarios involving different cell types or tissues. These rules and the methods used to derive them could be used for real-time feedback about ultrasound's biological effects. For example, it could assist clinicians during a drug delivery process, or could be imported into an implantable device inside the body for automatic drug delivery and monitoring.

Identification of Tumor Shape and Volume in Treatment of Sarcoma Reference [47] includes the detail analysis. This project involves the determination of tumor shape for adjuvant brachytherapy treatment of sarcoma, based on catheter images taken after surgery. In this application, the entities are overlapping consecutive triplets of catheter markings, each of which is used for determining the shape of the tumor contour. The triplets are to be classified into one of two groups: Group 1 = [triplets for which the middle catheter marking should be bypassed], and Group 2 = [triplets for which the middle marking should not be bypassed]. To develop and validate a classification rule, we used clinical data collected from fifteen soft tissue sarcoma (STS) patients. Cumulatively, this comprised 620 triplets of catheter markings. By careful (and tedious) clinical analysis of the geometry of these triplets, 65 were determined to belong to Group 1, the "bypass" group, and 555 were determined to belong to Group 2, the "do-not-bypass" group.

A set of measurements associated with each triplet is then determined. The choice of what attributes to measure to best distinguish triplets as belonging to Group 1 or Group

2 is non trivial. The attributes involved distance between each pair of markings, angles, curvature formed by the three triplet markings. Based on the selected attributes, our predictive model was used to develop a classification rule. The resulting rule provides 98% correct classification on cross-validation, and was capable of correctly determining/predicting 95% of the shape of the tumor on new patients' data. We remark that the current clinical procedure requires manual outline based on markers in films of the tumor volume. This study was the first to use automatic construction of tumor shape for sarcoma adjuvant brachytherapy [47, 107].

Discriminant Analysis of Biomarkers for Prediction of Early Atherosclerosis More detail on this work can be found in [50]. Oxidative stress is an important etiologic factor in the pathogenesis of vascular disease. Oxidative stress results from an imbalance between injurious oxidant and protective antioxidant events in which the former predominate [108, 109]. This results in the modification of proteins and DNA, alteration in gene expression, promotion of inflammation, and deterioration in endothelial function in the vessel wall, all processes that ultimately trigger or exacerbate the atherosclerotic process [110, 111]. It was hypothesized that novel biomarkers of oxidative stress would predict early atherosclerosis in a relatively healthy non-smoking population who are free from cardiovascular disease. One hundred and twenty seven healthy non-smokers, without known clinical atherosclerosis had carotid intima media thickness (IMT) measured using ultrasound. Plasma oxidative stress was estimated by measuring plasma lipid hydroperoxides using the determination of reactive oxygen metabolites (d-ROMs) test. Clinical measurements include traditional risk factors including age, sex, low density lipoprotein (LDL), high density lipoprotein (HDL), triglycerides, cholesterol, body-mass-index (BMI), hypertension, diabetes mellitus, smoking history, family history of CAD, Framingham risk score, and Hs-CRP.

For this prediction, the patients are first clustered into two groups: (Group 1: IMT $>=$ 0.68, Group 2: IMT $<$ 0.68). Based on this separator 30 patients belong to Group 1, and 97 belong to Group 2. Through each iteration, the classification method trains and learns from the input training set and returns the most discriminatory patterns among the 14 clinical measurements; ultimately resulting in the development of a prediction rule based on observed values of these discriminatory patterns among the patient data. Using all 127 patients as a training set, the predictive model identified age, sex, BMI, HDLc, Fhx CAD $<$ 60, hs-CRP and d-ROM as discriminatory attributes that together provide unbiased correct classification of 90% and 93%, respectively, for Group 1 (IMT $>=$ 0.68) and Group 2 patients (IMT $<$ 0.68). To further test the power of the classification method for correctly predicting the IMT status on new/unseen patients, we randomly selected a smaller patient training set of size 90. The predictive rule from this training set yields 80% and 89% correct rates for predicting the remaining 37 patients into Group 1 and Group 2, respectively. The importance of d-ROM as a discriminatory predictor for IMT status was confirmed during the machine learning process, this biomarker was selected in every iteration as the "machine" learned and trained to develop a predictive rule to correctly classify patients in the training set. We also performed predictive analysis using Framingham Risk Score and d-ROM; in this case the unbiased correct classification rates (for the 127 individuals) for Groups 1 and 2 are 77% and 84%, respectively. This is the first study to illustrate that this measure of oxidative stress can be effectively used along with traditional risk factors to generate a predictive rule that can potentially serve as an inexpensive clinical diagnostic tool for prediction of early atherosclerosis.

Fingerprinting Native and Angiogenic Microvascular Networks through Pattern Recognition and Discriminant Analysis of Functional Perfusion Data [51] describes the analysis anf findings. The cardiovascular system provides oxygen and nutrients to the entire body. Pathological conditions that impair normal microvascular perfusion can result in tissue ischemia, with potentially serious clinical effects. Conversely, development of new vascular structures fuels the progression of cancer, macular degeneration and atherosclerosis. Fluorescence-microangiography offers superb imaging of the functional perfusion of new and existent microvasculature, but quantitative analysis of the complex capillary patterns is challenging. We developed an automated pattern-recognition algorithm to systematically analyze the microvascular networks, and then apply our classification model herein to generate a predictive rule. The pattern-recognition algorithm identifies the complex vascular branching patterns, and the predictive rule demonstrates 100% and respectively 91% correct classification on perturbed (diseased) and normal tissue perfusion. We confirmed that transplantation of normal bone marrow to mice in which genetic deficiency resulted in impaired angiogenesis eliminated predicted differences and restored normal-tissue perfusion patterns (with 100% correctness). The pattern recognition and classification method offers an elegant solution for the automated fingerprinting of microvascular networks that could contribute to better understanding of angiogenic mechanisms and be utilized to diagnose and monitor microvascular deficiencies. Such information would be valuable for early detection and monitoring of functional abnormalities before they produce obvious and lasting effects, which may include improper perfusion of tissue, or support of tumor development.

The algorithm can be used to discriminate between the angiogenic response in a native healthy specimen compared to groups with impairment due to age, or chemical or other genetic deficiency. Similarly, it can be applied to analyze angiogenic responses as a result of various treatments. This will serve two important goals. First, the identification of discriminatory patterns/attributes that distinguish angiogenesis status will lead to a better understanding of the basic mechanisms underlying this process. Because therapeutic control of angiogenesis could influence physiological and pathological processes such as wound and tissue repairing, cancer progression and metastasis, or macular degeneration, the ability to understand it under different conditions will offer new insight in developing novel therapeutic interventions, monitoring and treatment, especially in aging, and heart disease. Thus, our study and the results form the foundation of a valuable diagnostic tool for changes in the functionality of the microvasculature and for discovery of drugs that alter the angiogenic response. The methods can be applied to tumor diagnosis, monitoring and prognosis. In particular, it will be possible to derive microangiographic fingerprints to acquire specific microvascular patterns associated with early stages of tumor development. Such "angioprinting" could become an extremely helpful early diagnostic modality, especially for easily accessible tumors such as skin cancer.

Prediction of Protein Localization Sites The protein localization database consists of 8 groups with a total of 336 instances (143, 77, 52, 35, 20, 5, 2, 2, respectively) with 7 attributes [96]. The 8 groups are 8 localization sites of protein, including cp (cytoplasm), im (inner membrane without signal sequence), pp (perisplasm), imU (inner membrane, uncleavable signal sequence), om (outer membrane), omL (outer membrane lipoprotein), imL (inner membrane lipoprotein), imS (inner membrane, cleavable signal sequence). However, the last 4 groups are taken out from our classification experiment since the population sizes are too small to ensure significance.

The 7 attributes include mcg (McGeoch's method for signal sequence recognition), gvh (von Heijne's method for signal sequence recognition), lip (von Heijne's Signal Peptidase II consensus sequence score), chg (Presence of charge on N-terminus of predicted lipoproteins), aac (score of discriminant analysis of the amino acid content of outer membrane and periplasmic proteins), alm1 (score of the ALOM membrane spanning region prediction program), and alm2 (score of ALOM program after excluding putative cleavable signal regions from the sequence).

In the classification we use 4 groups, 307 instances, with 7 attributes. Our classification model selected the discriminatory patterns mcg, gvh, alm1, and alm2 to form the predictive rule with unbiased correct classification rates of 89%, compared to the results of 81% by other classification models[112]

Pattern Recognition in Satellite Images for Determining Types of Soil The Satellite database consists of the multi-spectral values of pixels in 3x3 neighborhoods in a satellite image, and the classification associated with the central pixel in each neighborhood. The aim is to predict this classification, given the multi-spectral values. In the sample database, the class of a pixel is coded as a number. There are 6 groups with 4435 samples in the training dataset and 2000 samples in testing dataset; and each sample entity has 36 attributes describing the spectral bands of the image [96].

The original Landsat Multi-Spectral Scanner image data for this database was generated from data purchased from NASA by the Australian Centre for Remote Sensing. The Landsat satellite data is one of the many sources of information available for a scene. The interpretation of a scene by integrating spatial data of diverse types and resolutions including multispectral and radar data, maps indicating topography, land use etc. is expected to assume significant importance with the onset of an era characterized by integrative approaches to remote sensing (for example, NASA's Earth Observing System commencing this decade).

One frame of Landsat MSS imagery consists of four digital images of the same scene in different spectral bands. Two of these are in the visible region (corresponding approximately to green and red regions of the visible spectrum) and two are in the (near) infra-red. Each pixel is an 8-bit binary word, with 0 corresponding to black and 255 to white. The spatial resolution of a pixel is about 80m x 80m. Each image contains 2340 x 3380 such pixels.

The database is a (tiny) sub-area of a scene, consisting of 82 x 100 pixels. Each line of data corresponds to a 3x3 square neighborhood of pixels completely contained within the 82x100 sub-area. Each line contains the pixel values in the four spectral bands (converted to ASCII) of each of the 9 pixels in the 3x3 neighborhood and a number indicating the classification label of the central pixel. The number is a code for the following

6 groups: redsoil, cotton crop, grey soil, damp grey soil, soil with vegetation stubble, very damp grey soil. Running our classification model,17 discriminatory attributes were selected to form the classification rule, producing an unbiased prediction with 85% accuracy.

3.3. Further advances

Brooks and Lee 2007 [94, 113] devised other variations of the basic DAMIP model. They also showed that DAMIP is strongly universally consistent (in some sense) with very good rates of convergence from Vapnik and Chervonenkis Theory. A polynomial-time algorithm for discriminating between two populations with the DAMIP model was developed, and DAMIP was shown to be $\mathcal{NP}$-complete for a general number of groups. The proof demonstrating $\mathcal{NP}$-completeness employs results used in generating edges of the conflict graph [114, 115, 116, 117]. Exploiting the necessary and sufficient conditions that identify edges in the conflict graph is the central contribution to the improvement in solution performance over industry-standard software. The conflict graph is the basis for various valid inequalities, a branching scheme, and for conditions under which integer variables are fixed for all solutions. Additional solution methods are identified which include a heuristic for finding solutions at nodes in the branch-and-bound tree, upper bounds for model parameters, and necessary conditions for edges in the conflict hypergraph [118, 119]. Further, we have concluded that DAMIP is a computationally feasible, consistent, stable, robust, and accurate classifier.

4. PROGRESS AND CHALLENGES

We summarize below the mathematical programming techniques used in classification problems as reviewed in this chapter.

As noted by current research effort, multigroup classification remains $\mathcal{NP}$-complete and much work is needed to design effective models as well as to derive novel and efficient computational algorithms to solve these multigroup instances.

Table 2: Progress in mathematical programming-based classification models

Mathematical Programming Methods	**References (cite by published years**
Linear Programming	
Two-group classification:	
Separate data by hyperplanes	[10, 11]
Minimizing the sum of deviations (MSD), minimizing the maximum deviation (MMD), and minimizing the sum of interior distances (MSID)	[12, 7, 13, 14, 15, 20]
Hybrid model	[17, 20]
Review	[9, 19, 6]
Software	[21]
Issues about normalization	[22, 23, 24, 25, 29, 30, 27, 26, 28, 31]
Robust linear programming (RLP)	[32, 33]
Inclusion of second-order terms	[35, 36]
Effect of the position of outliers	[37]
Binary attributes	[38]
Multigroup classification:	
Single function classification	[13]
Multiple function classification	[39, 8]
Classification with reserved-judgment region using LP	[40, 41, 42, 43]
Mixed Integer Programming	
Two-group classification:	
Minimizing the number of misclassifications	[14, 56, 57, 52, 53, 58, 54, 55, 59]
Review	[9, 19, 6]
Software	[21]
Secondary goals	[60]
Binary attributes	[38]
Normalization and attribute selection	[61]
Dichotomous categorical variable formation	[62]
Multigroup classification:	
Multigroup classification	[63, 64]
Three-group classification	[65, 66, 67]
Classification with reserved-judgment region using MIP	[40, 41, 94, 49, 43]
Nonlinear Programming	
Two-group classification:	
L_p-norm criterion	[16]
Review	[9, 19, 6]
Piecewise-linear nonconvex discriminant function	[69]
Minimizing the number of misclassifications	[70, 71, 72]
Minimizing the sum of arbitrary-norm distances	[73]
Support vector machine	
Introduction and tutorial	[74, 75]
Generalized SVM	[76, 77]
Methods for huge-size problems	[78, 85, 83, 81, 82, 79, 84, 87, 80, 86]
Multigroup SVM	[40, 41, 91, 42, 92, 80, 94, 49, 43]

As noted by current research effort, multigroup classification remains $\mathcal{NP}$-complete and much work is needed to design effective models as well as to derive novel and efficient computational algorithms to solve these multigroup instances.

5. OTHER METHODS

While most classification methods can be described in terms of discriminant functions, some methods are not trained in the paradigm of determining coefficients or parameters for functions of a pre-defined form. These methods include *classification and regression trees (CART)*, *nearest-neighbor* methods, and *neural networks*.

Classification and regression trees [120] are nonparametric approaches to prediction. Classification trees seek to develop classification rules based on successive binary partitions of observations based on attribute values. Regression trees also employ rules consisting of binary partitions, but are used to predict continuous responses.

The rules generated by classification trees are easily viewable by plotting them in a tree-like structure from which the name arises. A test entity may be classified using rules in a tree plot by first comparing the entity's data with the root node of the tree. If the root node condition is satisfied by the data for a particular entity, the left branch is followed to another node; otherwise, the right branch is followed to another node. The data from the observation is compared to conditions at subsequent nodes until a leaf node is reached.

Nearest-neighbor methods begin by establishing a set of labeled prototype observations. The nearest-neighbor classification rule assigns test entities to groups according to the group membership of the nearest prototype. Different measures of distance may be used. The k-nearest-neighbor rule assigns entities to groups according to the group membership of the k nearest prototypes.

Neural networks are classification models that can also be interpreted in terms of discriminant functions, though they are used in a way that does not require finding an analytic form for the functions [1]. Neural networks are trained by considering one observation at a time, modifying the classification procedure slightly with each iteration.

6. CONCLUSIONS

In this chapter, we presented an overview of mathematical programming-based classification models, and analyzed their development and advances in recent years. Many mathematical programming methods are geared toward two-group analysis only, and performance is often compared to Fisher's linear discriminant, or Smith's quadratic discriminant. It has been noted that these methods can be used for multiple group analysis by finding $G(G-1)/2$ discriminants for each pair of groups ("one-against-one") or by finding G discriminants for each group versus the remaining data ("one-against-all"), but these approaches can lead to ambiguous classification rules [1].

Mathematical programming methods developed for multiple group analysis are described [39, 13, 40, 41, 63, 8, 42, 64]. Multiple group formulations for support vector machines have been proposed and tested [41, 79, 91, 92, 49, 43, 94], but are still considered computationally intensive [91]. The "one-against-one" and "one-against-all" methods with support vector machines have been successfully applied [91, 121].

We also discussed a class of multigroup general-purpose predictive models that we have developed based on the technology of large-scale optimization and support-vector machines [40, 41, 42, 49, 43, 94, 95]. Our models seek to maximize the correct classification rate while constraining the number of misclassifications in each group. The models incorporate the following features: 1) the ability to classify any number of distinct groups; 2) allow incorporation of heterogeneous types of attributes as input; 3) a high-dimensional data transformation that eliminates noise and errors in biological data; 4) constraining the misclassification in each group and a reserved-judgment region that provides a safeguard against over-training (which tends to lead to high misclassification rates from the resulting predictive rule); and 5) successive multi-stage classification capability to handle data points placed in the reserved judgment region. The performance and predictive power of the classification models is validated through a broad class of biological and medical applications.

Classification models are critical to medical advances as they can be used in genomic, cell, molecular, and system level analyses to assist in early prediction, diagnosis and detection of disease, as well as for intervention and monitoring. As shown in the CpG island study for human cancer, such prediction and diagnosis opens up novel therapeutic sites for early intervention. The ultrasound application illustrates its application to a novel drug delivery mechanism, assisting clinicians during a drug delivery process, or in devising implantable devices into the body for automated drug delivery and monitoring. The lung cancer cell motility offers an understanding of how cancer cells behave under different protein media, thus assisting in the identification of potential gene therapy and target treatment. Prediction of the shape of a cancer tumor bed provides a personalized treatment design, replacing manual estimates by sophisticated computer predictive models. Prediction of early atherosclerosis through inexpensive biomarker measurements and traditional risk factors can serve as a potential clinical diagnostic tool for routine physical and health maintenance, alerting doctors and patients to the need for early intervention to prevent serious vascular disease. Fingerprinting of microvascular networks opens up the possibility for early diagnosis of perturbed systems in the body that may trigger disease (e.g., genetic deficiency, diabetes, aging, obesity, macular degeneracy, tumor formation), identify target sites for treatment, and monitoring prognosis and success of treatment. Determining the type of Erythemato-Squamous disease and the presence/absence of heart disease helps clinicians to correctly diagnose and effectively treat patients. Thus classification models serve as a basis for predictive medicine where the desire is to diagnose early and provide personalized target intervention. This has the potential to reduce healthcare costs, improve success of treatment, and improve quality-of-life of patients.

ACKNOWLEDGMENTS

This research was partially supported by the National Science Foundation.

REFERENCES

1. R. O. Duda, P. E. Hart, and D. G. Stork, *Pattern Classification*, Wiley, New York, 2001.
2. C. Zopounidis, and M. Doumpos, *European Journal of Operational Research* **138**, 229–246 (2002).
3. A. O'Hagan, *Kendall's Advanced Theory of Statistics: Bayesian Inference*, vol. 2B, Halsted Press, New York, 1994.
4. R. Fisher, *Annals of Eugenics* 7, 179–188 (1936).
5. A. C. Rencher, *Multivariate Statistical Inference and Application*, Wiley, New York, 1998.
6. A. Stam, *Annals of Operations Research* **74**, 1–36 (1997).
7. N. Freed, and F. Glover, *Decision Sciences* **12**, 68–74 (1981).
8. W. Gochet, A. Stam, V. Srinivasan, and S. Chen, *Operations Research* **45**, 213–225 (1997).
9. E. Joachimsthaler, and A. Stam, *Multivariate Behavioral Research* **25**, 427–454 (1990).
10. O. Mangasarian, *Operations Research* **13**, 444–452 (1965).
11. O. Mangasarian, *IEEE Transactions on Information Theory* **14**, 801–807 (1968).
12. D. Hand, *Discrimination and classification*, John Wiley, New York, 1981.
13. N. Freed, and F. Glover, *European Journal of Operational Research* 7, 44–60 (1981).
14. S. Bajgier, and A. Hill, *Decision Sciences* **13**, 604–618 (1982).
15. N. Freed, and F. Glover, *Decision Sciences* **17**, 151–162 (1986).
16. A. Stam, and E. Joachimsthaler, *Decision Sciences* **20**, 285–293 (1989).
17. F. Glover, S. Keene, and B. Duea, *Decision Sciences* **19**, 269–280 (1988).
18. C. Smith, *Annals of Eugenics* **13**, 272–282 (1947).
19. S. Erenguc, and G. Koehler, *Managerial and Decision Economics* **11**, 215–225 (1990).
20. P. Rubin, *Decision Sciences* **21**, 373–386 (1990).
21. A. Stam, and D. Ungar, *European Journal of Operational Research* **86**, 374–388 (1995).
22. E. Markowski, and C. Markowski, *Decision Sciences* **16**, 237–247 (1985).
23. N. Freed, and F. Glover, *Decision Sciences* **17**, 589–595 (1986).
24. G. Koehler, *Decision Sciences* **20**, 239–257 (1989).
25. G. Koehler, *Decision Sciences* **20**, 844–848 (1989).
26. G. Koehler, *Decision Sciences* **25**, 331–333 (1994).
27. B. Xiao, *Decision Sciences* **24**, 699–712 (1993).
28. B. Xiao, *Decision Sciences* **25**, 335–336 (1994).
29. F. Glover, *Decision Sciences* **21**, 771–785 (1990).
30. P. Rubin, *Decision Sciences* **22**, 519–535 (1991).
31. B. Xiao, and Y. Feng, *Annals of Operations Research* **74**, 113–127 (1997).
32. K. Bennett, and O. Mangasarian, *Optimization Methods and Software* **1**, 23–34 (1992).
33. O. Mangasarian, W. Street, and W. Wolberg, *Operations Research* **43**, 570–577 (1995).
34. K. Bennett, "Decision tree construction via linear programming," in *Proceedings of the 4th Midwest Artificial Intelligence and Cognitive Science Society Conference*, edited by M. Evans, 1992, pp. 97–101.
35. A. D. Silva, and A. Stam, *European Journal of Operational Research* **72**, 4–22 (1994).
36. P. Wanarat, and R. Pavur, *European Journal of Operational Research* **93**, 582–601 (1996).
37. R. Pavur, *European Journal of Operational Research* **136**, 603–615 (2002).
38. O. Asparoukhov, and A. Stam, *Annals of Operations Research* **74**, 89–112 (1997).
39. K. Bennett, and O. Mangasarian, *Optimization Methods and Software* **3**, 27–39 (1994).
40. R. Gallagher, E. Lee, and D. Patterson, "An optimization model for constrained discriminant analysis and numerical experiments with iris, thyroid, and heart disease datasets," in *Proceedings of the 1996 American Medical Informatics Association*, 1996.
41. R. Gallagher, E. Lee, and D. Patterson, *Annals of Operations Research* **74**, 65–88 (1997).
42. E. Lee, R. Gallagher, and D. Patterson, *INFORMS Journal on Computing* **15**, 23–41 (2003).
43. E. Lee, *Annals of Biomedical Engineering, Systems Biology and Bioinformatics* **35**, 1095–1109 (2007).
44. J. P. Brooks, A. Wright, C. Zhu, and E. K. Lee, *Annals of Biomedical Engineering* (Submitted, 2007).
45. F. Feltus, E. Lee, J. Costello, C. Plass, and P. Vertino, *Proceedings of the National Academy of Sciences* **100**, 12253–12258 (2003).
46. F. Feltus, E. Lee, J. Costello, C. Plass, and P. Vertino, *Genomics* **87**, 572–579 (2006).

47. E. K. Lee, A. Y. Fung, J. P. Brooks, and M. Zaider, *Biology in Physics and Medicine* **47**, 1891–1910 (2002).
48. E. K. Lee, R. J. Gallagher, A. M. Campbell, and M. R. Prausnitz, *IEEE Transactions on Biomedical Engineering* **51**, 1–9 (2004).
49. E. Lee, "Discriminant analysis and predictive models in medicine," in *Interdisciplinary Research in Management Science, Finance, and HealthCare*, edited by S. Deng, Peking University Press, 2006, to appear.
50. E. Lee, T. Wu, S. Ashfaq, D. Jones, S. Rhodes, W. Weintrau, C. Hopper, V. Vaccarino, D. Harrison, and A. Quyyumi, Prediction of early atherosclerosis in healthy adults via novel markers of oxidative stress and d-ROMs (2007), working paper.
51. E. Lee, S. Jagannathan, C. Johnson, and Z. Galis, Fingerprinting native and angiogenic microvascular networks through pattern recognition and discriminant analysis of functional perfusion data (2006), submitted.
52. W. Banks, and P. Abad, *Decision Sciences* **22**, 1008–1023 (1991).
53. P. Abad, and W. Banks, *European Journal of Operational Research* **67**, 88–100 (1993).
54. A. D. Silva, and A. Stam, *Annals of Operations Research* **74**, 129–157 (1997).
55. P. Rubin, *Annals of Operations Research* **74**, 51–64 (1997).
56. A. Stam, and E. Joachimsthaler, *European Journal of Operational Research* **46**, 113–122 (1990).
57. G. Koehler, and S. Erenguc, *Decision Sciences* **21**, 63–85 (1990).
58. W. Banks, and P. Abad, *European Journal of Operational Research* **74**, 23–28 (1994).
59. N. Yanev, and S. Balev, *European Journal of Operational Research* **115**, 339–350 (1999).
60. R. Pavur, P. Wanarat, and C. Loucopoulos, *Annals of Operations Research* **74**, 173–189 (1997).
61. J. Glen, *Journal of the Operational Research Society* **50**, 1043–1053 (1999).
62. J. Glen, *Naval Research Logistics* **51**, 575–596 (2004).
63. W. Gehrlein, *Operations Research Letters* **5**, 299–304 (1986).
64. R. Pavur, *Annals of Operations Research* **74**, 37–50 (1997).
65. C. Loucopoulos, and R. Pavur, *Computers and Operations Research* **24**, 179–191 (1997).
66. C. Loucopoulos, and R. Pavur, *Annals of Operations Research* **74**, 191–209 (1997).
67. R. Pavur, and C. Loucopoulos, *Journal of the Operational Research Society* **52**, 896–904 (2001).
68. J. Anderson, *Journal of the Royal Statistical Society. Series B (Methodological)* **31**, 123–139 (1969).
69. O. Mangasarian, R. Setiono, and W. Wolberg, "Pattern recognition via linear programming: Theory and application to medical diagnosis," in *Large-Scale Numerical Optimization*, edited by T. Coleman, and Y. Li, SIAM, Philadelphia, Pennsylvania, 1990, pp. 22–31.
70. O. Mangasarian, *Journal of Global Optimization* **5**, 309–323 (1994).
71. O. Mangasarian, "Machine learning via polyhedral concave minimization," in *Applied Mathematics and Parallel computing – Festschrift for Klaus Ritter*, edited by H. Fischer, B. Riedmueller, and S. Schaeffler, Physica-Verlag, Germany, 1996, pp. 175–188.
72. C. Chen, and O. Mangasarian, *Advances in Computational Mathematics* **5**, 127–136 (1996).
73. O. Mangasarian, *Operations Research Letters* **24**, 15–23 (1999).
74. V. Vapnik, *The Nature of Statistical Learning Theory*, Springer Verlag, New York, 1995.
75. C. Burges, *Data Mining and Knowledge Discovery* **2**, 121–167 (1998).
76. O. Mangasarian, "Generalized support vector machines," in *Advances in Large Margin Classifiers*, edited by A. Smola, P. Bartlett, B. Schökopf, and D. Schuurmans, MIT Press, Cambridge, MA, 2000, pp. 135–146.
77. O. Mangasarian, and D. Musicant, "Data discrimination via nonlinear generalized support vector machines," in *Complementarity: Applications, Algorithms and Extensions*, edited by M. Ferris, O. Mangasarian, and J.-S. Pang, Kluwer Academic Publishers, Boston, MA, 2001, pp. 233–251.
78. O. Mangasarian, and D. Musicant, *IEEE Transactions on Neural Networks* **10**, 1032–1037 (1999).
79. G. Fung, and O. Mangasarian, "Proximal support vector machine classifiers," in *Proceedings KDD-2001*, San Francisco, 2001.
80. G. Fung, and O. Mangasarian, *Machine Learning* **59**, 77–97 (2005).
81. Y.-J. Lee, and O. Mangasarian, *Computational Optimization and Applications* **20**, 5–22 (2001).
82. Y.-J. Lee, and O. Mangasarian, "RSVM: Reduced support vector machines," in *Proceedings of the SIAM International Conference on Data Mining*, Chicago, 2001.
83. O. Mangasarian, and D. Musicant, *Journal of Machine Learning Research* **1**, 161–177 (2001).

84. G. Fung, and O. Mangasarian, "Incremental support vector machine classification," in *Proceedings of the Second SIAM International Conference on Data Mining*, edited by R. Grossman, H. Mannila, and R. Motwani, SIAM, Philadelphia, 2002, pp. 247–260.
85. P. Bradley, and O. Mangasarian, *Optimization Methods and Software* **13**, 1–10 (2000).
86. O. Mangasarian, *Optimization Methods and Software* **20**, 115–125 (2005).
87. O. Mangasarian, "Data mining via support vector machines," in *System Modeling and Optimization XX*, edited by E. Sachs, and R. Tichatschke, Kluwer Academic Publishers, Boston, 2003, pp. 91–112.
88. Y.-J. Lee, O. Mangasarian, and W. Wolberg, "Breast cancer survival and chemotherapy: A support vector machine analysis," in *DIMACS Series in Discrete Mathematical and Theoretical Computer Science*, American Mathematical Society, 2000, vol. 55, pp. 1–10.
89. Y.-J. Lee, O. Mangasarian, and W. Wolberg, *Computational Optimization and Applications* **25**, 151–166 (2003).
90. P. Luedi, A. Hartemink, and R. Jirtle, *Genome Research* **15**, 875–884 (2005).
91. C.-W. Hsu, and C.-J. Lin, *IEEE Transactions on Neural Networks* **13**, 415–425 (2002).
92. Y. Lee, Y. Lin, and G. Wahba, *Journal of the American Statistical Association* **99**, 67–81 (2004).
93. G. McLachlan, *Discriminant analysis and statistical pattern recognition*, Wiley, New York, 1992.
94. J. Brooks, and E. Lee, "Solving a mixed-integer programming formulation of a multi-category constrained discrimination model," in *Proceedings of the 2006 INFORMS Workshop on Artificial Intelligence and Data Mining*, Pittsburgh, PA, 2006.
95. J. Brooks, and E. Lee, "Mixed integer programming constrained discrimination model for credit screening," in *Proceedings of the 2007 Spring Simulation Multiconference, Business and Industry Symposium*, Norfolk, VA, 2007, pp. 1–6, aCM Digital Library.
96. P. Murphy, and D. Aha, UCI Repository of machine learning databases [http://www.ics.uci.edu/~mlearn/MLRepository.html], Department of Information and Computer Science, University of California, Irvine, California (1994).
97. E. Lee, T. Easton, and K. Gupta, *Annals of Operations Research, Operations Research in Medicine – Computing and Optimization in Medicine and Life Sciences* **148**, 167–187 (2006).
98. J. Costello, M. Fruhwald, D. Smiraglia, L. Rush, G. Robertson, X. Gao, F. Wright, J. Feramisco, P. Peltomaki, J. Lang, D. Schuller, L. Yu, C. Bloomfield, M. Caligiuri, A. Yates, R. Nishikawa, H. Su, N. Petrelli, X. Zhang, M. O'Dorisio, W. Held, W. Cavenee, and C. Plass, *Nature Genetics* **24**, 132–138 (2000).
99. J. Costello, C. Plass, and W. Cavenee, *Brain Tumor Pathology* **17**, 49–56 (2000).
100. M. Fruhwald, M. O'Dorisio, L. Rush, J. Reiter, D. Smiraglia, G. Wenger, J. Costello, P. White, R. Krahe, G. Brodeur, and C. Plass, *Journal of Medical Genetics* **37**, 501–509 (2000).
101. L. Rush, Z. Dai, D. Smiraglia, X. Gao, F. Wright, M. Fruhwald, J. Costello, W. Held, L. Yu, R. Krahe, J. Kolitz, C. Bloomfield, M. Caligiuri, and C. Plass, *Blood* **97**, 3226–3233 (2001).
102. G. Brock, T. Huang, C. Chen, and K. Johnson, *Nucleic Acids Research* **29**, e123 (2001).
103. P. Yan, C. Chen, H. Shi, F. Rahmatpanah, S. Wei, C. Caldwell, and T. Huang, *Cancer Research* **61**, 8375–8380 (2001).
104. P. Yan, M. Perry, D. Laux, A. Asare, C. Caldwell, and T. Huang, *Clinical Cancer Research* **6**, 1432–1438 (2000).
105. A. Zimmermann, and H. Keller, *Biomedicine & Pharmacotherapy* **41**, 337–344 (1987).
106. A. Raz, and A. Ben-Zéev, *Cancer and Metastasis Reviews* **6**, 3–21 (1987).
107. E. Lee, A. Fung, and M. Zaider, *International Journal of Radiation, Oncology, Biology and Physics* **51**, 391 (2001).
108. H. Sies, "Oxidative stress: introductory comments," in *Oxidative Stress*, edited by H. Sies, Academic Press, London, 1985, pp. 1–8.
109. J. McCord, *The American Journal of Medicine* **108**, 652–659 (2000).
110. M. Chevion, E. Berenshtein, and E. Stadtman, *Free Radical Research* **33(Suppl)**, S99–S108 (2000).
111. S. Tahara, M. Matsuo, and T. Kaneko, *Mechanisms of Ageing and Development* **122**, 415–426 (2001).
112. P. Horton, and K. Nakai, "A probablistic classification system for predicting the cellular localization sites of proteins," in *Proceedings of the Fourth International Conference on Intelligent Systems for Molecular Biology*, St. Louis, USA, 1996, pp. 109–115.

113. J. Brooks, and E. Lee, Analysis of the consistency of a mixed integer programming-based multi-category constrained discriminant model (2007), submitted.
114. R. E. Bixby, and E. K. Lee, *Operations Research* (1998).
115. E. K. Lee, *Solving a truck dispatching scheduling problem using branch-and-cut*, Ph.D. thesis, Computational and Applied Mathematics, Rice University, Houston, Texas (1993).
116. R. Borndörfer, *Aspects of set packing, partitioning and covering*, Ph.D. thesis, Technischen Universität Berlin, Berlin, Germany (1997).
117. A. Atamturk, *Conflict graphs and flow models for mixed-integer linear optimization problems*, Ph.D. thesis, School of Industrial and Systems Engineering, Georgia Institute of Technology, Atlanta, Georgia (1998).
118. T. Easton, K. Hooker, and E. K. Lee, *Mathematical Programming, Series B* **98**, 177–199 (2003).
119. E. K. Lee, and S. Maheshwary, Conflict hypergraphs in integer programming, Tech. rep., Georgia Institute of Technology (2006), submitted.
120. L. Breiman, J. H. Friedman, R. A. Olshen, and C. J. Stone, *Classification and Regression Trees*, Wadsworth and Brooks/Cole Advanced Books and Software, Pacific Grove, CA, 1984.
121. K.-R. Müller, S. Mika, G. Rätsch, K. Tsuda, and B. Schölkopf, *IEEE Transactions on Neural Networks* **12**, 181–201 (2001).

Employing optimization and sensitivity analyses tools to generate and analyze mathematical models of T cell signaling events

Maia M. Donahue[*], Wei Zhang[†], Marietta L. Harrison[¶], Jianghai Hu[†] and Ann E. Rundell[*]

[*] *Weldon School of Biomedical Engineering*
[†] *School of Electrical and Computer Engineering*
[¶] *Department of Medicinal Chemistry and Molecular Pharmacology*
Purdue University, West Lafayette, IN 47907, USA

Abstract. Approaches using optimization and sensitivity analysis to generate and analyze mathematical models relating to T cell signaling are reviewed and extended. Two approaches to rationally evolve a mathematical model of the signaling events associated with T cell receptor engagement are compared: one relies upon expert knowledge and the other is directed by sensitivity analyses. Both approaches extensively employ optimization to address model structure and parametric uncertainty through exogenous inputs and parameter identification. In isolation, the sensitivity analyses driven approach failed to converge to a model interpretable within the context of known signaling pathway information. This work suggests that expert knowledge supplemented with sensitivity analyses results may effectively guide model evolution through use of constrained optimization techniques. These methods are equally applicable to other cellular types even though they are presented within the context of T lymphocytes.

Keywords: Uncertain model structure and parameters, parameter identification, extended FAST sensitivity analysis, adjoint optimization methods, genetic algorithms, Erk MAPK pathway
PACS: 82.20.Wt, 82.39.Rt, 87.10.Ed, 87.17.Aa, 87.18.Mp

1. INTRODUCTION

Future advances in biomedicine will be directed, discovered, and supplemented through the use of optimization and systems analysis of mathematical models of cellular processes from biological systems. The current method of developing new drugs or therapies by pharmaceutical companies will eventually be cost prohibitive [1]. A recent study found that less than 0.02% of screened compounds become marketed drugs [1]. These drugs take an average of 15 years to develop and over 800 million dollars [1]. A new approach that utilizes a system-level understanding of the body, organs, tissues, and cells is required to efficiently design drugs with the highest impact and fewest side effects [2]. Important components of this approach are quantitative mathematical models. Some initial efforts have utilized models to pinpoint intervention points for drugs; future work will lead to their application for de novo drug design with a broader consideration of their systemic effects [2, 3].

CP953, *Data Mining, Systems Analysis, and Optimization in Biomedicine*
edited by O. Seref, O. E. Kundakcioglu, P. M. Pardalos

As our awareness of the inherent complexity associated with cellular systems increases, the limitations of intuition and exhaustive, exploratory experimental approaches become more obvious. Clearly, a quantitative approach would be beneficial to help systematically understand and engineer cellular processes. To date, numerous studies have focused on generating mathematical models of cellular processes and analyzing these models to understand behaviors. Many of these models have successfully predicted the response of cells to different stimuli. However, comparatively few have utilized these models for cellular engineering to manipulate the complex dynamics associated with cellular processes.

Mathematical models of the cellular processes are highly nonlinear and uncertain and may be time varying, hybrid and/or stochastic; these properties impede the systems analysis and to some extent prohibit the engineering design. To support systems analysis and engineering design, mathematical model genesis should consider the available analysis and optimization tools. For example, the construction of mathematical models for engineering applications routinely utilizes simple models and neglects unwarranted details. Although it is entirely possible to generate a very accurate model with hundreds of equations, the utility of the model may be limited. Analysis of overly complex models relies heavily upon numerous simulations to generate an understanding of the processes. Most analytical tools are not applicable to large, complex models. Simple models are preferred; in fact most engineering tools support the systems analysis and optimization for ordinary differential equation (ODE) models that are linear lumped-parameter mathematical models. Thus the genesis of the model must consider the desired application and evolve to provide sufficient accuracy for engineering design and accurate systems analysis.

This chapter describes a system analysis and optimization approach for generating mathematical models of T cell signaling that are suitable for supporting engineering analysis and design. The second section of this chapter briefly reviews the biological principles and events within the T cell that are initiated upon engagement of their receptors to provide a context for the rest of the document. The third section summarizes previously published modeling and analysis work related to these events. The subsequent section provides the theoretical background for the optimization and analysis techniques that support the genesis of appropriate ODE models. The fifth section provides an example of the evolution of a mathematical model of the early signaling events associated with T cell signaling utilizing embedded optimization and sensitivity analyses. The final section summarizes the findings and illustrates how the results and approaches are equally applicable to other cellular types and processes even though they are presented and discussed within the context of T lymphocytes.

2. T CELL SIGNALING AND ACTIVATION

The T cell (or T lymphocyte) will serve as the vehicle to illustrate and study the application of systems analysis and optimization techniques to mathematically model biological systems and cellular processes. The T cell is a crucial element of the human immune system that utilizes membrane bound receptors (TCRs) to identify and detect the presence of foreign particles (antigens), see Fig. 1. When antigenic peptides bind to these membrane-bound receptors, the T cells are activated by the biomechanical stimulus, initiating a sequence of biochemical events involving protein tyrosine kinase activation, and resulting in protein phosphorylation. This cascade of biochemical events is paralleled in many cells that are stimulated through membrane bound receptors. In fact, there is a nearly identical, analogous cascade, activated upon membrane-antibody binding on B cells, another element of the immune system. Portions of this T-cell signaling pathway, including the mitogen-activated protein kinases (MAPK), are ubiquitous to most cells, serving as a precursor to gene activation. Thus this document encompasses several highly conserved mechanisms that future mathematical models of the cell can build upon.

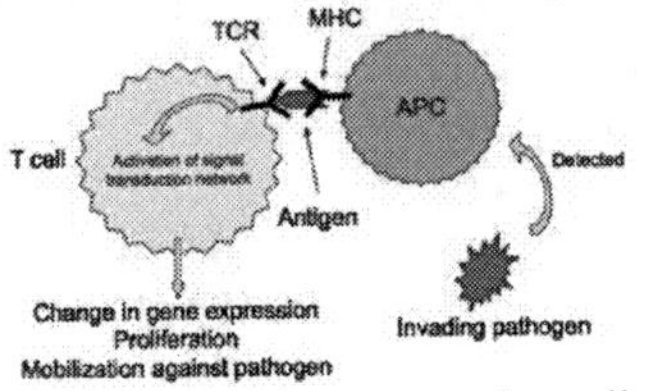

FIGURE 1. Activation of the T cell by an antigen-presenting cell (APC) in the immune system.

A biomechanical stimulus of the membrane receptors on the cell initiates a sequence of biochemical events (signal transduction) that serve to transmit, filter, and amplify the signal to the nucleus. Most signal transduction cascades consist of a sequence of reversible phosphorylation and dephosphorylation events. (Phosphorylation is the replacement of a hydroxl group by a phosphate group within the protein and dephosphorylation is the reversal of this process.) Each step of the cascade effectively integrates signals while transferring them to subsequent stages. Enzymes catalyze the chemical reactions to dramatically increase the rates and thereby increase the amount of substrate reacted upon in a short duration. Thus, enzymes ensure the signal is transmitted in a timely and amplified manner. Also, the specific nature of these enzymes enforces the stepwise process associated with the biochemical cascade. This specificity arises from protein kinases (enzymes that phosphorylate proteins in the presence of adenosine triphosphate, ATP) that act only upon a target serine, threonine, or tyrosine residues within a certain amino acid sequence in proteins. There are additional molecules involved in the biochemical cascade that are not enzymes, but are required for the signal transduction to proceed. These primarily serve as adaptor molecules providing binding sites (scaffolds) that generate the multi-protein complexes necessary for the signaling cascades to proceed.

The biochemical cascade is a carefully coordinated sequence of events, not simply a mixture of proteins that randomly collide. Proteins will not bind and react with just any other protein; there need to be some complementary features that promote binding between certain proteins. For the types of pathways under consideration, the dominant binding motifs identified thus far are the SH2 and SH3 domains on one protein binding to a phosphorylated tyrosine or a proline rich sequence, respectively, on a second protein. These pairings are critical for maintaining an ordered sequence of events and for promoting the next stage of the cascade. Extensive biochemical studies have identified many of the major molecular components and interactions of the T-cell signaling and activation pathways [4-8]. However, this molecular-level knowledge conveys little information on how the entire signaling network operates as a functional unit. Systems analysis, optimization, and modeling will help quantitatively understand the interactions and roles of the participating molecules.

Physiologic activation of the T cell signaling pathways begins with the binding of a specific antigen fragment complexed with a MHC (major histocompatibility complex) molecule to the TCR on the surface of the T cell and the activation of a protein kinase that initiates a sequence of biochemical events. (Each TCR is specific for its target antigen; different T cells express different TCRs). Simultaneous engagement of the T cell surface co-stimulatory molecule CD28 results in cell proliferation. Engagement of the TCR in the absence of CD28, results in cell anergy (inactivation) rather than cell proliferation. The Erk pathway is initiated when the TCR binds to its specific antigen. Upon engagement of the TCR, the constitutive receptor internalization, recycling, synthesis, and degradation process is augmented with active receptor phosphorylation and turnover during signaling [9, 10]. The bound TCR (TCRb) rapidly become phosphorylated receptors (TCRp). The phosphorylated receptors are internalized, and the internalized receptors (TCRi) are predominantly recycled back to the membrane while a fraction is degraded. Under saturating stimulation conditions, the recycled receptors are immediately bound and the cycle continues.

Fig. 2 presents a simplified view of the dominant biochemical interactions of the early signaling events and the MAPK pathway. The TCR engagement leads to the activation of Src-family protein tyrosine kinase (SFK) Lck/Fyn. Typically the SFK activation mechanism involves the coordinated interaction of a SFK-phosphorylated transmembrane scaffold protein Cbp, a SFK inhibitory kinase, Csk, and a membrane bound phosphatase, CD45. SFK is activated upon dephosphorylation at an inhibitory site followed by phosphorylation at an activation site. Activated SFK phosphorylates the TCR cytoplasmic tails. Phosphorylated TCR ζ binds to Zap70, allowing it to be phosphorylated by activated SFK at multiple tyrosine residues one of which is the activation site while another mediates its interaction with SFK [7]. A SFK-activated protein tyrosine phosphatase (PTP), SHP-1, can dephosphorylate SFK and Zap70 interfering with their association and activation. This early signal propagates through a number of membrane bound/recruited scaffold proteins including LAT and SLP76. Eventually, the small G protein Ras is activated by a number of factors, initiating the MAPK cascade, ending in the phosphorylation of Erk. The activation of Erk is necessary to activate transcription factors critical for T cell responses.

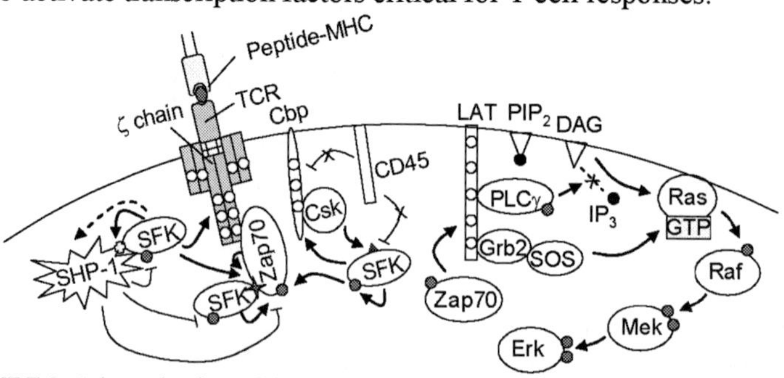

FIGURE 2. Schematic view of the TCR-activated Erk pathway. Arrows indicate protein catalyzed/facilitated phosphorylation or activation reactions while connectors with blunt ends indicate protein tyrosine phosphatase catalyzed dephosphorylation events. (Adapted from [11]).

In addition to the Erk MAPK pathway, other signaling cascades that serve major roles in transmitting extracellular signals to the nucleus but not described herein include the PLC-γ (phospholipase C) pathway, the Ca^{2+} activated pathway(s), and the PI3K (phosphoinositide 3-kinase) pathway (reviewed in [12-15]). The end result of successful signaling through these cascades is the regulation of nuclear transcription factors (e.g., AP-1, NF-AT, NF-κB, GATA-3) leading to gene expression (or in some cases repression). These pathways bifurcate and interconnect in complex patterns which mathematical models help to resolve and understand.

3. APPROACHES TO MODELING AND ANALYSIS OF T CELLS

Currently the field employs many different approaches to mathematically model cellular processes and phenomena: ordinary differential equations (ODEs), compartmental models, partial differential equations (PDEs), population balance (PB) models, stochastic models, Bayesian graphical models, Boolean/logical models, and hybrid systems. It has been noted that there is no obviously appropriate modeling framework for representing all cellular processes [16-18]; it is intimately linked with the application and the nature of the available experimental data. Despite attempts to base the models on the underlying biochemical and biophysical mechanisms, most of the models generated are phenomenological partially due to experimental limitations to simultaneously measure concentrations, rates, diffusion, etc. within a living intact cell [18-20]. A formal algorithmic method or process for generating families of models, or evolving models, that are compatible with prior knowledge and experimental data would benefit the community [18]. The modeling community agrees that utilizing black box models to fit the data serves little utility for cellular systems, as they do not provide insight into the underlying mechanisms and dynamics.

For a life scientist, the cell context is defined by biochemical and genetic pathways that map interactions between participating elements (similar to that in Fig. 2). Regardless of the modeling framework, these signaling diagrams constrain the mathematical model structure. The underlying biochemical and genetic signaling maps are continually revised as more interactions are discovered; many of these can be found at http://www.kegg.org, http://stke.sciencemag.org/, and http://doqcs.ncbs.res.in. These pathway diagrams have been painstakingly assembled primarily through protein interaction and knock-out studies. Engineering methods have been applied to help construct maps for unknown signaling pathways by utilizing network inference techniques to generate Bayesian graphical models where the nodes represent the measured signaling elements or events and the connections between the notes indicate the relationship between the measured elements [20, 21]. The drawback associated with this method is the quantity of data necessary to extract significant relationships, the computational expense, and its inability to identify feedback loops which are numerous and play significant regulatory roles in the signaling pathways.

ODEs and PDEs are widely used in modeling because they describe system time course dynamics and are easy to numerically simulate using differential equation solvers, such as Matlab, Mathematica, and Sundials. Although ODEs are preferred by the engineering community for systems analysis, optimization, and control studies, PDEs may more accurately reflect the signaling dynamics as they simulate the system as a function of time and space. Regardless of the approach, differential equations can over-simplify reactions and only approximate kinetic reactions. Stochastic models more realistically portray cellular processes by simulating random, discrete events. However, they are more computationally expensive to simulate and difficult to systematically analyze. Discrete spatial models are also stochastic, but add in equations that control the position of every element. These models require detailed biochemical information to be accurate and are the most complicated and computationally expensive. Logic networks do not use kinetic information, but instead calculate probable steady-state results based upon a predetermined tendency towards predefined events. These logical models require little computation but require significant a priori biochemical data to accurately determine probabilities of transitions. This chapter describes a method to derive ODE based models that are limited in complexity, retain fidelity to the underlying biochemical processes, and are fully compatible with the available experimental data.

The major published models related to T cell signaling dynamics are presented in Table 1. This table identifies the simulated cellular process, indicates the genesis of original experimental data to either generate or validate the model, states the role of optimization used in the model genesis or analysis, and summarizes any engineering or model analysis performed. Optimization occurrences in the table were utilized for parameter identification during the model construction or model structure minimization or variation. The optimizations for parameter identification were both global and local searches over pre-specified ranges. Optimization of model structure was performed through either model reduction techniques, using algorithms such as principal component analysis, flux analysis, and minimum cut sets, or varying reaction structures for optimal data fitting. In the case of Bayesian networks, optimization of model structure is built into the algorithms. Most of the model analyses performed were stability analyses, sensitivity analyses, and steady state analyses. Sensitivity analyses ranged from manual parameter manipulation to applications of global algorithms such as multi-parametric sensitivity analysis (MPSA).

TABLE 1. Summary of Published Optimization and Systems Analysis of Mathematical Models of T cells and Relevant Signaling Pathways

Source	Model Type	Process	Experimental results	Optimization performed	Model analysis
[22]	ODE	Receptor dynamics	None	None	Stability analysis
[5]	ODE	Signal transduction	Ca^{2+} with film-thinning apparatus	Local parameter ID	Local parameter sensitivity
[23]	ODE	Signal transduction	None	None	Stability analysis
[24]	ODE	Signal transduction	Sensitivity to small ΔSHP-1	None	Local parameter sensitivity
[25]	ODE	Signal transduction	Zap70 phosphorylation	Local parameter ID	Stability analysis, nonlinear dynamics
[11, 26]	ODE	Signal transduction	Receptor expression	Local and global parameter ID	GSA, stability analysis
[27]	ODE and Logical	Signal transduction	None	None	Stability analysis

[19]	ODE	Signal transduction	None	None	Stability analysis
[28]	ODE	Signal transduction	None	Hill coefficient estimation	Stability analysis
[29]	ODE	Signal transduction	None	None	Stability analysis
[30]	ODE	Cell differentiation	None	Parameter ID (method unpub)	Stability analysis
[31]	ODE	Signal transduction	Experimental validation of model	Parameter ID (method unpub)	Local parameter sensitivity analysis
[32]	ODE	Signal transduction	None	Model reduction and linearization	Steady state analysis
[33, 34]	ODE	Signal transduction	None	None	MPSA
[35]	ODE	Receptor kinetics	None	Local parameter ID, model structure variation	Local parameter sensitivity
[36]	ODE	Signal transduction	None	Model reduction	Global, time-varying parameter sensitivity analysis, principal component analysis, flux analysis
[37]	PDE	Receptor dynamics	None	None	Local parameter sensitivity
[38]	PDE	Receptor dynamics	Experimental validation of model	None	Local parameter sensitivity
[39]	PDE	Signal transduction	None	None	Control coefficients
[40]	Stochastic	Signal transduction	None	None	Local parameter sensitivity
[41]	Stochastic	Receptor dynamics	None	None	Local parameter sensitivity
[42]	Stochastic	Receptor dynamics	TCRp and pZap in knockout mice	None	Local parameter sensitivity
[43]	Logical	Signal transduction	None	None	Logical analysis
[44]	Logical	Signal transduction	None	Model structure variation	Logical analysis
[45]	Logical	Signal transduction	None	Minimal cut sets	Logical steady-state analysis, dependency matrix
[46]	Algebraic	Signal transduction	None	None	Local parameter sensitivity
[47]	Algebraic	Receptor dynamics	None	None	Local parameter sensitivity
[48]	Algebraic	Receptor dynamics	None	None	Local parameter sensitivity
[20]	Graphical	Signal transduction	Phosphorylation data	Bayesian network formation	None
[49]	Graphical	Signal transduction	None	None	Stability analysis
[50]	Graphical	Gene regulation	Their published cDNA data	Bayesian network formation	None
[21]	Graphical	Cell differen-tiation	Proliferation and phosphorylation data	Bayesian network formation	None

4. OPTIMIZATION AND SENSITIVITY ANALYSIS FUNDAMENTALS

As demonstrated in the previous section, mathematical model development has relied upon optimization techniques primarily for parameter identification and systems analysis predominantly for sensitivity analyses or stability analyses. This section generally describes constrained optimization and sensitivity analyses in an effort to illustrate how they direct and support model development in a systematic manner.

The mathematical models employed in the study of cellular processes, including the T-cell signaling and activation process studied in this chapter, can be generally categorized as dynamical systems. In such a system, a finite dimensional variable $x \in \Re^n$, called the state variable, is identified as a minimal set of variables needed to represent the entire state of the system at any given time; its time-course trajectory will thus summarize the dynamics of the underlying biological process. In addition to the state variable, other types of variables can also be identified. For example, input variable $i \in R^l$ consists of those variables that one can control to influence the evolution of the state variable, such as those modeling external stimuli and, in the case of the example studied in this chapter, the exogenous input introduced to account for the incompleteness of the model structure. The output variable $y \in \Re^q$ is a set of variables whose values can be measured through experiments. The progress of the state x in time can be described by an ODE called the dynamics equation:

$$\frac{dx}{dt} = f(p,i,x,t), \tag{1}$$

and the output can be determined from the state through a transformation:

$$y = g(p,x,t). \tag{2}$$

Here, $p \in \Re^m$ is a set of parameters that can affect both the state dynamics and the way the output y is determined. In stochastic models, random perturbations may also be introduced to the above two equations to accommodate the various uncertainty inherent in the system, resulting in stochastic differential equations (SDEs).

In this framework, after the state, the input and the output variables are identified, the goal of developing dynamical system models for a cellular process can be typically decomposed into two stages. In the first stage, the model structure, namely, the specific forms of the functions f and g, is determined by taking into consideration the nature of the underlying biochemical reaction processes. Some of the commonly used forms of f for this purpose are linear, bilinear, and rational functions [51]; thus the model structure is specified by the way that these forms are combined in the dynamics equation. Once a model structure has been settled, in the second stage, the value of the parameter p is then adjusted so that the system output y produced under this particular model structure and parameter value agrees with the experimentally measured data $\hat{y}$. It is possible that these two stages may be repeated multiple times before a satisfactory model can be reached. Optimization and sensitivity analysis can play a large role in directing this model generation as described in this section and further illustrated in the following example.

4.1. Sensitivity analysis

The mathematical models of cellular signaling processes typically have largely unknown parameter values as the models are not mechanistic and the parameters often account for the effects of multiple events. For example, many ODE model parameters account for diffusion without explicit representation of the process. Revisions, extensions, or adaptations of previously published models may require entirely new parameter values as the explicit addition or removal of a process or element may alter the interpretation of parameters. Hence, there is considerable uncertainty in all model parameter values. Given this uncertainty, it is important to understand how sensitive the model output is to variations in parameter values.

Conducting a sensitivity analysis not only quantifies the sensitivity of the output with regards to the model parameter values but also provides information for directing optimization (as described in the next section), model development and experiment design [26, 52]. The output of a sensitivity analysis helps to identify dominant processes or elements, recognizes events/elements that can be considered negligible, and suggests areas where future model modifications or additional experimental data would be valuable. Typically sensitivity analyses have been used for model reduction [34, 36]; we investigate their use in model genesis and parameter identification.

For large highly nonlinear models, the sensitivity of a model output to variations in a single parameter value is not apparent. This problem becomes more complex when considering that large models typically have numerous parameters, many of which may be uncertain. While a local sensitivity analysis determines the model sensitivity to parameter variations over a localized region around the nominal parameter values, a global sensitivity analysis (GSA) investigates the sensitivity over the entire parameter space. For a more complete review of sensitivity analysis of complex kinetic systems applicable towards chemical systems, see [53].

The effects from small changes in parameter values can be expressed by a Taylor series expansion about those changes with first order, second order, and higher order terms. For simplicity consider an algebraic expression $h(p)$ where $p \in \Re^m$, the expected value of the function for a small change in p is given by:

$$h(p+\Delta p) = h(p) + \sum_{j=1}^{m} \frac{\partial h}{\partial p_j}\Delta p_j + \frac{1}{2}\sum_{j=1}^{m}\sum_{i=1}^{m} \frac{\partial^2 h}{\partial p_j \partial p_i}\Delta p_i \Delta p_j + \cdots \tag{3}$$

This expansion is only valid in a small region of the parameter space about the nominal values. The local sensitivity can be computed exactly using deterministic methods that differentiate the model (reviewed in [53, 54]). The decoupled direct method, often applied to chemical systems, differentiates the mathematical model of the system with respect to each parameter and then solves the fully differentiated system [55]. The first order local sensitivities, namely the slope of the model output for small variations in one parameter about the nominal parameter values, can be estimated using a finite difference approximation. This "brute force" method computes the model output at least twice: once at the nominal parameter values and again with a small perturbation at one parameter value. The sensitivity is estimated as the change in the model output divided by the change in the parameter value:

$$\frac{\partial h}{\partial p_i} = \frac{h(p+\Delta p_i)-h(p)}{\Delta p_i} \tag{4}$$

This estimate can be inaccurate if the perturbation chosen is too large and so violates the local linearity assumption.

Statistical methods are generally used to estimate global sensitivities for models of cellular processes by varying numerous model parameters simultaneously to help understand how the model output varies over the entire parameter space and elucidate parametric interactions. These methods can be categorized as either regression or variance based [56]. Regression based methods are most applicable when each parameter is independent of the others; a rank-transformed regression method can accommodate nonlinear models but is only valid for monotonic models. Variance based methods do not make assumptions about the model type. However, these variance based models are computationally expensive. The variance methods can estimate the main-effect index, the contribution from a single parameter, and the total effects sensitivity index to account for interactions between parameters. A comparison of the relative parameter sensitivity rankings for the T-cell Erk MAPK pathway model [26] indicates that the regression based method, partial rank correlation coefficient (PRCC) [57-59], and variance based methods such as the Sobol's method [60] and extended Fourier amplitude sensitivity test (FAST) [61] methods give similar results. Not surprisingly, the extended FAST method was the most computationally efficient of the variance-based methods.

The extended FAST computes the main-effect and total-effect indices from numerous model outputs evaluated over the m-dimensional parameter space sampled along curves that vary in angular frequency for each parameter. This is defined by the following transformation function:

$$p_i = \frac{1}{2} + \frac{1}{\pi}\arcsin(\sin(\omega_i s + \varphi_i)) \tag{5}$$

where each parameter varies periodically at an angular frequency ω_i and random phase shift φ_i, with a scalar variable $s \in (-\pi, \pi)$. (The angular frequencies are chosen to be linearly independent from one another.) With this transformation, the model output function can be expressed as a Fourier series with respect to s. The FAST sensitivity indices are computed from the Fourier series coefficients [61]. The extended FAST sensitivity analysis is applied in the following example to quantify the parameter sensitivities and direct model development.

4.2. Optimization

For a given model structure, the problem of parameter identification can be solved by a method outlined in the following, which makes use of the sensitivity analysis results described in the previous section. The parameter identification problem is often formulated as an optimization problem, namely, finding the parameter values p for which the simulated model output y fits the experimental data $\hat{y}$ the "best". For example, in one possible model of the T cell signaling diagram, the system output consists of $q = 4$ components, corresponding to the four relevant signaling elements: percentages of TCR receptors expressed on the surface and internalized, and percentages of Zap70 and Erk phosphorylated. Let j be an index of these signals, and assume that measurements of the j-th output have been taken at n_j different times $t_{j,1}, t_{j,2}, \ldots, t_{j,n_j}$ to yield the data $\hat{y}_{j,1}, \hat{y}_{j,2}, \ldots, \hat{y}_{j,n_j}$. Then an objective function to measure how well the simulated outputs of the model fit the experimental data can be chosen as

$$J(p) = \sum_{j=1}^{q} \sum_{i=1}^{n_j} w_{j,i} [y(p, t_{j,i}) - \hat{y}_{j,i}]^2, \tag{6}$$

which is the weighted least squares error (WLSE) between the simulated outputs and the experimentally measured time course data with properly chosen weights $w_{j,i}$. The notation $y(p, t_{j,i})$ is adopted to highlight the dependency of the simulated output on the choice of the parameter value p: the output y depends on the state x, whose evolution is in turn affected by p. With the objective function defined above, the parameter identification problem can be formulated as

$$\begin{aligned} &\min && J(p) \\ &s.t. && \frac{dx}{dt} = f(p, i, x, t), y = g(p, x, t) \end{aligned} \tag{7}$$

which is a constrained optimization problem. The constraints are specified by state dynamics and the output function. Additional constraints may exist that bound the range of the parameter variations.

Many existing methods can be used to solve the above constrained optimization problem. However, most of them suffer from local minimum trap and poor convergence rate [62]. Since the generic algorithm (GA) excels at finding (rough) global minimum and gradient based methods can provide fast convergence rate, the combination of these two are adopted in this chapter: first a good initial guess is obtained using GA and then the solution is further improved using a gradient descent algorithm. To expedite the search in the second steps, the true gradient vector can be computed using the sensitivities of the states with respect to the parameters. Define $s_{jk}(p,t) = \partial x_j(p,t)/\partial p_k$ as the sensitivity of the state x_j with respect to parameter p_k at time t under parameter p. Taking the derivative of the equation $\dot{x}_j = f_j(p, x, t)$ with respect to p_k, leads to the following ODE:

$$\frac{ds_{jk}}{dt}(p,t) = \frac{\partial f_j}{\partial p_k}(p,x,t) + \sum_{i=1}^{n} \frac{\partial f_j}{\partial x_i}(p,x,t) \cdot s_{ik}(p,t). \tag{8}$$

Solutions to the above equations for all j and k yield the state sensitivity functions at the given parameter value p. The gradient of the objective function $J(p)$ with respective to the k-th parameter p_k can then be obtained as:

$$\frac{\partial J}{\partial p_k} = 2 \sum_{\alpha=1}^{q} \sum_{i=1}^{n_\alpha} w_{\alpha,i} [y(p, t_{\alpha,i}) - \hat{y}_{\alpha,i}] \cdot \left[\frac{\partial g(p, x, t_{\alpha,i})}{\partial p_k} + \sum_{j=1}^{n} \frac{\partial g(p, x, t_{\alpha,l})}{\partial x_j} \cdot s_{jk}(p,t) \right] \tag{9}$$

In the gradient descent algorithm, the sensitivity variables are first obtained from the initial guess p^{old}. The gradient of the objective function with respect to p is then computed using the above expression. The parameter is then updated as:

$$p^{new} = p^{old} + \delta \cdot \frac{\partial J}{\partial p}, \tag{10}$$

where δ is the adjustable step size. This process is repeated until certain convergence criterion is met (e.g., no significant improvement in the objective function).

Note that this approach of sensitivity analysis is a special case of the much more powerful technique called the adjoint method. The adjoint method, introduced by Lions in [63] and further developed by Jameson [64] in the context of aerodynamic design, computes the gradient of an objective function whose variables are subject to PDE constraints. One of its advantages is the flexibility with which optimization problems can be formulated. In the context of parameter identification of the mathematical modeling of cellular processes, several generalized optimization problems can be formulated and solved. For example, the system dynamics equation can admit time-varying parameter $p(t)$; thus for any given model structure, a much better fitting of the experimental data may be achieved by the optimized time-varying parameter than that achievable by a constant one. Another generalization is that the system dynamics may allow mode switching, resulting in hybrid system models [65]. For instance, as the concentration of certain signaling element varies, some reaction rates may change abruptly. We are currently exploring these directions.

5. MODEL EVOLUTION FOR T CELL SIGNALING EVENTS

This section provides an example of the generation of experimental data and the application of constrained optimization and sensitivity analyses to direct the evolution of an ODE model of T cell activation.

5.1. Data generation

All mathematical models of cellular processes require experimental data, whether knock-out experiments or detailed proteomic data. When feasible, previously published data is utilized to help determine interactions, concentrations, and kinetic parameters. More often than not, the necessary data to support model development is not available. A benefit of generating new data is self-consistency: each data point can be produced under the same conditions and with similar degrees of reliability.

To generate a model of T cell activation, detailed time-course data was required. Time-course data determines the degree of activation of an element over time. Ideally, data would be generated for each element of the model. However, investigators are limited by instrumentation, reagents, and indicators. For instance, the phosphorylation state of a protein can be determined using either western blots (which average the results over thousands of cells) or flow cytometry (which produces single-cell data);but both methods require reliable, phospho-specific antibodies.

For this example, the phosphorylation of two kinases, Zap70 and Erk, and TCR location were measured. Western blots were used to determine the percent phosphorylation of Zap70 and Erk over time, while flow cytometry was used to monitor the surface expression of TCR and the accumulation of a fluorescent antibody controlled by TCR internalization. Time points were chosen so that the majority of data was of early times but still covered 90 minutes.

Specifically, the human leukemia Jurkat E6.1 T cell line was utilized. Prior to stimulation, cells were resuspended in serum free media at 1×10^{8} cells/ml and incubated for 30 minutes at 37°C. Cells were then stimulated through the TCR by incubation with 10 μg/ml anti-CD3ε (UTCH1, BD Bioscience) on ice for 5 minutes. Cells were transferred to 37°C and incubated for different time durations up to 90 minutes. For quantitative SDS-PAGE and immunoblotting, cell lysates were analyzed on 12% SDS-PAGE gels followed by transfer to a nitrocellulose membrane. The membrane was stained with antibodies against human phospho-Zap70 Y319 (rabbit polyclonal, Cell Signaling Technology), phospho-Erk1/2 pT202/pY204 (rabbit polyclonal, Cell Signaling Technology) and α-tubulin (mouse monoclonal, Sigma), followed by incubation with Odessey® anti-mouse Alexa Fluor 680 and anti-rabbit IRDye 800 fluorescent secondary antibodies (LI-COR Biosciences). Treated membranes were scanned and fluorescent signals were quantified using an Odessey® infrared imager. For flow cytometry, following stimulation as above, cells were washed and incubated for 1hr at 4°C with FITC conjugated goat anti-mouse IgG (Santa Cruz Biotechnology) in staining buffer. Stained cells were washed once and resuspended in staining buffer at 2×10^6 cells/ml for flow cytometry analysis using a Beckman-Coulter Cytomics FC500 flow cytometer (Purdue Cancer Center Analytical Cytology facility). For intracellular TCR experiments, cells were stimulated as above with 10 ug/mL Alexa Fluor 700 conjugated anti-CD3 antibody (Biologic). At each time point, cells were then washed in stripping buffer, fixed, and analyzed on the FC500. Total surface expression levels were obtained by stimulating with conjugated antibody on ice and washing away excess label. The fluorescent antibody is assumed to neither leave the cell nor be degraded; therefore, the intracellular MFI represents the cumulative internalization of TCR.

5.2. ODE model development

A minimal signaling diagram was constructed to contain only the elements whose time course has been measured. This signaling diagram was a gross simplification that lumped many signaling processes (transport, multiple phosphorylation events, etc.) into individual events but still captured the information transfer between the signaling elements considered. The diagram was translated into a mathematical model with rates defined by the Law of Mass Action and simple chemical kinetics. This generated nonlinear ordinary differential equations with only a few rate parameters. The model could be represented as in Eqs. (1) and (2) without an exogenous input. The rate parameters associated with this minimal structure were largely unknown as they were not defined mechanistically. In the initial iteration, time course data existed to define the dynamics of each model variable so the dimension of the parameter space, m, was significantly less than the number of experimental data points, N. Thus, the parameter identification problem was well posed [66] and any inability to fit the data suggested the omission of important processes.

The discrepancies between the simulated and the experimental were addressed in the next model generation either by adding an omitted process or through the use of a piecewise linear exogenous input, $i \in R^l$. This exogenous input was temporarily utilized to mediate intermediate reactions and events to fit the available data. The variable time nodes of the exogenous input nodes were defined as (t_0, t_1,...$t_{l/2}$) with their corresponding magnitudes denoted by (i_0, i_1,...$i_{l/2}$) where $\frac{l}{2} \leq \left\lfloor \frac{(N-1)/2-m}{2} \right\rfloor$. Linear interpolation between time nodes computed the exogenous input value in the model. If the maximum allowable $l < 2$ and/or an acceptable fit was not obtainable with the model using the exogenous input, additional experimental data was obtained either from literature or the laboratory to further develop the model.

After fitting with an exogenous input, the function of the exogenous input was interpreted within the known signaling pathway and replaced by signaling element(s) or event(s) that serve a similar function. This process was iterated until the data was well fit and all possible exogenous inputs were interpreted as known signaling elements or processes. In some instances, the current state of knowledge was not capable of recreating the action of the exogenous input, thereby suggesting the presence of an unknown signaling event to investigate via future experiments. This approach encouraged the mathematical translation of the signaling pathways to be developed in a stepwise manner with well defined inputs and outputs, consistent with the approaches suggested in [7, 8]. The unknown model parameters and definition of the exogenous input in terms of node times and magnitudes were identified using local and global optimization techniques as described in the optimization section.

Model development evolved in two manners with the first guided by expert knowledge of the signaling pathway details and the second being directed through a sensitivity analysis. The sensitivity analysis identified the most sensitive portions of the model to address in order to remedy the inadequacies in the model fitness. The initial model was the same for both approaches.

5.3. Results

The process began with the creation of a basic diagram, Fig. 3(a), containing the measured elements: active Zap70 (Zap*), active Erk (Erk*), and three forms of TCR: internalized TCR, TCRi, bound TCR, TCRb, and phosphorylated TCR, TCRp. (TCRb and TCRp comprise the TCR surface expression data.) Translation of the signaling diagram created an initial model with unknown parameters. The fitted simulation results, shown in Fig. 4, matched the surface expression data well. However, after the initial time period, the simulation results deviated from the experimental data for the internalized receptors and the activated Zap70 and Erk dynamics. Receptor internalization, Zap70 and Erk activation were inappropriately sustained and required downregulation. Subsequently, two parallel approaches were taken and compared: evolution directed by expert knowledge and evolution directed by sensitivity analyses.

5.3.1. Evolution directed by expert knowledge

The evolution of the minimal models as directed by expert knowledge is illustrated in Fig. 3 (b) and (c). As receptor phosphorylation initiates and sustains receptor internalization and Zap70 activation, signaling mechanisms to mediate this process were reviewed. In the second iteration, an exogenous input, i, was incorporated to modify the TCRp dynamics and degradation of internalized receptors. (The degradation rate is affected by protein tyrosine kinase (PTK) activity during stimulation [67] while the recycling rate remains relatively constant.) A GA optimization problem was solved to define the exogenous input and determine the best fit model parameters. The resulting model dramatically improved the fitted dynamics (see Fig. 4). With the action of the exogenous input mediating the phosphorylation of the TCR, internalization ceased and Zap70 and Erk returned to basal activity levels following their initial activation.

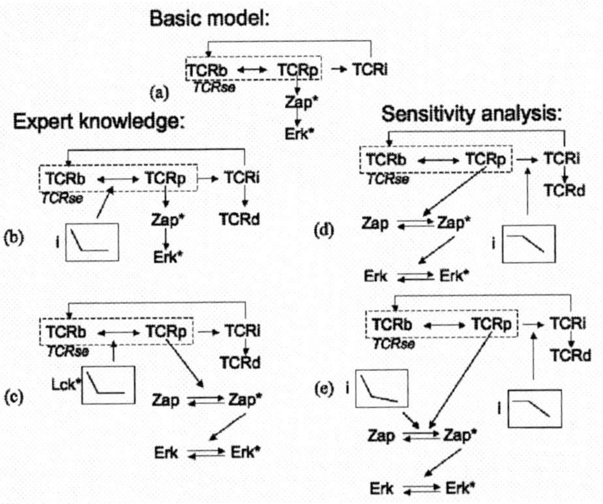

FIGURE 3. Evolution of system diagrams of minimal models. (a) Original, most basic model. (b),(c) Model evolution directed by expert knowledge only. (d), (e) Model evolution directed by sensitivity analysis only.

Not surprisingly, the fitted time course of the exogenous input qualitatively matched the time course of Lck, a known TCR kinase [68]. Therefore, in the next generation, Fig. 3 (c), the model was altered to more closely resemble the known signaling pathway elements and dynamics. Namely, the exogenous input was replaced by an Lck waveform with dynamics consistent with that reported in literature and the phosphorylation and dephosphorylation process of Zap70 and Erk activation and inactivation were incorporated. The resulting simulations after fitting by both the GA and the adjoint method are shown in Fig. 5. The parameters for this system were first obtained by GA, dashed lines, which have noticeable deviations from the experimentally measured data (points with error bars) (see Fig. 5). The bold lines are the simulated responses using the parameters obtained by the local gradient method. As can be seen, the latter represent a much better fit for the experimental data, demonstrating the utility of combining GA and gradient algorithms as described in the optimization section.

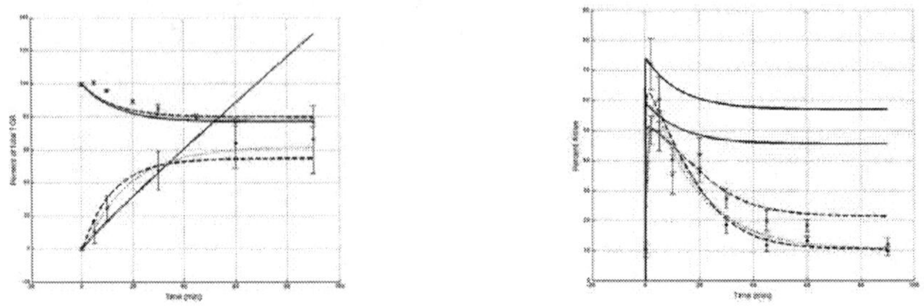

(a) (b)

FIGURE 4. Evolution directed by expert knowledge. Comparison of fitted minimal model A (continuous lines), B (dotted lines), and C (dashed lines) simulations with experimental data (points with error bars). (a) TCR dynamics. (b) Percent activation of Zap70 and Erk.

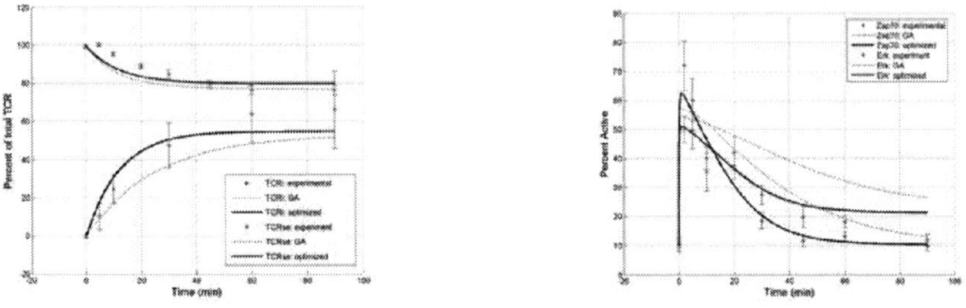

(a) (b)

FIGURE 5. Results from model C. Comparison of the simulated responses using the initial guess obtained by GA (dashed lines), using the parameters obtained by the local gradient descent algorithm (bold lines), and experimental data (points with error bars). (a) TCR kinetics, (b) Percent activation of Erk and Zap70.

5.3.2. Evolution directed by sensitivity analysis

A parallel approach utilized sensitivity analyses to direct model iterations. With this approach, illustrated in Fig. 3 (d) and (e), the next step was determined by the results of an extended FAST analysis as described in the sensitivity analysis section. Analysis of model A revealed it to be most sensitive to the receptor internalization rate parameter. The second generation model employed an exogenous input, *i*, to control internalization with physiological activation of Zap70 and Erk. GA optimization defined the exogenous input and found the model parameters. The exogenous input, i, achieved a closer fit to surface and cumulative internalized TCR without a significant improvement to the fitness of the Zap70 and Erk dynamics (Fig. 6, dotted lines). Ideally, the exogenous input would be replaced in the next generation model by a known signaling element or event, but this could not be achieved. An extended FAST sensitivity analysis of model D was performed to improve the Zap70 (and thereby Erk) fit. This analysis found Zap70 dynamics were slightly more sensitive to its phosphorylation parameter than all others. Therefore, a second exogenous input was added to phosphorylate Zap70 in Fig. 3 (e) with results shown in Fig. 6, dashed lines. This iteration required a more complicated input than used previously, and also decreased the dependence of Zap70 on TCRp by an order of magnitude from model D.

5.3.3. Comparison of approaches

Although both approaches to guiding model modifications resulted in reasonable fits to experimental data, the model evolution pathway diverged immediately. Expert knowledge directed the placement of the first exogenous input to the TCR phosphorylation process, while a sensitivity analysis indicated that manipulating the internalization process would most significantly alter the output. The simulations from the two approaches exhibited similar surface and cumulative internalization kinetics; however, as shown in Fig. 7, the simulated TCR phosphorylation differed dramatically. The expert knowledge approach downregulated TCRp while the sensitivity analysis approach allowed TCRp to remain elevated. Additional experimental data for TCRp is necessary to definitively determine which simulation was correct for TCRp, if either.

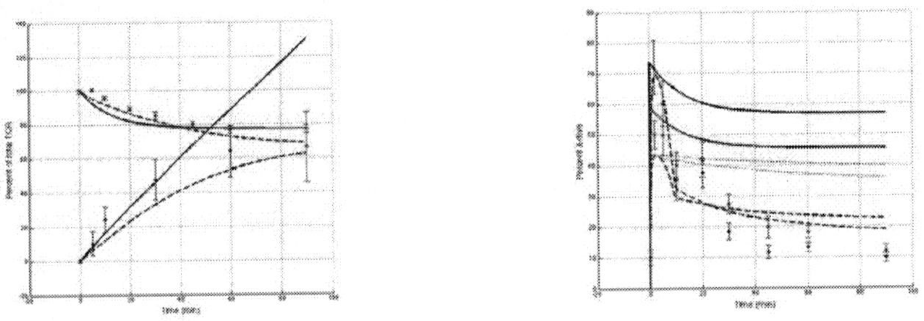

(a) (b)

FIGURE 6. Evolution directed by sensitivity analysis. Comparison of fitted minimal model A (continuous lines), D (dotted lines), and E (dashed lines) simulations with experimental data (points with error bars). (a) TCR dynamics, which did not vary between D and E. (b) Percent activation of Zap70 and Erk.

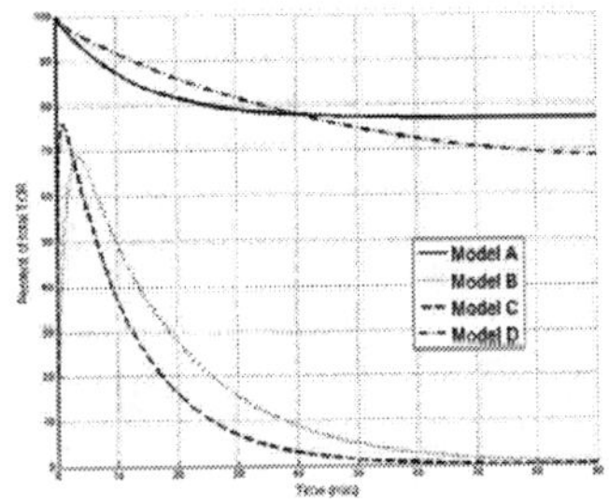

FIGURE 7. Comparison of evolution approaches. TCRp dynamics for each model.

The sensitivity analysis directed approach resulted in a more complicated model without better fits to experimental data. Furthermore, it was not possible to replace or interpret the identified exogenous inputs by known signaling elements or events as in the first approach. From this simple example, it is clear that sensitivity analyses should not be used in isolation to direct model genesis. Instead, expert knowledge supplemented by sensitivity analysis is more likely to succeed.

6. SUMMARY

The majority of published systems analysis and optimization methods conducted on mathematical models relating to T cell signaling dynamics were for parameter identification, model reduction, and stability, sensitivity and steady state analyses. Herein, a method to address model structural uncertainty through the extensive use of optimization techniques with exogenous inputs was proposed and evaluated. Under the guidance of expert knowledge, the resulting model captures the dominant biochemical events to explain the experimental observations. Although sensitivity analyses can identify dominant processes or elements, its direction of model modifications was not productive for these complex biological systems. Nevertheless, sensitivity analysis in support of expert knowledge-driven model evolution has significant potential. Thus, optimization and sensitivity analysis can help systematically evolve mathematical models that are appropriate for advancing biomedicine. The proposed process limits the model complexity, retains sufficient model accuracy, and utilizes expert knowledge to ensure the model structure is consistent with the known biochemical pathways. Although this methodology was developed for studying T cell signaling pathways, it is broadly applicable to other cell types and can accommodate a wide range of data types.

ACKNOWLEDGMENTS

This work was supported by NSF Grant CNS-0643805 and a NSF Graduate Research Fellowship.

REFERENCES

1. DiMasi, J.A., R.W. Hansen, and H.G. Grabowski, *The price of innovation: new estimates of drug development costs.* J Health Econ, 2003. **22**(2): p. 151-85.
2. Butcher, E.C., E.L. Berg, and E.J. Kunkel, *Systems biology in drug discovery.* Nat Biotechnol, 2004. **22**(10): p. 1253-9.
3. Gorska, M.M. and R. Alam, *Signaling molecules as therapeutic targets in allergic diseases.* J Allergy Clin Immunol, 2003. **112**(2): p. 241-50; quiz 251.
4. Germain, R.N. and I. Stefanova, *The dynamics of T cell receptor signaling: complex orchestration and the key roles of tempo and cooperation.* Annu Rev Immunol, 1999. **17**: p. 467-522.
5. Kim, S., S.M. Patrick, N.S. Braunstein, J.L. Thomas, and E.F. Leonard, *Modeling of early events in T cell signal transduction after controlled T cell activation by peptide major histocompatibility complex.* Ann Biomed Eng, 2001. **29**(5): p. 373-83.
6. Nel, A.E., *T-cell activation through the antigen receptor. Part 1: signaling components, signaling pathways, and signal integration at the T-cell antigen receptor synapse.* J Allergy Clin Immunol, 2002. **109**(5): p. 758-70.
7. Mustelin, T. and K. Tasken, *Positive and negative regulation of T-cell activation through kinases and phosphatases.* Biochem J, 2003. **371**(Pt 1): p. 15-27.
8. Veillette, A., S. Latour, and D. Davidson, *Negative regulation of immunoreceptor signaling.* Annu Rev Immunol, 2002. **20**: p. 669-707.

9. Naramura, M., et al., *c-Cbl and Cbl-b regulate T cell responsiveness by promoting ligand-induced TCR down-modulation.* Nat Immunol, 2002. **3**(12): p. 1192-9.
10. Alcover, A. and B. Alarcon, *Internalization and intracellular fate of TCR-CD3 complexes.* Crit Rev Immunol, 2000. **20**(4): p. 325-46.
11. Zheng, Y., *A systems biology study to delineate the T cell receptor activated ERK-MAPK signaling pathway*, in *BME*. 2005, Purdue University: West Lafayette. p. 200.
12. Leo, A. and B. Schraven, *Adapters in lymphocyte signalling.* Curr Opin Immunol, 2001. **13**(3): p. 307-16.
13. Ward, S.G. and D.A. Cantrell, *Phosphoinositide 3-kinases in T lymphocyte activation.* Current Opinion in Immunology, 2001. **13**(3): p. 332-338.
14. Latour, S., et al., *Bidirectional negative regulation of human T and dendritic cells by CD47 and its cognate receptor signal-regulator protein-alpha: Down-regulation of IL-12 responsiveness and inhibition of dendritic cell activation.* Journal of Immunology, 2001. **167**(5): p. 2547-2554.
15. Latour, S. and A. Veillette, *Proximal protein tyrosine kinases it immunoreceptor signaling.* Current Opinion in Immunology, 2001. **13**(3): p. 299-306.
16. Viswanathan, S. and P.W. Zandstra, *Towards predictive models of stem cell fate.* Cytotechnology, 2003. **41**(2-3): p. 75-92.
17. Sontag, E.D., *Molecular systems biology and control.* European Journal of Control, 2005. **11**(4-5): p. 396-435.
18. Aldridge, B.B., J.M. Burke, D.A. Lauffenburger, and P.K. Sorger, *Physicochemical modelling of cell signalling pathways.* Nature Cell Biology, 2006. **8**(11): p. 1195-1203.
19. Wolkenhauer, O., M. Ullah, P. Wellstead, and K.H. Cho, *The dynamic systems approach to control and regulation of intracellular networks.* Febs Letters, 2005. **579**(8): p. 1846-1853.
20. Sachs, K., O. Perez, D. Pe'er, D.A. Lauffenburger, and G.P. Nolan, *Causal protein-signaling networks derived from multiparameter single-cell data.* Science, 2005. **308**(5721): p. 523-529.
21. Woolf, P.J., W. Prudhomme, L. Daheron, G.Q. Daley, and D.A. Lauffenburger, *Bayesian analysis of signaling networks governing embryonic stem cell fate decisions.* Bioinformatics, 2005. **21**(6): p. 741-753.
22. Hlavacek, W.S., A. Redondo, C. Wofsy, and B. Goldstein, *Kinetic proofreading in receptor-mediated transduction of cellular signals: receptor aggregation, partially activated receptors, and cytosolic messengers.* Bull Math Biol, 2002. **64**(5): p. 887-911.
23. Chan, C., J. Stark, and A.J. George, *Feedback control of T-cell receptor activation.* Proc Biol Sci, 2004. **271**(1542): p. 931-9.
24. Altan-Bonnet, G. and R.N. Germain, *Modeling T cell antigen discrimination based on feedback control of digital ERK responses.* Plos Biology, 2005. **3**(11): p. 1925-1938.
25. Zheng, Y.N., et al., *Modeling and analysis of early events in T-lymphocyte antigen-activated intracellular-signaling pathways.* Journal of Computational and Applied Mathematics, 2005. **184**(1): p. 320-341.
26. Zheng, Y. and A. Rundell, *Comparative study of parameter sensitivity analyses of the TCR-activated Erk-MAPK signalling pathway.* Iee Proceedings Systems Biology, 2006. **153**(4): p. 201-211.
27. Mendoza, L., *A network model for the control of the differentiation process in Th cells.* Biosystems, 2006. **84**(2): p. 101-114.
28. Saez-Rodriguez, J., A. Kremling, H. Conzelmann, K. Bettenbrock, and E.D. Gilles, *Modular analysis of signal transduction networks.* Ieee Control Systems Magazine, 2004. **24**(4): p. 35-52.
29. Kholodenko, B.N., *Negative feedback and ultrasensitivity can bring about oscillations in the mitogen-activated protein kinase cascades.* European Journal of Biochemistry, 2000. **267**(6): p. 1583-1588.
30. Mariani, L., M. Lohning, A. Radbruch, and T. Hofer, *Transcriptional control networks of cell differentiation: insights from helper T lymphocytes.* Progress in Biophysics & Molecular Biology, 2004. **86**(1): p. 45-76.
31. Babu, C.V.S., S. Yoon, H.S. Nam, and Y.S. Yoo, *Simulation and sensitivity analysis of phosphorylation of EGFR signal transduction pathway in PC12 cell model.* Systems Biology, 2004. **1**(2): p. 213-221.
32. Conzelmann, H., et al., *Reduction of mathematical models of signal transduction networks: simulation-based approach applied to EGF receptor signalling.* Syst Biol (Stevenage), 2004. **1**(1): p. 159-69.
33. Zi, Z.K., et al., *In silico identification of the key components and steps in IFN-gamma induced JAK-STAT signaling pathway.* Febs Letters, 2005. **579**(5): p. 1101-1108.
34. Cho, K.-H., S.-Y. Shin, W. Kolch, and O. Wolkenhauer, *Experimental design in systems biology, based on parameter sensitivity analysis using a Monte Carlo method: a case study for the TNFalpha-mediated NF-kappaB signal transduction pathway.* SIMULATION, 2003. **79**(12): p. 726-739.
35. Sousa, J. and J. Carneiro, *A mathematical analysis of TCR serial triggering and down-regulation.* Eur J Immunol, 2000. **30**(11): p. 3219-27.
36. Liu, G., M.T. Swihart, and S. Neelamegham, *Sensitivity, principal component and flux analysis applied to signal transduction: the case of epidermal growth factor mediated signaling.* Bioinformatics, 2005. **21**(7): p. 1194-1202.

37. Coombs, D., A.M. Kalergis, S.G. Nathenson, C. Wofsy, and B. Goldstein, *Activated TCRs remain marked for internalization after dissociation from pMHC.* Nat Immunol, 2002. **3**(10): p. 926-31.
38. Gonzalez, P.A., et al., *T cell receptor binding kinetics required for T cell activation depend on the density of cognate ligand on the antigen-presenting cell.* Proceedings of the National Academy of Sciences of the United States of America, 2005. **102**(13): p. 4824-4829.
39. Kholodenko, B.N., G.C. Brown, and J.B. Hoek, *Diffusion control of protein phosphorylation in signal transduction pathways.* Biochemical Journal, 2000. **350**: p. 901-907.
40. Eide, J.L. and A.K. Chakraborty, *Effects of quenched and annealed macromolecular crowding elements on a simple model for signaling in T lymphocytes.* J Phys Chem B, 2006. **110**(5): p. 2318-24.
41. Chan, C., A.J. George, and J. Stark, *Cooperative enhancement of specificity in a lattice of T cell receptors.* Proc Natl Acad Sci U S A, 2001. **98**(10): p. 5758-63.
42. Lee, K.H., et al., *The immunological synapse balances T cell receptor signaling and degradation.* Science, 2003. **302**(5648): p. 1218-22.
43. Kaufman, M., F. Andris, and O. Leo, *A logical analysis of T cell activation and anergy.* Proc Natl Acad Sci U S A, 1999. **96**(7): p. 3894-9.
44. Sarkar, A. and B.R. Franza, *A logical analysis of the process of T cell activation: different consequences depending on the state of CD28 engagement.* J Theor Biol, 2004. **226**(4): p. 455-66.
45. Klamt, S., J. Saez-Rodriguez, J.A. Lindquist, L. Simeoni, and E.D. Gilles, *A methodology for the structural and functional analysis of signaling and regulatory networks.* BMC Bioinformatics, 2006. **7**: p. 56.
46. Heinrich, R., B.G. Neel, and T.A. Rapoport, *Mathematical models of protein kinase signal transduction.* Mol Cell, 2002. **9**(5): p. 957-70.
47. Rabinowitz, J.D., C. Beeson, D.S. Lyons, M.M. Davis, and H.M. McConnell, *Kinetic discrimination in T-cell activation.* Proc Natl Acad Sci U S A, 1996. **93**(4): p. 1401-5.
48. Wofsy, C., D. Coombs, and B. Goldstein, *Calculations show substantial serial engagement of T cell receptors.* Biophys J, 2001. **80**(2): p. 606-12.
49. Angeli, D., J.E. Ferrell, Jr., and E.D. Sontag, *Detection of multistability, bifurcations, and hysteresis in a large class of biological positive-feedback systems.* Proc Natl Acad Sci U S A, 2004. **101**(7): p. 1822-7.
50. Beal, M.J., F. Falciani, Z. Ghahramani, C. Rangel, and D.L. Wild, *A Bayesian approach to reconstructing genetic regulatory networks with hidden factors.* Bioinformatics, 2005. **21**(3): p. 349-56.
51. Fall, C.P., *Computational cell biology.* Interdisciplinary applied mathematics ; v. 20. 2002, New York: Springer. xx, 468 p.
52. Yue, H., et al., *Insights into the behaviour of systems biology models from dynamic sensitivty and identifiability analysis: a case study of an NF-kappa B signalling pathway.* Molecular Biosystems, 2006. **2**(12): p. 640-649.
53. Turanyi, T., *Sensitivity Analysis of Complex Kinetic Systems - Tools and Applications.* Journal of Mathematical Chemistry, 1990. **5**(3): p. 203-248.
54. Ionescu-Bujor, M. and D.G. Cacuci, *A comparative review of sensitivity and uncertainty analysis of large-scale systems - I: Deterministic methods.* Nuclear Science and Engineering, 2004. **147**(3): p. 189-203.
55. Dunker, A.M., *The Decoupled Direct Method for Calculating Sensitivity Coefficients in Chemical-Kinetics.* Journal of Chemical Physics, 1984. **81**(5): p. 2385-2393.
56. Frey, H.C. and S.R. Patil, *Identification and review of sensitivity analysis methods.* Risk Analysis, 2002. **22**(3): p. 553-578.
57. Draper, N. and H. Smith, *Applied Regression Analysis.* 2nd ed. 1981, New York: Wiley.
58. Blower, S.M. and H. Dowlatabadi, *Sensitivity and uncertainty analysis of complex-models of disease transmission - an HIV model, as an example.* International Statistical Review, 1994. **62**(2): p. 229-243.
59. Mckay, M.D., R.J. Beckman, and W.J. Conover, *Comparison of 3 methods for selecting values of input variables in the analysis of output from a computer code.* Technometrics, 1979. **21**(2): p. 239-245.
60. Sobol, I.M., *Global sensitivity indices for nonlinear mathematical models and their Monte Carlo estimates.* Mathematics and Computers in Simulation, 2001. **55**(1-3): p. 271-280.
61. Saltelli, A., S. Tarantola, and K.P.S. Chan, *A quantitative model-independent method for global sensitivity analysis of model output.* Technometrics, 1999. **41**(1): p. 39-56.
62. Esposito, W.R. and C.A. Floudas, *Global optimization for the parameter estimation of differential-algebraic systems.* Industrial & Engineering Chemistry Research, 2000. **39**(5): p. 1291-1310.
63. Lions, J.L., *Optimal control of systems governed by partial differential equations.* 1971, Berlin, New York,: Springer-Verlag. xi, 396 p.
64. Jameson, A., *Aerodynamic design via control theory.* J. of Scientific Computing, 1988. **3**: p. 322-260.
65. Raffard, R.L., J.H. Hu, and C.J. Tomlin, *Adjoint-based optimal control of the expected exit time for stochastic hybrid systems.* Hybrid Systems: Computation and Control, 2005. **3414**: p. 557-572.
66. Sontag, E.D., *For differential equations with r parameters, 2r+1 experiments are enough for identification.* Journal of Nonlinear Science, 2002. **12**(6): p. 553-583.

67. Geisler, C., *TCR trafficking in resting and stimulated T cells.* Critical Reviews in Immunology, 2004. **24**(1): p. 67-85.
68. Filipp, D., et al., *Regulation of Fyn through translocation of activated Lck into lipid rafts.* Journal of Experimental Medicine, 2003. **197**(9): p. 1221-1227.

Automated MR image processing and analysis of malignant brain tumors: enabling technology for data mining

Shishir Dube[*], Jason J. Corso[¶], Timothy F. Cloughesy[†], Suzie El-Saden[*], Alan L. Yuille[¥] and Usha Sinha[£]

[*] *Department of Biomedical Engineering, Medical Imaging Informatics Group, UCLA, 924 Westwood Blvd, Suite 420, Los Angeles, CA, 90024*
[¶] *Department of Computer Science, University at Buffalo SUNY, 201 Bell Hall, North Campus, Buffalo, NY, 14260*
[†] *Department of Neurology, UCLA, Box 951769, 1-230 RNRC, Los Angeles, CA, 90024*
[¥] *Department of Statistics, UCLA, Box 951563, 8967 MSB, Los Angeles, CA, 90024*
[£] *Department of Physics, San Diego State University, 5500 Campanile Drive, San Diego, CA, 92182*

Abstract. Glioblastoma multiforme (GBM) is a malignant brain cancer with poor patient prognosis (i.e. time to survival, time to tumor progression). A number of clinical trials are underway evaluating novel therapeutic strategies and magnetic resonance imaging is the most routinely performed procedure for accurate serial monitoring of patients. The electronic availability of the comprehensive data collected as part of the clinical trials provides an unprecedented opportunity to discover new relationships in complex diseases such as GBM. However, imaging data, which is the most accurate non-invasive assessment of GBMs, is not directly amenable for data mining. The focus of this chapter is on image analysis techniques including image spatial and intensity standardization, novel methods for robust tumor and edema segmentation, and quantification of tumor intensity, texture, and shape characteristics. The chapter concludes with an application of discovering the relationship between these quantitative image-derived features and time to survival in GBM patients; the data is part of a comprehensive large electronically accessible archive at UCLA (UCLA Neuro-oncology database).

Keywords: Data Mining, Image Segmentation, Classification, Pattern Recognition
PACS: 87.57.-s

1. INTRODUCTION

The high volume of brain imaging data that is routinely acquired clinically requires sophisticated tools in order to discover novel associations between imaging findings and other measures such as cognitive function, prognosis, therapeutic response, and genetic profiles. The field of data mining with an application to brain imaging is an area of considerable research efforts.

There has been a tremendous effort in developing robust data mining applications in brain imaging in the hopes of discovering novel associations that can give a better

CP953, *Data Mining, Systems Analysis, and Optimization in Biomedicine*
edited by O. Seref, O. E. Kundakcioglu, P. M. Pardalos

understanding of structure-function relationships. The important criteria for any large scale data mining application remains the automatic processing of image data including robust segmentations of the regions of interest. Feature extraction (location, size, surface area, volume, and shape) is fairly straight-forward, though some effort has to be directed at identifying relevant features. We focus, in this chapter, on the image processing techniques required for robust data mining of brain tumor MR imaging studies. We conclude the chapter with a report on quantitative image derived features of the tumor and edema and their correlation with prognosis.

2. BACKGROUND

2.1. Overview of existing brain image data mining studies

There has been much research done on discovering novel patterns in functional imaging and structural imaging. For functional brain mapping, there have been two main approaches: seeking associations between lesioned structures and connected neurological or neuropsychological deficits (i.e. brain tumor and deficits such as partial visual impairment); measuring brain activation in subjects as they perform certain tasks. Data mining on structural images primarily involves extracting features about the potential shape variability of brain structures and abnormalities that can occur, such as tumors or Multiple Sclerosis lesions [1].

Within data mining of functional MRI, there are several approaches that have been taken in the field. The scale-space method builds a multi-resolution representation of functional data and the voxels are organized in clusters that could be of different shapes and sizes at a variety of scales. The clusters are then hierarchically organized, and scales higher than the voxel level corresponds to the convolution of the original n-dimensional signal with a Gaussian kernel of width equivalent to the scale [2-4]. The multi-filtering approach performs analysis on smoothed and unsmoothed data to identify large regions of relatively weak activations [5]. Another analysis technique that does not involve smoothing of the data and directly operates on the image information is of the wavelet-domain type [6-8]. Hemodynamic response functions have also been convolved with the signal series at each voxel and used in temporal smoothing of fMRI. Usually the hemodynamic response functions are approximated by a Poisson function, Gamma function, a linear combination of the Gamma function and its temporal derivatives, or the Gaussian function. The hemodynamic response function approach identifies activated voxels, but the modeling of the function must be accurate and is an implicit limitation of the analysis [9,10].

Several analysis approaches exist in identifying significant activated voxels in functional imaging. One commonly used method is called statistical parametric mapping (SPM); it independently calculates the voxel changes and generates a map of statistic values for each voxel. To determine the significance of specific voxel activations, SPM can apply a variety of tests that include: t-test, F-test, correlation coefficient, or any univariate statistical parametric test [11,12]. Other analysis techniques involve correlation of observed signal changes with experimental blocks, analyzing voxel groups with an overall mean change in signal, incorporating prior information, Expectation-Maximization to estimate labeling parameters in a Markov random field model, or by using structural equation modeling to relate functional neuroimaging signals to underlying neurobiological behavior [1].

A data mining technique developed by Pokrajac et al. consisted of classifying three dimensional medical images, such as lesion-deficit data and fMRI activation data, by analyzing spatial distributions to model and characterizes the arrangement of the regions of interest. The probability distributions of the regions of interest were estimated by semi-parametric methods, such as the Expectation-Maximization algorithm. The two types of classification metrics used were common similarity measures (Kullback-Leibler, Mahalanobis, etc), and the maximum likelihood principle. The classification was performed at the voxel level and did not take into account local to global context explicitly. Another drawback is the dependency upon an accurate segmentation, and if a region has a fuzzy delineation for a diseased region, the classification techniques would not perform well [13]. Nielsen et al. developed an image retrieval system that would find similar functional neuroimaging images by converting the three dimensional locations in Talairach space, in a given experiment from the BrainMap Database, to a voxel-volume representation by convolving the location with a Gaussian kernel. A distance metric was applied to the voxelized feature set to find whether outliers exist; thus, the system could be used as a potential error check to ensure reproducibility of experiments [14]. Kontos et al. derived morphological features in the form of concentric spheres radiating out of a region's center of mass. The morphological features are advantageous because they were faster to calculate than other more computationally expensive mathematical morphological methods (i.e. three dimensional shape descriptors based on three dimensional Fourier transform). The classification of the regions on artificial and real data was computed by neural networks, and the results proved to be encouraging [15].

Mining methods, from structural imaging data, find structure-function relations through the study of lesions and associated deficits. The two general types of analysis in data mining of structural imaging data are: atlas-based analysis and voxel-based analysis [16].

For atlas-based analysis, the anatomical structures are considered functional units and provide a higher sensitivity than voxel-based analysis since the atlas contains prior knowledge. To do an exhaustive analysis (assuming there is no knowledge about the association of certain variables), one can apply a pairwise or multivariate technique [1].

Since atlas-based techniques are heavily dependent upon the integrity of the atlas, an alternative approach could be to analyze the data on a voxel-level basis. In a typical setting, voxels are labeled as normal and abnormal, and a method of analysis can be applied to determine whether voxels belonging to a certain region are associated with an outcome under investigation. One technique uses voxel-based regression to relate lesions in a sphere of a given radius and location to a deficit, which can then be used to determine regions of the brain that have the strongest association with the deficit under study. However, this technique's main weaknesses lie in its computational expense and lack of accounting for multifocal associations [16].

A common mining technique used in structural imaging is extracting morphological characteristics of brain structures to understand variability among a population. The morphological features, such as location, size, surface area, volume, and shape of brain regions can potentially result in understanding normal brain organization and quantifying pathological changes (i.e. ventricles deforming in presence of brain tumor). Typically, voxel-based morphometry has been used to identify changes in brain tissue after large shape differences have been accounted for by atlas alignment or other spatial normalization methods [17,18]. Deformable registration has also been a popular technique for obtaining information about a subject's brain structures [19,20]. Classification is a significant issue in data mining applications, and can be used in further aiding other applications such as content based image retrieval systems. For breast tumor classification, Korn et al. applied morphological concepts to extract features and define the distance metric [21]. One of the weaknesses was that the method was applied on synthetic tumor shapes, so the accuracy could not be assessed on real images. Other types of methods, such as fractal descriptors and texture parameters, have been extracted in attempting to quantitatively describe and recognize brain tumors in volumetric MR images [22].

2.2. Survey of automated techniques for brain tumor segmentation

Image segmentation of malignant brain tumors is challenging given the variable nature of tumor presentation. Tumor segmentation is however a critical step in the automated extraction of tumor features; these features can then be used in data mining tasks in a manner similar to text data. The area of image segmentation of brain tumors is ongoing and the research community has directed a significant amount of effort to develop automated algorithms to accurately segment brain tumors from MR imaging.

One of the most popular methods involves voxel-based fuzzy clustering, and it has been evaluated on different kinds of brain tumors. Fuzzy clustering is an unsupervised pattern recognition technique and is similar to the k-means algorithm. The fuzzy clustering approach allows for a pixel to be part of several classes, and the clustering occurs by tissue type that is based on separated distributions. An early study used fuzzy clustering for brain tumor by computing on the raw multi-sequence data, and it was visually shown that the intensity distributions for tumor and normal tissue overlap. Since the intensity distributions could not discriminate between tumor and normal tissues accurately, researchers incorporated additional knowledge into the segmentation process [23].

Clark et al. integrated knowledge-based techniques to remove extracranial tissues and multi-spectral histogram analysis to segment GBM tumors in a multi-channel setting. However, the major drawback of this technique was that it resulted in a low accuracy in the segmentation [24]. A knowledge-based fuzzy clustering approach with three dimensional connected components for building of tumor shape on a multi-channel image set was proposed by Fletcher-Heath et al. An initial segmentation is computed by fuzzy clustering, and clusters are labeled by shape, location, and intensity features. In the method by Fletcher-Heath et al., the accuracy of tumor segmentation was highly variable and did not provide consistent results that could be clinically implemented [25].

A knowledge-based segmentation method based on learning intensity distributions for normal brain matter and detecting outliers (considered tumor) was shown to have good accuracy. The intensity distributions of the T1 and T2 channels are learned from with kernel-based density estimation methods. The initial outlier detection is subsequently followed by a region competition algorithm. The disadvantage of the technique was that the time to completion for tumor segmentation was approximately ninety minutes; thus, it was not clinically practical [26].

Another common method of brain tumor segmentation involves voxel-based statistical classification. One of the techniques involves using the adaptive template-moderated classification algorithm to segment the MR image into five different tissue classes: background, skin, brain, ventricles, and tumor. The technique starts as an iterative sequence of spatially varying classification and non-linear registration and it resulted in a high accuracy for segmenting low grade gliomas. The efficacy of the adaptive template-moderated technique had not been tested on malignant gliomas, however [27].

Prastawa et al. presented a technique for segmenting tumor and edema of malignant gliomas by a spatial atlas that was modified to include prior probabilities of tumor and edema. The abnormal tissue probabilities were generated from a parametric distribution across the T1 and T2 channels, and an Expectation-Maximization algorithm was used to perform segmentation and iteratively adapt the model parameters. One of the main drawbacks of the approach was that the whole segmentation process took approximately 100 minutes to complete [28].

Although the voxel-based statistical classification and fuzzy clustering methods have shown some promising results, they are limited by the fact that no local or global context is taken into account. Within the more complex and varied GBM subgroup, the success of segmentation methods has been limited. A particular technique of significance is by Lee et al.; they use the context-sensitive discriminative random fields model with support vector machines for classification [29]. However, the method resulted in variable accuracy. Potentially, energy and shape models could be useful in malignant brain tumor segmentation but are typically iterative and as a result sensitive to initialization; thus, some level of user interaction is required to assist in making the segmentation more optimal.

2.3. Image derived features relevant in characterizing brain tumors

The characterization of brain tumors, via image derived features, for diagnosis, prognosis, and treatment response has been an active research area. Many research groups are investigating what radiographic aspects of brain tumors can potentially give a more robust understanding of the nature of the malignancy of primary brain tumors.

Diagnosing brain tumors non-invasively gives a tremendous benefit to patients. The ability to use radiographic findings to characterize the type of brain tumor has an advantage in that it can be used to avoid giving the patient a biopsy, which includes risks associated with anesthesia and surgery. Several studies have attempted to harness the power of conventional MR to accurately classify brain tumors into distinct grades for diagnostic use.

An early study evaluated MR characteristics to determine if classification of brain tumors could be done with low-grade astrocytomas, anaplastic astrocytoma, and GBM. The characteristics that were found to be statistically significant were mass effect and cyst formation or necrosis [30].

Another study evaluated the relationship between the degree of contrast enhancement and the volume of peritumoral edema in meningiomas and malignant gliomas. Meningiomas have a variable extent of edema and enhance prominently. Thus, differentiating between meningiomas and malignant gliomas via MR characteristics would be very useful. For gliomas, a direct linear relationship existed between the degree of contrast enhancement and volume of peritumoral edema. However, for meningiomas, a statistically significant mathematical relationship could not be established [31].

One group studied the MR characteristics indicative of a meningioma that produces brain edema. The key characteristics found to positively correlate with a statistically significantly higher incidence of edema were an irregular tumor margin, lack of a peritumoral rim, and signal intensity of tumor (hyperintense) on T2 weighted images [32]. An investigation was done to see whether tumor contrast enhancement could be used as a characteristic for differentiating between different grades of oligodendrogliomas. However, there was no statistical significant evidence that tumor contrast enhancement could discriminate between low-grade and anaplastic oligodendrogliomas [33].

Correlating imaging features with survival in patients with malignant gliomas has been studied in the literature for several years. The malignant gliomas appear as heterogeneous tumors and survival times vary significantly between grade III and IV gliomas (where grade III tumors have significantly higher survival times). Several groups had results showing the degree of necrosis and edema negatively correlated with survival times. It was shown that tumors with a significantly high amount of necrosis resulted in a median survival time of 12 months. Also, the intensity of the enhancement of the tumor nodule yielded a negative correlation with survival time where the highest enhancement group yielded in a median survival time of 13.5 months [34].

One study showed that the presence of large tumor cysts (96% of the cysts involved eloquent or near eloquent areas of the brain) resulted in a positive correlation with survival (22 patients). Patients exhibiting cystic GBM had a median survival time of 18.2 months with 43% of the patients surviving at 2 years, and patients with non-cystic GBM had a median survival time of 14.3 months with 16% of the patients surviving at 2 years [35].

There have been mixed results on whether extensive surgical resection prolongs survival. In one study, the measurement of the preoperative and post-resection tumor volumes (considered to be the contrast enhancing area including any central necrotic region) was used to calculate the percentage of resection and volume of residual disease. The percentage of resection and volume of residual disease were found to have a significant effect on time to tumor progression and survival [36].

A more recent study performed an exhaustive analysis of imaging variables for correlation with prognosis on approximately 100 patients. The imaging variables were: necrosis, enhancement, enhancing rim, enhancing margin, non-enhancing margin, solid enhancement, non-contrast enhancing tumor, edema, cyst, multifocal, satellites, extent of resection, tumor crossing midline, edema crossing midline, location, size, and side. For each study, the presence of necrosis, enhancement, non-contrast enhancing tumor, cyst(s), multifocality, satellites, tumor crossing midline, edema crossing midline (if applicable) was observed. The location variable was assigned as either frontal, parietal, temporal, occipital, corpus callosum, or thalamus [37]. The results showed that the extent of resection did not have a significantly statistical correlation with survival. [37, 38].

The molecular subtype of brain tumors has also being studied as a possible indicator of response to treatment. One study investigated whether the allelic loss of 1p and 19q was evident via conventional MR imaging characteristics. The MR imaging characteristics observed was the following: sharp vs. indistinct tumor border, homogeneous vs. heterogeneous tumor signal intensity on T1 and T2, paramagnetic susceptibility effect (hyperintense on T1 and hypointense on T2), contrast enhancement, and percent cortical involvement of tumor. Their findings showed that the loss of 1p and 19q was associated with an indistinct border on T1 images and mixed signal intensity on T1 and T2. The loss of 1p and 19q also had an association with paramagnetic susceptibility effect and with calcification [39].

Another study looked at the correlation of genotype in oligodendroglial tumors with MR imaging characteristics, since oligodendroglial neoplasms with the –1p/-19q genotype are more indolent with longer survival and increased therapeutic responsiveness than ones with intact 1p/19q. The tumor imaging features used were: border type (sharp vs. distinct), contour type (smooth vs. irregular), signal intensity type (homogeneous vs. heterogeneous), contrast enhancement, and paramagnetic susceptibility effect. Tumor cases where the T1 or T2 signal intensity was homogeneous resulted in intact 1p/19q. Also, tumors with sharp/smooth borders had a higher likelihood of intact 1p/19q than those without [40].

The studies discussed above were mostly qualitative in nature, or simple quantitative measurements (i.e. bi- or tri-linear measurements) were calculated to characterize the state of the brain tumor. A significant effort has been underway in quantifying imaging features of tumors to establish correlations with molecular subtype, time to survival, and time to tumor progression.

Parametric values, such as T1, T2, and proton density, have been experimented with to see whether normal and abnormal tissue could be differentiated. One studied obtained the parametric values for T1, T2, and proton density on approximately 200 patients with brain tumors. Regions of interest were chosen to be representative of the tumor, and white matter regions were chosen in the contralateral hemisphere with a similar location to the lesion. It was shown that normal brain tissues showed small inter-individual variations of tissue parameters, whereas brain tumor tissues showed significantly large variations of the tissue parameters (that were significantly different from normal tissue) [41].

The morphological characteristics of the tumor were addressed in one study by determining the T1 and T2 border sharpness and ratio of T2-bright volume to enclosed T1-contrast enhanced volume in order to find a correlation to molecular subtype. The border sharpness was found by calculating a border sharpness coefficient via the rates of change in grayscale intensity of adjacent 0.02 cm^2 voxels sampled at the anterior, posterior, and lateral borders on T1-enhanced and T2 images. The mean volume ratio was found to be 4.7 for EGFR-amplified GBMs (classified as primary), and the mean volume ratio was 2.3 for TP53-mutated GBMs. The GBM cases with T2 border sharpness coefficient less than 30.8 were EGFR-amplified [42].

Spherical harmonics (SH) has been used in approximating brain tumor shape for surgical planning. SH are an orthonormal basis of functions defined on a unit sphere, and the SH expansion defines a three dimensional surface in a spherical coordinate system in terms of the coefficients of the corresponding SH basis functions [43]. SH were used to quantitatively define three dimensional geometric features in multiple sclerosis lesions and complicated anatomic shapes (i.e. cerebral ventricles) [44,45].

Another method employed volumetric texture analysis in differentiating between different types of normal and tumor tissue. The texture parameters were extracted from volumes of interest regions chosen on T1-weighted sequences. The regions consisted of: solid tumor, necrosis, edema, peritumoral white matter, far homo-lateral white matter, and contra-lateral white matter. Classification between necrosis and solid tumor, edema and solid tumor, peritumoral white matter and far homo-lateral white matter resulted in higher accuracy with the volumetric gray level co-occurrence matrix approach [46].

A recent work assessed the variables that affect the extent of resection and what makes an incomplete surgery in low grade gliomas. It was found that the main tumor characteristics that resulted in incomplete tumor resection were: diffuse tumor margin on T2-weighted images, oligodendroglioma or oligoastrocytoma histopathologic type, and large tumor volume. Also, tumor involvement with corpus callosum, corticospinal tract, insular lobe, middle cerebral artery, motor cortex, optic radiation, visual cortex, and basal ganglia resulted in incomplete resection. The initial and residual tumor volume was computed from the initial and final intraoperative MR images [47].

The use of advanced imaging techniques with diffusion weighted imaging (ADC coefficients), MR spectroscopy (peak integration values), and MR perfusion (rCBV) are starting to become utilized more for their complementary role in further characterizing tumor type, response to therapy, and prognosis [48,49].

3. IMAGE DATA MINING FOR MALIGNANT BRAIN TUMOR

3.1. Infrastructure of the data mining application

A key component of effective data mining of brain tumors is a database that has comprehensive information on subjects including clinical, histological, therapeutic and imaging assessments. One of the most extensive collections of brain tumor imaging in the nation is the UCLA Neuro-oncology database. In the next sections, we outline the infrastructure to perform image data mining of brain tumor data. Our focus is on creating a pipeline that will automatically extract imaging features from the MR images to be used subsequently for data mining applications. Automatic feature extraction proceeds in several stages:

1) Normalization of imaging data to reduce the variability of raw image intensities and subject orientation; this is important both for the segmentation and for consistent feature values,
2) Segmentation of tumor and tumor sub-regions (i.e. edema, necrotic tissue, etc)
3) Feature extraction
4) A data mining application determining the correlation between imaging features and prognosis (i.e. time to survival)

3.1.1. Brain tumor database

The UCLA Neuro-oncology brain tumor database obtains approximately a hundred new patients every year exhibiting malignant primary brain tumors. Over the past several years, data from these patients has been collected which has resulted in a current database size of over 700 malignant primary tumor cases. The imaging protocol in the database includes: axial pre- and post-contrast T1-weighted image series; volumetric 3D post-contrast T1-weighted series (intended for pre-surgical evaluation); axial proton density (PD)/T2-weighted double echo image series; and axial FLAIR image series.

3.2. Image study normalization

Routine clinical image acquisitions exhibit high spatial and intensity variability within a patient's image dataset as well as between different patient's images, making serial/inter-subject analysis problematic. To obtain accurate quantitative measures to aid in brain tumor data mining applications, it is desirable to have image studies with standard intensity values.

Atlas-based registration is one of the most common techniques used for spatial normalization. For affine alignment based on intensity, the registration process varies the alignment parameters of rotation, translation, and scale and determines the most ideal values which will minimize a cost function. The cost function essentially quantifies the goodness of the alignment. Aligning the image set to a pre-determined atlas reduces the spatial variation between image sets and it also allows one to determine the location of a brain structure (i.e. tumor area is approximately in right temporal lobe).

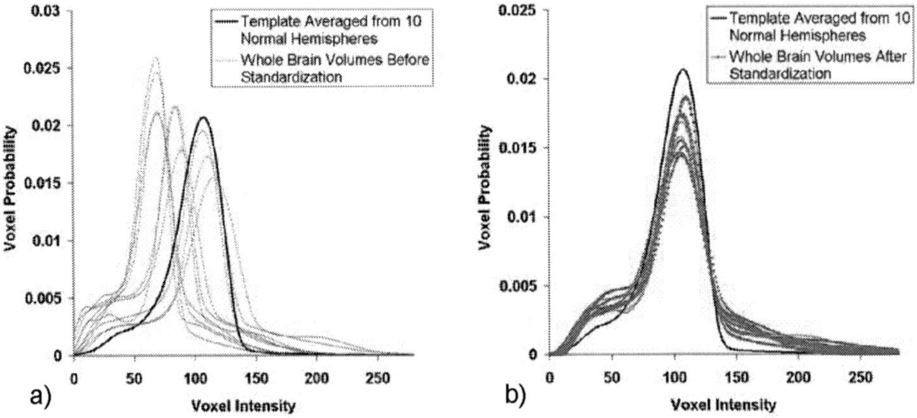

FIGURE 1. a) Histogram plots of $h_{template}$ and patient data before standardization b) Histogram plots of $h_{template}$ and patient data after standardization.

There have been past attempts made to calibrate MR signal at acquisition time using MR visible markers scanned along with the subjects, but an efficient and accurate post-processing method applied to the images would need no specific acquisition requirements [50]. One possible method for intensity standardization is an extension of the work done by Nyul et al [51], based on a histogram matching method that is applicable to normal or abnormal subjects with small lesions. Once the image studies are aligned to a normal atlas, the contra-lateral side is then used to derive intensity histogram and the resulting second order polynomial transformation mapping parameters (based on histogram landmarks of the template and subject) to standardize the image intensities. The difference in the intensity histograms before and after standardization can be seen quantitatively in Fig. 1a and 1b, and qualitatively in Fig. 2.

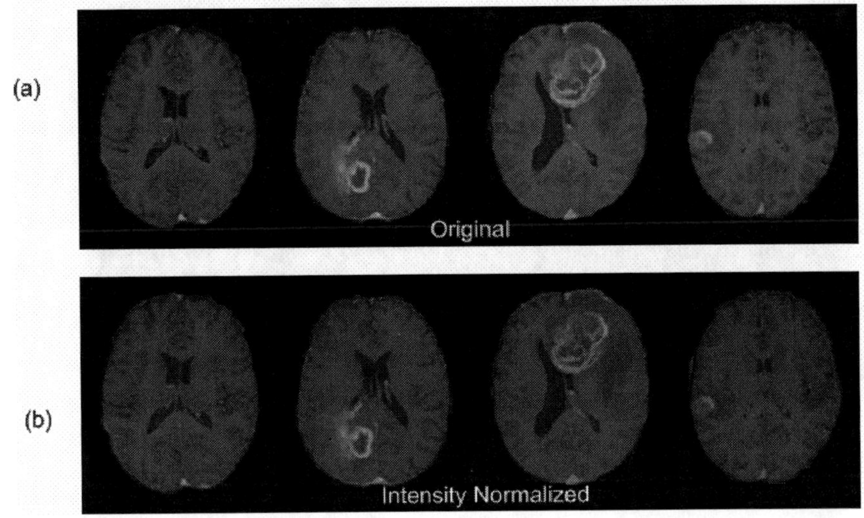

FIGURE 2. a) Example of images from different subjects before and b) after intensity standardization at the same windowing level for all images.

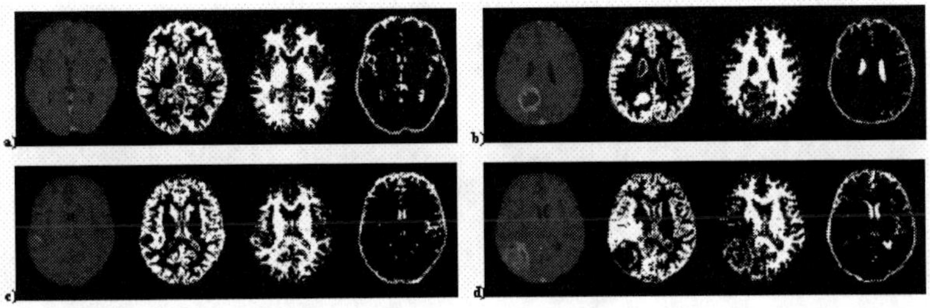

FIGURE 3. Four different patient images (a, b, c, and d). Each patient set shows the standardized T1 post contrast image followed by images segmented into binary gray, white and csf regions using the same tissue-specific threshold ranges for all patients.

As a result of the standardization, a standard threshold obtained from the template can be applied to roughly segment gray, white and cerebrospinal (csf) tissue types. The results of applying the same thresholds to roughly segment four different patient sets for visualization of the efficacy of the method are shown in Fig.3.

Image study normalization is an important component in assisting effective image data mining for malignant brain tumors. A post-processing method to obtain image standardization to account for spatial and intensity variations within a given database is highly attractive because of the potential application to image studies derived from different protocols. In addition to reducing variability, post-processing methods for image study normalization assist in robust segmentation.

3.3. Segmentation of brain tumors

Once an image study has been normalized, it must be segmented into meaningful regions. These regions can then be quantified by a variety of mathematical operations and associations can be determined between tumor features and tumor type, survival time, and response to a particular treatment protocol. Tumor and edema size are important predictors in patient outcome. Manual methods of segmenting an image are cumbersome and prone to error. Thus, a robust segmentation algorithm has clear benefits to brain tumor data mining applications. There have been numerous attempts at fully automated segmentation of brain tumors, but a majority of them take a long time to process and do not produce clinically viable results. One of the more promising developments in brain tumor segmentation involves hierarchical analysis of an image and is an extension of the segmentation by weighted aggregation (SWA) algorithm originally developed by Sharon et al [52,53].

The multilevel segmentation by weighted aggregation with integrated Bayesian model classification unifies two effective approaches to segmentation:

- explicit representation of phenomenon via class modeling
- affinity-based graph cuts

Explicit representation of distinct phenomenon allows for segmentation and classification to be solved jointly by having a learned statistical distribution from training data for a given model. This approach is accurate since the segmentation and classification results are guaranteed from the learned distribution, but it is not practical since the algorithms used to obtain the segmentation are significantly slow and the choice of the model is problematic.

The second approach of affinity-based graph cuts is based on the concept of normalized cuts, where the input image initializes a sparse graph and each edge in the graph is given an affinity measurement. The conventional affinity, w_{uv}, for nodes u and v is defined as follows:

$$w_{uv} = \exp(-D(s_u, s_v; \theta) \tag{1}$$

where D is a non-negative distance measure, s_u and s_v are the statistics at nodes u and v respectively (typically average intensity), and θ are predetermined parameters learned from the training data. The objective of the affinity-based methods is to find salient regions which are defined as having low values of the following saliency function:

$$\Gamma(C) = \frac{\sum_{u \in C, v \notin C} w_{uv}}{\sum_{u, v \in C} w_{uv}} \tag{2}$$

As can be seen in Fig. 4, a cluster, C, that exhibits a small $\Gamma(C)$ have a low affinity across their boundaries and high affinity within their interior, which is more commonly known as the normalized cut criterion.

The affinity measurement quantifies similarity between two neighboring nodes, and the actual cuts made in the graph are groups of edges that partition the graph into a number of subsets (Fig. 4). SWA recursively coarsens the initial graph using an adapted algebraic multigrid algorithm, and it has been shown to approximate the normalized cut measure. Once the recursive coarsening process has finished, a multilevel set of segmentations (each node in the hierarchy representing a potential segment) exists. The advantage of the hierarchy in the brain tumor example is that at a given level, different components of the tumor emerge as independent nodes which then group together as a single node in a higher level; thus, allowing for more detailed quantification. The multiscale properties are captured effectively and rapidly in the SWA framework, but do not provide a way to determine a final segmentation.

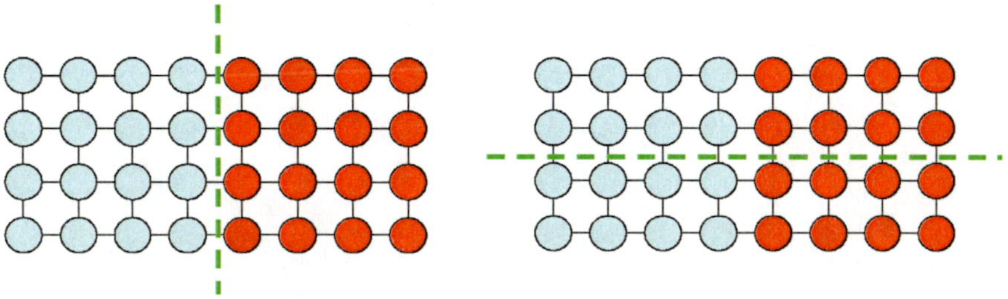

FIGURE 4. Illustration of a cut resulting in low (left image) and high (right image) saliency.

In the multilevel SWA with integrated Bayesian model classification, the two segmentation approaches are unified by incorporating models into the calculation of the affinities (model-aware affinity) on the graph, and then using the models to determine a final classification from the hierarchy of segments.

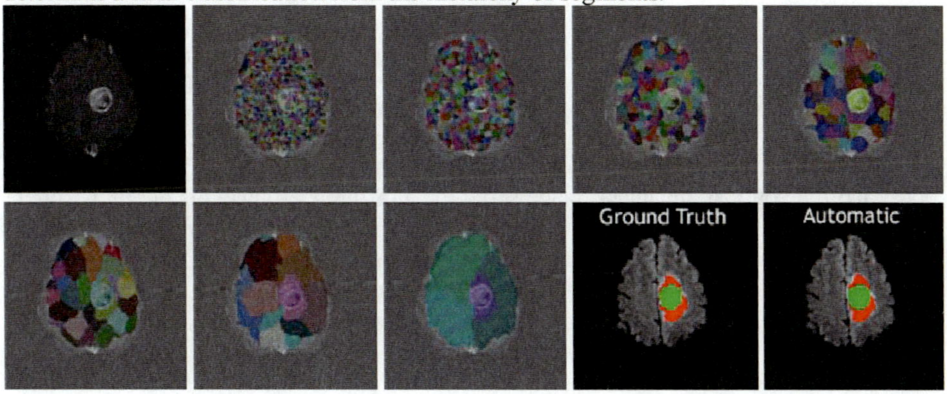

FIGURE 5. A depiction of the model-aware segmentation process showing the different levels of the hierarchy formation process and the ground truth and final automated segmentation result for a patient exhibiting malignant glioma. Reprinted from [53].

The learned statistical distribution of tumor and edema regions is computed by a Gaussian mixture model on the multi-channel labeled dataset consisting of T1-pre/post contrast, T2, and FLAIR sequences. During the recursive process of building the hierarchy of segments, the model-aware affinities influence the aggregation process in the higher levels of the pyramid by adaptively modifying newly formed aggregates with the model information included in the process.

For the final extraction, a majority vote of the class labels of a voxel at each level was taken to determine the final class assignment (Fig. 5). The results obtained for tumor (Jaccard score of 0.69) and edema segmentation (Jaccard score of 0.62) were encouraging [53].

We are currently investigating methods to improve the accuracy by incorporation of knowledge into the model-aware SWA. Likelihood maps are generated from knowledge-based information of tumor [28], necrosis, and edema presentation in the different MR channels.

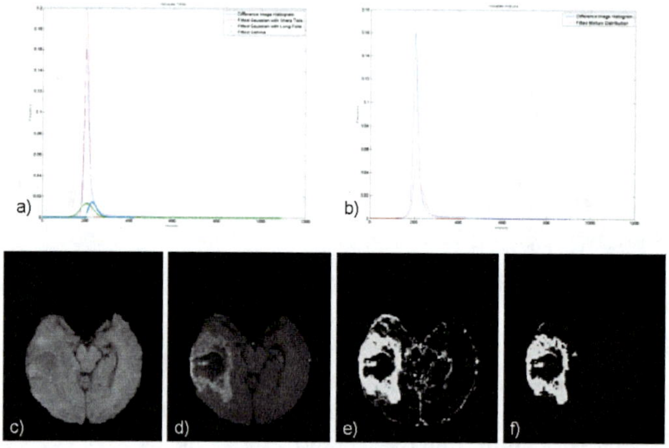

FIGURE 6. a) Histogram image analysis of the two Gaussian fitted functions (red and green), Gamma fitted function (cyan), and original difference image histogram (blue). b) Sum of all three fitted distributions (red) and original difference image histogram). c) T1 pre image. d) T1 post image. e) Initial likelihood map of contrast enhancing tumor. f) Refined likelihood map with shape constraints.

Contrast enhancing tumor region is extracted from the difference image of the T1 post and pre contrast enhanced image series. The difference image histogram is fitted to the sum of two Gaussians for the noise component and a gamma distribution corresponding to the contrast enhancing regions [28]. The gamma probability density function is converted to a cumulative density function, and this cdf is used as a mapping function to convert the difference image intensities to probabilities (Fig. 6a and 6b). Since enhancing blood vessels result in high probability values, a shape constraint can be imposed (such as eccentricity) to remove linearly shaped objects and thereby reduce false positives of tumor regions. The refined likelihood map of contrast enhancing tumor (Fig. 6f) could potentially be used as a prior probability feature in the model-aware SWA framework and significantly reduce false positives and improve the segmentation accuracy. Likelihood maps for necrosis and edema were built in a similar fashion based on the presentation of these structures in the images.

Since intensity measurements overlap between tissue types, we are exploring methods to incorporate texture information along with the intensities. A technique for obtaining texture information is the gray level co-occurrence matrix (GLCM) method. The GLCM can be calculated for different distance separation between voxel pairs (d), direction (θ), and neighborhood window size [54]. From the resultant matrix, one can calculate texture parameters (i.e. entropy, angular second momentum, etc). Fig. 7 shows an example of the GLCM method applied on an image and the resultant texture feature maps that were computed for potential use in tumor discrimination.

Incorporating texture features into the multilevel segmentation framework can be potentially helpful. An example of segmentation with non-linear SVM classification with texture features can be seen in Fig. 8. The potential benefits of incorporating additional features that quantify visual appearance are significant.

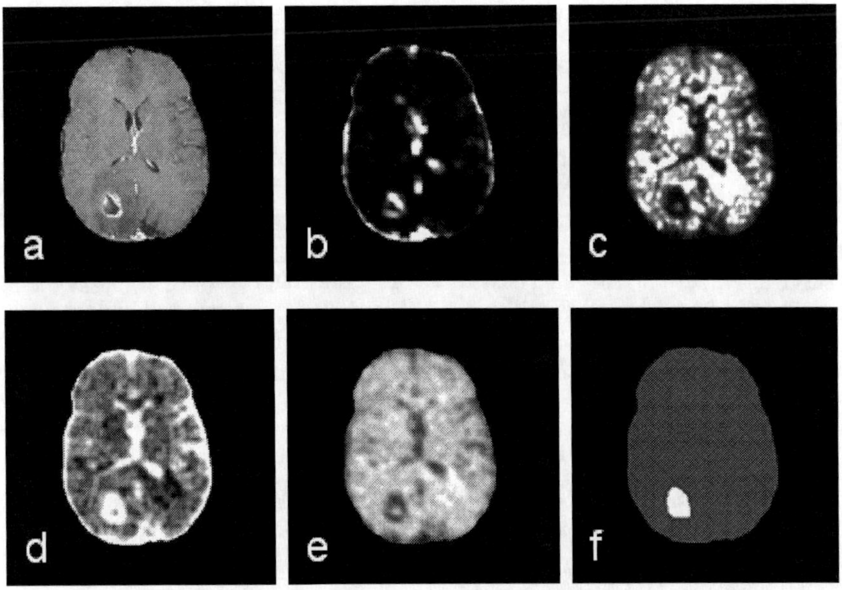

FIGURE 7. The T1 post-contrast enhanced intensity image (a) for a given patient as well as the respective contextual filter response images and ground truth at the corresponding slice: (b) contrast, (c) energy, (d) entropy, (e) homogeneity, (f) ground truth.

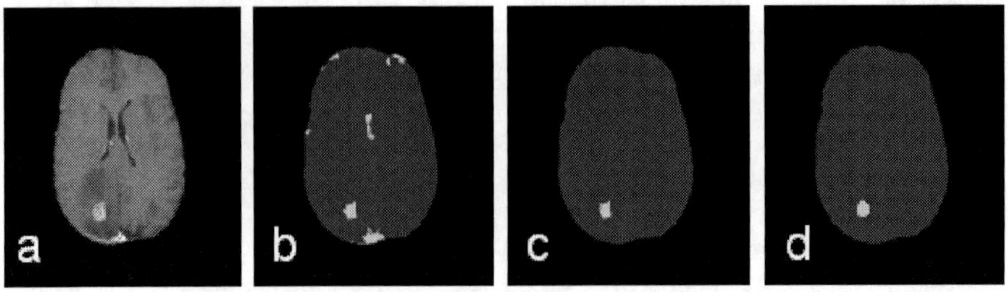

FIGURE 8. Images showing final segmentation of tumor (green) and brain (blue) tissues. (a) A T1 post contrast enhanced image with an outline of the tumor segmentation resulting from SVM classification on all features, (b) segmentation resulting from LDA classification on all features, (c) segmentation resulting from SVM classification on all features, (d) ground truth.

3.4. Feature extraction and correlations

With an automated system for accurate segmentation in place, tumor features can be extracted for potential data mining applications. Since prognostic outcomes within the GBM group are highly variable, discovering robust MR imaging markers correlated with prognosis is highly significant.

A recent example highlighting the significant potential of quantitative tumor feature correlation to prognosis involves deriving qualitative features from the radiology report and quantitative features from T1 post contrast enhanced images only [55,56]. The data mining application was to find the MR imaging features correlated with time to survival. The current analysis used imaging data (prior to surgery) from approximately forty patients exhibiting GBM, and all the patients had undergone gross total resection with subsequent radiation and chemotherapy. Prior to any processing, all the imaging studies were normalized. For feature extraction, three visually distinct regions that are readily identifiable by an expert radiologist were used for analysis: edema, contrast enhancing tumor, and necrosis (Fig.9) [55].

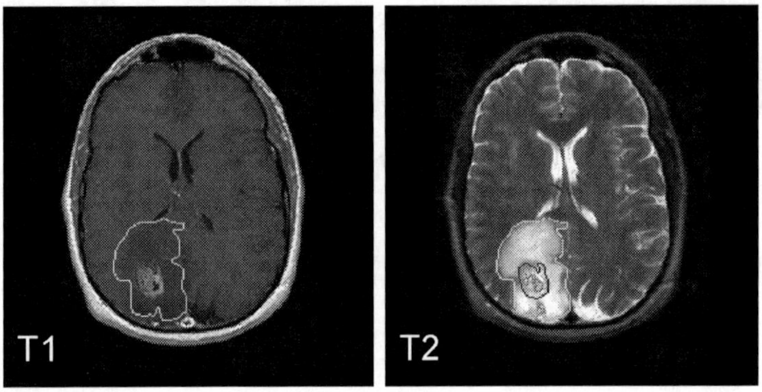

FIGURE 9. The left and right images are the T1 post contrast and respective T2 with the labeled regions that were segmented for quantitative analysis. The edematous boundary is outermost boundary, contrast enhancement is middle boundary, and necrosis by innermost boundary.

TABLE 1. List of the different features extracted for each sub-region. The features consisted of texture parameters derived from 3D co-occurrence matrices, 3D shape moments, and histogram parameters.

Category	Features
3D Texture via Co-occurrence Matrix	Correlation Contrast Homogeneity Energy
3D Shape Moments	Order 0 Invariant Moment (Volume) Order 1 Invariant Moments Order 2 Invariant Moments Order 3 Invariant Moments
Histogram	Mean Variance Skewness Kurtosis

Table 1 is a list of the shape, texture, and intensity histogram features that were extracted from each of the sub-regions that were segmented from each patient image study. A stepwise multivariate linear regression model was applied to the tumor features to determine which descriptors were most significant in predicting time to survival.

Applying the stepwise multivariate linear regression model to the data found resulted in four quantitative descriptors of contrast enhancing tumor and one quantitative descriptor of edema and necrosis were significant. The resulting multivariate model had an adjusted correlation coefficient of 0.59.

The shape and intensity properties of contrast enhancing tumor, and shape features of edema and necrotic regions had significant correlation with time to survival. The shape features chosen were found to have a negative contribution to survival time because they characterized the uniformity of the shape of the contrast enhancing tumor region. Higher values of the selected moment invariant features of contrast enhancing tumor corresponded to a more irregular appearance of the contrast enhancing tumor as well as a larger presence of necrotic tissue. The contrast enhancing tumor intensity histogram features found significant were kurtosis and mean. The kurtosis of the contrast enhancing tumor intensity histogram primarily reflects peakedness of the histogram.

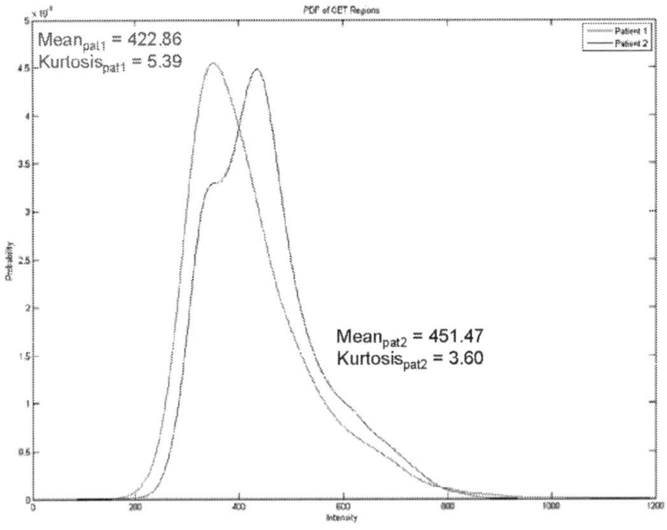

FIGURE 10. Histograms contrast enhancing tumor regions of two patients with high (Pat 1) and low (Pat 2) survival.

Fig. 10 shows two patient intensity histograms and their corresponding kurtosis and mean, where patient one has a significantly long survival time (700 days) and patient two had a short survival time (100 days). The kurtosis of the contrast enhancing tumor as a significant feature would imply that uniform contrast uptake (resulting in a narrow histogram and larger kurtosis values) is correlated with longer time to survival as opposed to a spread in contrast uptake (more rounded histogram with lower kurtosis) which is correlated to lower time to survival. The novel kurtosis finding is in line with the general observation that higher heterogeneity of contrast uptake is correlated with malignancy, and consequently to survival time. The other significant predictor of mean intensity of contrast enhancing tumor, which was found to have a negative correlation to time to survival, implies that higher mean intensities of the contrast enhancing region were associated with a lower survival time estimate. Since the mean intensity is related to the degree of uptake of the contrast agent, it could be potentially correlated to malignancy.

The selected invariant shape moment descriptor of edema as a significant predictor is meaningful since edema preferentially travels along white matter tracts. Essentially, the selected edema shape feature implied that a more irregularly shaped edematous region with a larger presence of tumor resulted in a lower time to survival estimate. A higher irregularity of the edema shape coupled with a larger presence of tumor could result in higher infiltration of the tracts and could potentially be captured in the shape indices as structures of high morphological anisotropy. The significant necrosis shape feature captured the irregularity of necrotic regions, and higher irregularity of necrotic tissue correlated with lower time to survival [55].

In addition to the quantitative features computed from the segmented sub-regions, qualitative features from the radiology report could be included as binary features to improve the correlation and obtain better predictors of prognostic indices. Image-derived qualitative features such as multifocality, satellites, and non-contrast enhancing tumor have been found to be significant in literature. Features such as non-contrast enhancing tumor are not easily segmentable currently, and a combination of qualitative and quantitative indicators could further improve survival time estimates as well as other prognostic indicators (i.e. time to tumor progression).

4. CONCLUSIONS

In this chapter, the focus was image processing for data mining brain tumor datasets. Several earlier studies have shown the potential of imaging features obtained from routine images to correlate with diagnosis, prognosis, and therapeutic response. An automated method of extracting image features will enable data mining large databases with the exciting possibilities of uncovering complex relationships between image features and other variables. The infrastructure to enable integration of image information is discussed in this chapter with emphasis on image normalization, accurate segmentation, relevant feature extraction and concluding with a potential data mining application. Potential future uses of these imaging-derived features for data mining applications include: fast and effective real-time differentiation between tumor abnormality and less serious events (i.e. bleeding); tracking changes serially in genetic sub-type of tumor after intervention; discovering pre-surgical image-derived feature groupings with respect to treatment response; and identifying the most effective treatment for a given tumor sub-type.

ACKNOWLEDGMENTS

This research was funded by NLM Grant # LM07356.

REFERENCES

1. V. Megalooikonomou, J. Ford, L. Shen, F. Makedon, and A. Saykin, "Data mining in brain imaging," in *Statistical Methods in Medical Research*, 2000, pp. 359-394.
2. T. Lindberg, P. Lidberg, and P.E. Roland, "Analysis of brain activation patterns using a 3-D scale-space primal sketch," in *Human Brain Mapping*, 1999, pp. 166-194.
3. O. Coulon, J.F. Mangin, V. Frouin, and I. Bloch, "Structural group analysis of functional maps," in Kuba MSaAT-P, *16th International Conference on Information Processing in Medical Imaging*, Visegrad, 1999, pp. 448-453.
4. J.K. Worsley, S. Marrett, and P. Neelin, AC. Evans, "Searching scale space for activation in PET images," in *Human Brain Mapping*, 1996, pp. 74-90.
5. J.B. Poline and BM. Mazoyer, "Enhanced detection in brain activation maps using a multifiltering approach," in *Journal of Cerebral Blood Flow Metabolism*, 1994, pp. 639-642.
6. M.J. Brammer, "Multidimensional wavelet analysis of functional magnetic resonance images," in *Human Brain Mapping*, 1998, pp. 378-382.
7. M. Jansen, G. Uytterhoeven, and A. Bultheel, "Image de-noising by integer wavelet transforms and generalized cross validation," in *Medical Physics*, 1999, pp. 622-630.
8. UE. Ruttimann, M. Unser, RR. Rawlings, D. Rio, NF. Ramsey, VS. Mattay, DW. Hommer, JA. Frank, and DR. Weinberger, "Statistical analysis of functional MRI data in the wavelet domain," in *IEEE Transactions on Medical Imaging*, 1998, pp. 142-154.

9. F. Kruggel and D.Y. Cramon, "Physiologically oriented models of the hemodynamic response in functional MRI," in *Lecture Notes in Computer Science*, Berlin: Springer, 1999.
10. F. Kruggel and D.Y. Cramon, "Modeling the hemodynamic response in single-trial functional MRI experiments," in *Magnetic Resonance in Medicine*, 1999, pp. 787-797.
11. K.J. Friston, AP. Holmes, K.J. Worsley, J.P. Poline, C.D. Frith, and R.S.J. Frackowiak, "Statistical parametric maps in functional imaging: a general linear approach," in *Human Brain Mapping*, 1995, pp. 189-210.
12. K. Friston, "Statistical parametric mapping and other analyses of functional imaging data," in *Brain mapping: the methods*, edited by A. Toga, J. Mazziotta, San Diego, CA: Academic Press, 1996.
13. D. Pokrajac, V. Megalooikonomou, A. Lazarevic, D. Kontos, and Z. Obradovic, "Applying spatial distribution analysis techniques to classification of 3D medical images," in *Artificial Intelligence in Medicine*, 2005, pp. 261-280.
14. F.A. Nielse and L.K. Hansen, "Finding related functional neuroimaging volumes," in *Artificial Intelligence in Medicine*, 2004, pp. 141-151.
15. D. Kontos and V. Megalooikonomou, "Fast and effective characterization for classification and similarity searches of 2D and 3D spatial region data," in *Pattern Recognition*, 2005, pp. 1831-1846.
16. V. Megalooikonomou, C. Davatzikos, and E. Herskovits, "Mining lesion-deficit associations in a brain image database," in *ACM SIGKDD International Conference on Knowledge Discovery and Data Mining*, San Diego, CA, 1997, pp. 347-351.
17. C.J. Mummery, K. Patterson, C.J. Price, J. Ashburner, R.S. Frackowiak, and J.R. Hodges, "A voxel-based morphometry study of semantic dementia: relationship between temporal lobe atrophy and semantic memory," in *Annals in Neurology*, 2000, pp. 36-45.
18. J. Ashburner and K.J. Friston, "Voxel-based morphometry-the methods," in *NeuroImage*, 2000, pp. 805-821.
19. C. Davatzikos, M. Vaillant, S.M. Resnick, J.L. Prince, S. Letovxky, and R.N. Bryan, "A computerized approach for morphological analysis of the corpus callosum," in *Journal of Computer Assisted Tomography*, 1996, pp. 88-97.
20. J. Csernansky, S. Joshi, L. Wang, J.W Haller, M. Gado, J.P. Miller, U. Grenander, and M.I. Miller,"Hippocampal morphometry in schizophrenia by high dimensional brain mapping," in *Proceedings of the National Academy of Sciences of the United States of America*, 1998, pp. 11406-11411.
21. F. Korn, N. Sidiropoulos, C. Faloutsos, E. Siegel, and Z. Protopapas,"Fast and effective similarity search in medical tumor databases using morphology," in *SPIE Proceedings*, Boston, MA, 1996.
22. C. Rossmanith, H. Handels, S. Poppl, E. Rinast, and D. Weiss, "Characterisation and classification of brain tumors in three-dimensional MR image sequences," in *Fourth International Conference on Visualization in Biomedical Computing*, 1996, pp. 429-438.
23. W.E. Phillips, R.P. Velthuizen, S. Phupanich, L.O. Hall, L.P. Clarke, and M.L. Silbiger, "Applications of Fuzzy C-Means Segmentation Technique for Tissue Differentiation in MR Images of a Hemorrhagic Glioblastoma Multiforme," in *Journal of Magnetic Resonance Imaging*, 1995, pp. 277-290.
24. M.C. Clark, L.O. Hall, D.B. Goldgof, R. Velthuzien, R. Murtagh, and M.L. Silbiger, "Automatic tumor segmentation using knowledge-based techniques," in *IEEE Transactions on Medical Imaging*, 1998, pp. 187-201.
25. L.M. Fletcher-Heath, L.O. Hall, D.B. Goldgof, and F.R. Murtagh, "Automatic segmentation of non-enhancing brain tumors in magnetic resonance images," in *Artificial Intelligence in Medicine*, 2001, pp. 172-180.
26. M. Prastawa, E. Bullitt, S. Ho, and G. Gerig, "A brain tumor segmentation framework based on outlier detection," in *Medical Image Analysis*, 2004, pp. 275-283.
27. M. Kaus, S. Warfield, A. Nabavi, P.M. Black, F.A. Jolez, and R. Kikinis, "Automated segmentation of MR images of brain tumors," in *Radiology*, 2001, pp. 586-591.
28. M. Prastawa, E. Bullitt, N. Moon, K.V. Leemput, and G. Gerig, "Automatic brain tumor segmentation by subject specific modification of atlas priors," in *Academic Radiology*, 2003, pp. 1341-1348.
29. C.H. Lee, M. Schmidt, A. Murtha, A. Bistritz, J. Sander, and R. Greiner, "Segmenting brain tumor with conditional random fields and support vector machines," in *Proceedings of Workshop on*

Computer Vision for Biomedical Image Applications at International Conference on Computer Vision, 2005.

30. B.L Dean, B.P Drayer, C.R Bird, R.A. Flom, J.A. Hodak, S.W. Coons, and R.G Carey, "Gliomas: Classification with MR Imaging," in *Radiology*, 1990, pp. 411-415.
31. A.I. Holodny, A.O. Nusbaum, S. Festa, I.N. Pronin, H.J. Lee, and A.J. Kalnin, "Correlation between the degree of contrast enhancement and the volume of peritumoral edema in meningiomas and malignant gliomas," in *Neuroradiology*, 1999, pp. 820-825.
32. T. Nakano, K. Asano, H. Miura, S. Itoh, and S. Suzuki, "Meningiomas with brain edema. Radiologic characteristics on MRI and review of the literature," in *Journal of Clinical Imaging*, 2002, pp. 243-249
33. M.L. White, Y. Zhang, P. Kirby, and T.C. Ryken, "Can Tumor Contrast Enhancement be used as a Criterion for Differentiating Tumor Grades of Oligodendrogliomas," in *American Journal of Neuroradiology*, 2005, pp. 784-709.
34. M.A. Hammoud, R. Sawaya, W. Shi, P.F. Thall, and N.E. Leeds, "Prognostic significance of preoperative MRI scans in glioblastoma multiforme," in *Journal of Neurooncology*, 1996, pp. 65–73.
35. M.V. Maldaun, D. Suki, F.F. Lang, S. Prabhu, W. Shi, G.N. Fuller, D.M. Wildrick, and R. Sawaya, "Cystic glioblastoma multiforme: survival outcomes in 22 cases," in *Journal of Neurosurgery*, 2004, pp. 61–67.
36. G.E. Keles, B. Anderson,and M.S. Berger, "The effect of extent of resection on time to tumor progression and survival in patients with glioblastoma multiforme of the cerebral hemisphere," in *Surgical Neurology*, 1999, pp. 371–379.
37. WB Pope, J Sayre, A Perlina, J.P. Villablanca, P.S. Mischel, and T.F. Cloughesy, "MR Imaging Correlates of Survival in Patients with High-Grade Gliomas," in *American Journal of Neuroradiology*, 2005, pp. 2466-2474.
38. B. Donahue, C.B. Scott, J.S. Nelson, M. Rotman, K.J. Murray, D.F. Nelson, F.L. Banker, J.D. Earle, J.A. Fischbach, S.O. Asbell, L.E. Gaspar, A.M. Markoe, and W. Curran, "Influence of an oligodendroglial component on the survival of patients with anaplastic astrocytomas: a report of Radiation Therapy Oncology Group 83-02," in *International Journal of Radiation Oncology, Biology, and Physics*, 1997, pp. 911–914.
39. J.F. Megyesi, E. Kachur, D.H. Lee, C.Z. Magdalena, R.A. Betensky, P.A. Forsyth, Y. Okada, H. Sasaki, M. Mizoguchi, D.N. Louis, and J.G. Cairncross, "Imaging Correlates of Molecular Signatures in Oligodendrogliomas," in *Clinical Cancer Research*, 2004, pp. 4303-4306
40. M.D. Jenkinson, D.G. Du Plessis, T.S. Smith, K.A. Joyce, P.C. Warnke, and C. Walker, "Histological growth patterns and genotype in oligodendroglial tumors: correlation with MRI features," in *Brain*, 2006, pp. 1884-1891.
41. M. Just and M. Thelen, "Tissue Characterization with T1, T2, and Proton Density Values: Results in 160 Patients with Brain Tumors," in *Radiology*, 1988, pp. 779-785.
42. M. Aghi, P. Gaviani, J.W. Henson, T.T. Batchelor, D.N. Louis, and F.G. Barker, "Magnetic Resonance Imaging Characteristics Predict Epidermal Growth Factor Receptor Amplification Status in Glioblastoma," in *Clinical Cancer Research*, 2005, pp. 8600-8605.
43. D. Goldberg-Zimring, I.F. Talos, J.G. Bhagwat, S.J. Haker, P.M. Black, and K.H Zou, "Statistical validation of brain tumor shape approximation via spherical harmonics for image-guided neurosurgery," in *Academic Radiology*, 2005, pp. 459-466.
44. D. Goldberg-Zimring, B. Shalmon, K.H. Zou, H. Azhari, D. Nass, and A. Achiron, "Assessment of multiple sclerosis lesions with spherical harmonics: comparison of MR imaging and pathologic findings," in *Radiology*, 2005, pp. 1036-1044.
45. M. Styner, G. Gerig, J. Lieberman, D. Jones, and D. Weinberger, "Statistical shape analysis of neuroanatomical structures based on medial models," in *Medical Image Analysis*, 2003, pp. 207-220.
46. D. Mahmoud-Ghoneim, G. Toussaint, J.M. Constans, and J.D. de Certaines, "Three dimensional texture analysis in MRI: a preliminary evaluation in gliomas," in *Magnetic Resonance Imaging*, 2003, pp. 983-987.

47. I.F. Talos, K.H. Zou, L. Ohno-Machado, J.G. Bhagwat, R. Kikinis, P.M. Black, and F.A. Jolesz, "Supratentorial Low-Grade Glioma Resectability: Statistical Predictive Analysis Based on Anatomic MR Features and Tumor Characteristics, in *Neuroradiology*, 2006, pp. 506-513.
48. J.M.Provenzale, S. Mukundan, and D.P. Barboriak, "Diffusion-weighted and Perfusion MR Imaging for Brain Tumor Characterization and Assessment of Treatment Response," in *Radiology*, 2006, pp. 632-649.
49. A. Devos, A.W. Simonetti, M. van der Graaf, L. Lukas, J.K. Suykens, L. Vanhamme, L.C. Buydens, A. Heerschap, and S. Van Huffel, "The use of multivariate MR Imaging intensities versus metabolic data from MR spectroscopic imaging for brain tumor classification," in *Journal of Magnetic Resonance*, 2005, pp. 218-228.
50. L.G. Nyul and J.K. Udupa, "On Standardizing the MR Image Intensity Scale," in *Magnetic Resonance in Medicine*, 1999, pp. 1072-81.
51. L.G. Nyul, J.K. Udupa, X. Zhang, "New variants of a method of MRI scale standardization," in *IEEE Transactions on Medical Imaging*, 2000, pp. 143-50.
52. E. Sharon, A. Brandt, and R. Basri, "Fast Multiscale Image Segmentation," in *Proceedings of IEEE Conference on Computer Vision and Pattern Recognition*, 2000, pp. 70-77.
53. J.J. Corso, E. Sharon, S. Dube, S. El-Saden, U. Sinha, and A. Yuille, "Efficient Multilevel Brain Tumor Segmentation with Integrated Bayesian Model Classification," in *IEEE Transactions on Medical Imaging, 2007*, in press.
54. R.M. Haralick, K. Shanmugam, and I. Dinstein, "Texture features for image classification," in *IEEE Transactions on Systems, Man, and Cybernetics*, 1973, pp. 610-621.
55. S. Dube, J. Sayer, T.F. Cloughesy, and U. Sinha, "Multivariate Linear Modeling of Quantitative MR Features for Prognosis in Patients with Glioblastoma Multiforme," in *Proceedings of the Joint Annual International Conference of ISMRM-ESMRMB*, Berlin, Germany, 2007, in press.
56. S. Dube, T.F. Cloughesy, and U. Sinha, "Content Based Image Retrieval of Glioblastoma Multiforme," in *Proceedings of the International Medical Informatics Association (MedInfo)*, Brisbane, Australia, 2007, in press.

Nonparametric smoothing and its applications in biomedical imaging

John D. Carew* and Ming Yuan†

*Departments of Biostatistics and Radiology, Emory University, 1518 Clifton Rd NE, Atlanta, GA 30322

†School of Industrial and Systems Engineering, Georgia Institute of Technology, 755 Ferst Dr. NW, Atlanta, GA 30332

Abstract. This chapter provides a brief review of smoothing splines, a powerful nonparametric smoothing tool that have found applications in a variety of biomedical imaging problems. It involves an optimization problem in a reproducing kernel Hilbert space that balances the tradeoff between the fidelity to the data and the smoothness of the estimate. After reviewing the basic theory, we then provide two examples where these methods have been applied to magnetic resonance imaging (MRI) data. The first example involves smoothing a functional MRI (fMRI) time series to induce a known correlation structure. The second example shows how to model arterial blood flow and shear.

Keywords: Magnetic resonance imaging, Reproducing kernel Hilbert space, Smoothing splines
PACS: 87.85.Pq

1. INTRODUCTION

Biomedical imaging systems can be used to investigate complex biological problems and their associated pathologies. The image data are realizations of a complicated spatiotemporal process that is governed by the underlying biological system, the imaging device (which usually observes the biological system indirectly), and experimentally manipulated variables in the case of designed experiments.

Recent advances in imaging systems now allow clinicians and scientists to acquire data at high spatial and temporal resolutions. Consequently, there has been rapid growth in the number of studies that use imaging as a primary tool. As the costs decrease and the equipment becomes even more accessible, the number and frequency of biomedical imaging studies will continue to grow.

Investigators who employ imaging methods are faced with a number of challenges in the design, analysis, and interpretation of their studies. One challenge is the large quantities of data that are collected during an imaging study. Such large data sets require special considerations for processing and statistical inference. Perhaps the most difficult challenge is to identify an appropriate model for the data.

Nonparametric smoothing has been widely applied to biomedical image data in various contexts. This method provides a flexible model for complicated data without the need to explicitly specify a parametric form. For this reason, nonparametric smoothing is appropriately advantageous for biomedical images. In a common situation, we observe a set of pixel or voxel (volume element) intensities $y_i \in \mathbb{R}$, for $i = 1, \ldots, n$, at

CP953, *Data Mining, Systems Analysis, and Optimization in Biomedicine*
edited by O. Seref, O. E. Kundakcioglu, P. M. Pardalos

corresponding image locations $\mathbf{t}_i \in \mathbb{R}^d$. The basic nonparametric model for our data is

$$y_i = f(\mathbf{t}_i) + \varepsilon_i, \tag{1}$$

where f is some smooth function that depends on location in the image and ε_i is an additive error. We will require that f lies in a function space that is sufficiently rich to provide a satisfactory model of the imaged phenomenon. Later, we will introduce an approach that puts f in a reproducing kernel Hilbert space (RKHS).

In this chapter, we begin with begin with a review of the basic theory of the RKHS approach to smoothing splines. We then provide two examples where these methods have been applied to magnetic resonance imaging (MRI) data. The first example involves smoothing a functional MRI (fMRI) time series to induce a known correlation structure. The second example shows how to model arterial blood flow and shear. The biased choice of examples reflects the authors' own experience and is not intended to be a comprehensive review.

2. SMOOTHING SPLINES

2.1. Reproducing kernel Hilbert spaces and roughness penalty

In parametric regression, $f(\cdot)$ of (1) is assumed to take the form $f(\mathbf{t};\beta)$, which is known up to the parameters β. In particular, in the classical linear regression, f is assumed to be linear and β represents the unknown regression coefficient. To avoid potential model misspecification, in nonparametric regression, f can vary in a RKHS which may be of infinite dimension.

A RKHS is a function space in which point-wise evaluation is a continuous linear functional. They can also be characterized by reproducing kernels. A reproducing kernel, or kernel for simplicity, is a positive definite bivariate function. Let $\mathcal{T}$ be an index set, i.e., the domain for $\mathbf{t}_i$. A bivariate function $K(\cdot,\cdot) : \mathcal{T} \times \mathcal{T} \mapsto \mathbb{R}$ is called (strictly) positive definite if and only if

$$\sum_{i,j=1}^{n} a_i a_j K(t_i, t_j)$$

for any n, $t_1, \ldots, t_n \in \mathcal{T}$, and $a_1, \ldots, a_n \in \mathbb{R}$. It is known [1] that there is a one-to-one correspondence between a positive definite kernel $K(\cdot,\cdot)$ and a RKHS $\mathcal{H}$. Moreover, for any $\mathbf{t} \in \mathcal{T}$ and $f \in \mathcal{H}$, $< K_{\mathbf{t}}, f >= f(\mathbf{t})$ where $K_{\mathbf{t}}(\cdot) = K(\mathbf{t},\cdot)$. Furthermore, it can be shown [1] that $< K_{\mathbf{t}}, K_{\mathbf{s}} >= K(\mathbf{t},\mathbf{s})$ defines an inner product and consequently a norm on $\mathcal{H}$.

As an example, consider the function space that consists of all periodic functions on the unit interval $[0,1]$ which integrate to 0 and have square integrable second derivatives. It is well known that the kernel corresponds to the norm $\int_0^1 [f''(s)]^2 ds$ is $K(s,t) = B_2(s)B_2(t)/4 - B_4(|s-t|)/24$ where $s,t \in [0,1]$, and B_m is the mth Bernoulli polynomial. Another class of popular kernels is the Matérn kernels defined for $\mathbb{R}^d$:

$$K(\mathbf{s},\mathbf{t}) = \frac{1}{\Gamma(\nu)2^{\nu-1}} \left(\frac{\sqrt{2\nu}}{\ell} \|\mathbf{s}-\mathbf{t}\| \right)^{\nu} B_{\nu} \left(\frac{\sqrt{2\nu}}{\ell} \|\mathbf{s}-\mathbf{t}\| \right),$$

where parameters ν and ℓ are positive, and B_ν is a modified Bessel function. Functions from a RKHS with Matérn kernel is $\nu - 1$ times differentiable, and when $\nu \to \infty$,

$$K(\mathbf{s},\mathbf{t}) \to \exp\left(-\frac{||\mathbf{s}-\mathbf{t}||^2}{2\ell^2}\right)$$

which is the popular Gaussian kernel.

2.2. Smoothing splines and representer lemma

Assume that $t \in [0,1]$. A popular approach to the nonparametric estimation of f in (1) is the smoothing splines which minimizes the penalized residual sum of squares

$$\frac{1}{n}\sum_{i=1}^{n}(y_i - f(t_i))^2 + \lambda \int_0^1 [f''(u)]^2 du,$$

where $\lambda \geq 0$ is a tuning parameter that balances the tradeoff between the fidelity to the data measured by the first term and the smoothness of the estimate that is measured by $\int (f'')^2$. When $\lambda = 0$, it becomes interpolation; whereas when λ approaches $+\infty$, the smoothing spline estimate must satisfy $\int (f'')^2 = 0$ and therefore reduces to linear regression.

More generally, given some particular basis functions $\{\phi_1, \ldots, \phi_M\}$ on $\mathcal{T}$ and a kernel K, we estimate f by minimizing

$$\frac{1}{n}\sum_{i=1}^{n}(y_i - f(t_i))^2 + \lambda ||h||^2_{\mathcal{H}_K}$$

over all functions $f = \sum_\nu d_\nu \phi_\nu + h$ where $h \in \mathcal{H}_K$ and $\mathcal{H}_K$ is the RKHS equipped with reproducing kernel K. The flexibility of this nonparametric approach is in that f is only known to come from a possibly infinitely dimensional function space. On the other, this flexibility also poses problem in computation because the minimization is now taken over a possibly infinitely dimensional space. This problem, however, can be overcome thanks to the so-called representer lemma [2].

Lemma 1 *Consider the following problem: find a function f of the form $\sum_\nu d_\nu \phi_\nu + h$ where $h \in \mathcal{H}_K$ to minimize*

$$\frac{1}{n}\sum_{i=1}^{n}(y_i - f(t_i))^2 + \lambda ||h||^2_{\mathcal{H}_K}. \qquad (2)$$

Then the minimizer has a representation of the form

$$\hat{f}(\cdot) = \sum_\nu d_\nu \phi_\nu(\cdot) + \sum_{i=1}^{n} c_i K(\mathbf{t}_i, \cdot). \qquad (3)$$

This result shows that although the original minimization problem is of infinite dimension, its solution can be characterized by a finite number of coefficients $d = (d_1, \ldots, d_M)$ and $c = (c_1, \ldots, c_n)$ and these coefficients can be computed by substituting (3) into the first term of (2) and using the fact that

$$\|\sum_{i=1}^{n} c_i K(\mathbf{t}_i, \cdot)\|^2_{\mathcal{H}_K} = c' K_n c$$

where K_n is a $n \times n$ matrix whose (i, j)th entry is $K(\mathbf{t}_i, \mathbf{t}_j)$.

2.3. Tuning with GCV

The performance of the smoothing spline estimate critically depends on the choice of the tuning parameter λ. Clearly when λ is too small, the estimate can be too wiggly because of the overfitting. But if λ is too big, the estimate may be too smooth and fail to capture the main characteristics of f. A good choice of the tuning parameter aims at minimizing the mean squared error (MSE) $E[\hat{f}(\mathbf{t})_\lambda - f(\mathbf{t})]^2$ where the expectation is taken over a randomly selected $\mathbf{t}$. Not knowing the true function f, one may rely on a proxy of the MSE, the cross-validation score:

$$CV(\lambda) = \frac{1}{n} \sum_{i=1}^{n} [y_i - \hat{f}_\lambda^{[-i]}(\mathbf{t}_i)]^2$$

where $\hat{f}_\lambda^{[-i]}$ is the smoothing spline estimate with the ith observation $(\mathbf{t}_i, y_i)$ removed. The cross-validation score mimics the prediction error for a future observation and we can choose a value for the tuning parameter that minimizes $CV(\lambda)$. Evaluating $CV(\lambda)$ naively, however, can be computationally expensive since a total n smoothing splines, $\hat{f}_\lambda^{[-1]}, \ldots, \hat{f}_\lambda^{[-n]}$, need to be fitted. Fortunately, it has been shown [3] that

$$y_i - \hat{f}_\lambda^{[-i]}(\mathbf{t}_i) = \frac{y_i - \hat{f}_\lambda(\mathbf{t}_i)}{1 - a_{ii}(\lambda)}$$

where $\hat{f}_\lambda$ is the smoothing spline estimate with all the data and $a_{ii}(\lambda)$ is the (i, i)th entry of its hat matrix $A(\lambda)$. The hat matrix is a $n \times n$ matrix such that

$$\begin{pmatrix} \hat{f}_\lambda(\mathbf{t}_1) \\ \hat{f}_\lambda(\mathbf{t}_2) \\ \vdots \\ \hat{f}_\lambda(\mathbf{t}_n) \end{pmatrix} = A(\lambda) \begin{pmatrix} y_1 \\ y_2 \\ \vdots \\ y_n \end{pmatrix}.$$

Therefore, $CV(\lambda)$ can be computed using only $\hat{f}_\lambda$ as opposed to $\hat{f}_\lambda^{[-1]}, \ldots, \hat{f}_\lambda^{[-n]}$,

$$CV(\lambda) = \frac{1}{n} \sum_{i=1}^{n} \left(\frac{y_i - \hat{f}_\lambda(\mathbf{t}_i)}{1 - a_{ii}(\lambda)} \right)^2.$$

Craven and Wahba [3] further observed that $CV(\lambda)$ may exhibit some abnormal behavior and can be stabilized by replacing $a_{ii}(\lambda)$ with trace$(A(\lambda))/n$, resulting in what they called generalized cross validation (GCV) score:

$$GCV(\lambda) = \frac{n^{-1}\sum_{i=1}^{n}\left(y_i - \hat{f}_\lambda(\mathbf{t}_i)\right)^2}{\left[n^{-1}\text{trace}\left(I - A(\lambda)\right)\right]^2}.$$

3. SPLINE SMOOTHING OF FMRI TIME SERIES

Functional MRI (fMRI) measures the blood oxygen level dependent signal in tissue. When applied to the brain, the goal is to measure brain function where changes in blood oxygenation are surrogates for changes in neuronal function. One method for analyzing fMRI time series involves smoothing the data to induce a known autocorrelation structure. The goal of smoothing is to reduce the bias in estimates of linear model parameters.

3.1. Introduction

Linear parametric regression models of fMRI time series have autocorrelated residual errors [4]. Two general approaches to deal with these autocorrelated residuals are temporal smoothing [5, 6] and whitening [7]. Data are whitened by first modeling the intrinsic autocorrelations and then removing them from the data. Provided that the model of the autocorrelations is correct, whitening yields the minimum variance estimate among all unbiased estimates of the linear regression model parameters [7]. Smoothing conditions the autocorrelation structure of an fMRI time series. Appropriate smoothing can minimize the bias in variance estimators for a contrast of a linear model parameter and, thus, the difference between the applied autocorrelation structure and the intrinsic autocorrelations [8]. Friston *et al.* [8] argue that it is preferable to smooth with the goal of minimizing bias rather than whiten data since it is difficult to obtain an accurate estimate of the intrinsic autocorrelations. Thus, it is prudent to investigate the appropriateness of various smoothing methods for fMRI time series. Here, we focus on spline smoothing of fMRI time series where the appropriate degree of smoothing is selected by GCV. We compare the variance and the bias of the variance estimator for a contrast of a regression model parameter under GCV-spline smoothing to those under the low pass filter (SPM-HRF) implemented in SPM99 (Wellcome Department of Cognitive Neurology, London) and no smoothing. We begin by introducing the general linear model for fMRI time series.

3.2. Linear models for fMRI time series

One popular model of fMRI time series is a general linear model (GLM) [5, 6],

$$\mathbf{y} = \mathbf{X}\beta + \mathbf{K}\varepsilon, \tag{4}$$

where $\mathbf{y} = (y_1, \ldots, y_n)'$ is an $n \times 1$ matrix of equally-spaced samples of the time series, $\mathbf{X}$ is the model (design) matrix with columns that contain signals of interest and nuisance signals, β is an unknown parameter, $\mathbf{K}$ is an unknown convolution matrix which characterizes the autocorrelations, and $\varepsilon \sim \mathcal{N}(\mathbf{0}, \sigma^2\mathbf{I})$. The autocorrelation matrix is given by $\mathbf{V} \propto \mathbf{KK}'$. Let $\mathbf{S}$ be a linear transformation. The matrix $\mathbf{S}$ is applied to the model (4) to give

$$\mathbf{Sy} = \mathbf{SX}\beta + \mathbf{SK}\varepsilon. \tag{5}$$

If $\mathbf{K}$ is known, the approach of whitening is to choose $\mathbf{S} = \mathbf{K}^{-1}$. This transformation will allow a minimum variance, unbiased ordinary least squares (OLS) estimate of β in equation (5) given by $\hat{\beta} = (\mathbf{SX})^{+}\mathbf{Sy}$ where $^{+}$ denotes the pseudoinverse, i.e., $\mathbf{X}^{+} = (\mathbf{X}'\mathbf{X})^{-1}\mathbf{X}'$. Suppose that $\mathbf{S} \neq \mathbf{K}^{-1}$. Then the assumed autocorrelation $\mathbf{V}_a \propto \mathbf{S}^{-1}(\mathbf{S}^{-1})'$ will differ from the actual autocorrelation $\mathbf{V}$ and result in biased OLS estimates of the variance of a contrast of β. The amount of bias depends on the accuracy of the approximation of $\mathbf{K}^{-1}$. Friston *et al.* [8] found that computing $\mathbf{V}_a$ with a low order autoregressive (AR) model or a $1/f$ model [9] results in unacceptable bias for fMRI inference. An alternative approach to whitening is to smooth model (4) with $\mathbf{S}$ such that the assumed autocorrelation $\mathbf{V}_a \propto \mathbf{SS}' \approx \mathbf{SVS}'$, the true autocorrelation under smoothing. Even when $\mathbf{V}$ is unknown, bias can be minimized [8]. Since the bias as a function of $\mathbf{S}$ is difficult to directly minimize it is more practical to determine if a method for computing $\mathbf{S}$ gives acceptable levels of bias. Our approach is to take $\mathbf{S}$ as the influence matrix of a cubic smoothing spline.

Given $\mathbf{S}$ computed with GCV-spline, the variance of a contrast of $\hat{\beta}$ and the bias of the variance estimator can be computed with equations given in [8]:

$$\mathrm{var}(\mathbf{c}'\hat{\beta}) = \sigma^2\mathbf{c}'(\mathbf{SX})^{+}\mathbf{SVS}'(\mathbf{SX})^{+T}\mathbf{c} \tag{6}$$

and

$$\begin{aligned} \mathrm{bias}(\mathbf{S},\mathbf{V}) &= \frac{\mathrm{var}(\mathbf{c}^T\hat{\beta}) - \mathrm{E}[\widehat{\mathrm{var}(\mathbf{c}'\hat{\beta})}]}{\mathrm{var}(\mathbf{c}^T\hat{\beta})} \\ &= 1 - \frac{\mathrm{tr}[\mathbf{LSVS}']\mathbf{c}'(\mathbf{SX})^{+}\mathbf{SV}_a\mathbf{S}'(\mathbf{SX})^{+'}\mathbf{c}}{\mathrm{tr}[\mathbf{LSV}_a\mathbf{S}']\mathbf{c}'(\mathbf{SX})^{+}\mathbf{SVS}'(\mathbf{SX})^{+'}\mathbf{c}}, \end{aligned} \tag{7}$$

where $\mathbf{L} = \mathbf{I} - \mathbf{SX}(\mathbf{SX})^{+}$ is the residual forming matrix and $\mathbf{c}$ is a contrast vector for hypothesis testing of the components of $\hat{\beta}$. An estimate of $\mathrm{var}(\mathbf{c}'\hat{\beta})$ is obtained by replacing $\mathbf{V}$ with its assumed value, $\mathbf{V}_a$, and σ^2 with its estimate

$$\hat{\sigma}^2 = \frac{(\mathbf{LSy})'\mathbf{LSy}}{\mathrm{tr}(\mathbf{LV}_a)}, \tag{8}$$

given in [6].

3.3. fMRI experiment

One hundred thirty-two brain volumes were acquired from a healthy volunteer with a 1.5T scanner running a gradient echo EPI pulse sequence for BOLD contrast. During the fMRI experiment, four symmetric blocks of photic stimulation and darkness were presented to the volunteer. This stimulus was designed to activate the visual cortex. The raw data were spatially smoothed in the frequency domain with a Hamming filter to increase SNR [10]. An intensity mask was applied to the images to exclude voxels located outside of the brain. Head motion was corrected with the algorithm in SPM99.

The data were analyzed with the GLM in equation (5) with $\mathbf{S}$ determined by GCV-spline. The design matrix contained one column with a boxcar function that was convolved with the SPM99 HRF to model the expected BOLD response to the fMRI experiment. Polynomial terms up to order three were also included to achieve high pass filtering following the motivation of [11]. Moreover, the design matrix $\mathbf{X} = [\mathbf{s}\ \mathbf{1}\ \mathbf{t}\ \mathbf{t}^2\ \mathbf{t}^3]$, where $\mathbf{s}$ is the convolved boxcar, $\mathbf{1}$ is a $n \times 1$ column of 1s, and $\mathbf{t} = (t_1, t_2, \cdots, t_n)'$. An optimal spline smoothing matrix was computed for each time series. The optimal $\lambda =: \lambda_{\text{opt}}$ was determined by a grid search for the minimum of the GCV score over $\lambda \in [10^{-3}, 10^6]$ on a $\log_{10}$ scale with steps of size 0.1. The data were also analyzed with $\mathbf{S}$ given by the SPM-HRF and $\mathbf{S} = \mathbf{I}$ (no smoothing). Images of voxel t-statistics from a test of the null hypothesis $\mathbf{c}'\beta = 0$ for the contrast $\mathbf{c}' = [1\ 0\ 0\ 0\ 0]$ were created for each of the three smoothing strategies with the statistic given by

$$t = \frac{\mathbf{c}'\hat{\beta}}{\sqrt{\widehat{\text{var}}(\mathbf{c}'\hat{\beta})}}. \tag{9}$$

3.4. Bias computations with simulated data

Bias of the variance estimator can be computed with equation (7). However, this demands knowledge of the true variance. For an observed fMRI time series, the true variance is unknown. It is possible to generate simulated a time series with a known autocorrelation structure to allow direct calculation of bias. To generate a set of reasonable simulated fMRI time series, autocorrelation structures can be estimated from real data. The autocorrelation estimates can be used to induce correlations in pseudo-random numbers independently sampled from a Gaussian density with a known variance. Finally, a signal can be added to the correlated Gaussian samples.

Autocorrelation structures were estimated from the residuals of the fit of the time series from the fMRI experiment with the smoothing matrix $\mathbf{S} = \mathbf{I}$ (i.e., no smoothing) with an AR(8) model. The residuals for a single time series $\mathbf{y}$ are

$$\mathbf{r} = \mathbf{L}\mathbf{y}. \tag{10}$$

The AR(8) model of the i-th residual is

$$r_i = b_1 r_{i-1} + \cdots + b_8 r_{i-8} + \zeta_i, \tag{11}$$

where $b_1, \cdots, b_8$ are the AR coefficients and $\zeta_i \sim \mathcal{N}(0, \sigma_r^2)$. The more convenient matrix representation of equation (11) is, following [8],

$$\mathbf{r} = (\mathbf{I} - \mathbf{B})^{-1}\zeta, \tag{12}$$

where $\mathbf{B}$ is a matrix of AR model coefficients and $\zeta \sim \mathcal{N}(\mathbf{0}, \sigma_r^2\mathbf{I})$. The AR coefficients were estimated with the least squares procedure [12]. Then, the estimated convolution matrix is

$$\tilde{\mathbf{K}} = (\mathbf{I} - \hat{\mathbf{B}})^{-1}. \tag{13}$$

A simulated time series $\tilde{\mathbf{y}}$ is constructed by

$$\tilde{\mathbf{y}} = 0.15 \cdot \mathbf{s} + \tilde{\mathbf{K}}\mathbf{e}, \tag{14}$$

where $\mathbf{s}$ is the BOLD signal and $\mathbf{e}$ is a sample from $\mathcal{N}(\mathbf{0}, \mathbf{I})$. The coefficient of the signal, namely 0.15, and the unit variance error term $\mathbf{e}$ were selected to match the simulated data parameters used by [13].

For each voxel in the masked data set, a separate $\tilde{\mathbf{K}}$ was estimated and a separate $\mathbf{e}$ was generated to produce a simulated time series with equation (14). The simulated time series were assigned to the spatial location where the $\tilde{\mathbf{K}}$ was estimated. This preserves the spatial variability of autocorrelation structures in the simulated data. A subset of the simulated data was selected based on 100 voxels with the largest t-statistics of a test of the null hypothesis $\mathbf{c}'\beta = 0$ for the contrast $\mathbf{c}' = [1\ 0\ 0\ 0\ 0]$ in the real data under $\mathbf{S} = \mathbf{I}$. This subset contained simulated time series with autocorrelation structures from regions that were presumably related to the stimulus in the fMRI experiment.

The variance and bias of the variance estimator were calculated for each time series in the entire simulated data set with three different smoothing matrices: $\mathbf{S} = \mathbf{I}$, $\mathbf{S}$ equal to the SPM-HRF filter matrix, and the spline smoothing (influence) matrix. The SPM-HRF smoothing matrix is generated with the `spm_make_filter` function with the option for no high pass filtering and "hrf" for the low pass filter. This function is part of the SPM99 package. For the spline smoothing, the optimal λ was determined with the same parameters as used with the real data, i.e., for each time series a GCV search over $\lambda \in [10^{-3}, 10^6]$ on a $\log_{10}$ scale with steps of size 0.1. The variance of $\mathbf{c}'\beta$ with $\mathbf{c}' = [1\ 0\ 0\ 0\ 0]$ and the bias of the variance estimator were computed for each of the time series under each smoothing method with equations (6) and (7), respectively. The known variance-covariance matrix $\mathbf{V} = \tilde{\mathbf{K}}\tilde{\mathbf{K}}'$ and the assumed variance-covariance matrix $\mathbf{V}_a = \mathbf{S}\mathbf{S}'$ with $\mathbf{S}$ given by the particular smoothing matrix under investigation.

3.5. Real data results

To illustrate the GCV-spline smoothing the real fMRI data, we considered two example time series that have statistically significant signals, but require different amounts of smoothing. An example of a time series that required little smoothing and one that required more smoothing are included in Fig. 1 (a) and (b), respectively. The corresponding plots of the GCV score demonstrate the numerical stability of GCV for these two

TABLE 1. Bias for Three Smoothing Strategies (Whole Brain)

	GCV-Spline	SPM-HRF	No Smoothing
Mean Bias	0.0200	0.0701	0.4019
Median Bias	0.0037	0.1608	0.5399

TABLE 2. Bias for Three Smoothing Strategies (100 Significant Voxels)

	GCV-Spline	SPM-HRF	No Smoothing
Mean Bias	-1.800×10^{-5}	0.1801	0.5964
Median Bias	-0.0332	0.2312	0.6627

time series (Fig. 1 (c) and (d)). These plots are similar in the relative amount of curvature to the GCV scores from the other time series. The GCV-spline method performs, on average, more smoothing than the SPM-HRF smoothing kernel. Fig. 1 (e) and (f) shows the equivalent smoothing spline kernels and the SPM-HRF kernel for the time series in Fig. 1 (a) and (b), respectively. For $\log_{10}(\lambda) \approx 0.78$, both methods provide the same amount of smoothing (Fig. 1 (e)).

3.6. Simulated data

The amount of bias in the variance estimates for the simulated data differed between the three smoothing strategies. The mean and median bias with the simulated data for the three methods are shown in Table 1. The mean and median bias were positive for all methods. GCV-spline smoothing had bias that was closest to zero, whereas the bias was greatest for no smoothing. Histograms of the bias show that, on average, GCV-spline is unbiased and that SPM-HRF and no smoothing are biased (Fig. 2). The 100 voxel subset of the simulated data show similar trends in the bias that are summarized in Table 2.

3.7. Conclusions

We have shown that that cubic spline smoothing with generalized cross-validation provides a method to determine the optimal amount of smoothing for an fMRI time series. This method not only conditions the autocorrelation structure on the data, but it does so in a way to optimally separate the underlying signal from noise. By selecting the λ that minimizes the GCV score, the smoothing spline estimator for the signal will minimize the predictive mean-square error [3]. The empirical results from the spline smoothing of the fMRI data show that GCV is a sufficient method to automatically choose the appropriate amount of smoothing.

4. NONPARAMETRIC ESTIMATION OF FLUID VELOCITY AND SHEAR

In this section, we describe flexible, nonparametric estimators for fluid velocity and shear. The estimation is based on penalized least squares over a reproducing kernel Hilbert space (RKHS). A Bayesian confidence interval is derived for the shear estimator. Our methods are applied to three phase contrast MRI (PC-MRI) data sets that were acquired from vascular “phantoms.” The phantoms are idealized models of human arteries.

4.1. Introduction

The level of cardiovascular plaque formations is related to the risk of stroke and heart disease. Atherosclerosis preferentially forms in areas of reduced wall shear stress (WSS) [14, 15, 16]. Thus, arterial WSS measurements may provide a diagnostic indicator for atherosclerosis. Most of the evidence supporting this connection is from studies of arteries that were dissected from cadavers. In vivo methods are beneficial for studying the formation and progression of the disease. One method for noninvasively estimating WSS is through post-processing of phase contrast magnetic resonance images. PC-MRI is capable of measuring blood velocity along the direction of arterial blood flow. The WSS is proportional to the component normal to the wall of the gradient of the blood velocity at the arterial wall. The post-processing step involves fitting a surface to the PC-MRI velocity images over the cross-section of the artery and computing the derivatives at the boundary. It is difficult to find a parametric surface that fits the velocity data well. In fact, only for the simplest vessel geometries are there closed-form analytical descriptions of fluid flow [17]. The general flow of fluid is governed by solutions to the Navier-Stokes partial differential equations. Since vascular geometries are complex, it is most appropriate to fit a nonparametric function to the velocity data.

4.2. Theory and methods

When a fluid flows along one axis, WSS[1] at a point $\mathbf{t} := (x,y)' \in \mathbb{R}^2$ along the wall with unit normal vector $\mathbf{n}(\mathbf{t}) := (n_x, n_y)'$ is

$$\begin{aligned} W(\mathbf{t}) &:= \mu \nabla f(\mathbf{t}) \cdot \mathbf{n}(\mathbf{t}) & (15) \\ &= \mu \left(\frac{\partial f(\mathbf{t})}{\partial x} n_x + \frac{\partial f(\mathbf{t})}{\partial y} n_y \right). & (16) \end{aligned}$$

The function $f(\mathbf{t})$ is the spatially dependent fluid velocity where $\{\mathbf{t}\}$ lie in a plane that is perpendicular to the axis of fluid flow. The constant μ is the dynamic viscosity of

[1] Shear can be calculated at any point within the vessel and not only at the wall. It is because of the connection to plaque formation that we are particularly interested in shear at the wall.

the fluid. We can estimate $W(\mathbf{t})$ by first estimating $f(\mathbf{t})$ and then differentiate $\hat{f}(\mathbf{t})$. Substituting derivatives of $\hat{f}(\mathbf{t})$ for the derivatives of the velocity function in equation (16) gives an estimator for $W(\mathbf{t})$ that is

$$\hat{W}(\mathbf{t}) = \mu\left(\frac{\partial \hat{f}(\mathbf{t})}{\partial x}n_x + \frac{\partial \hat{f}(\mathbf{t})}{\partial y}n_y\right). \tag{17}$$

The objective is to estimate a function from a set of fluid velocity observations that are contaminated by noise. Consider data $\{(\mathbf{t}_i, z_i)\}$, $i = 1, \ldots, n$, where $\mathbf{t}_i = (x_i, y_i)' \in \mathbb{R}^2$ is the point where velocity z_i is observed[2]. In our particular application, the $\mathbf{t}_i$ are pixel locations in a PC-MR image, and the z_i are the corresponding pixel intensities, which represent fluid velocity. The data are modeled as

$$z_i = f(\mathbf{t}_i) + \varepsilon_i, \tag{18}$$

where f is assumed to be a smooth function and the errors $\varepsilon_i \sim_{iid} \mathcal{N}(0, \sigma^2)$.

Model (18) is fit from data by selecting an appropriate f in a RKHS. Let $\mathcal{H}_R$ be a RKHS with reproducing kernel R. An estimator for $f \in \mathcal{H}_R$ is given from the solution to the variational problem

$$\hat{f}(\mathbf{t}) := \arg\min_{f \in \mathcal{H}_R}\left(\sum_{i=1}^{n}(z_i - f(\mathbf{t}_i))^2 + \lambda\|f\|^2_{\mathcal{H}_R}\right). \tag{19}$$

The regularization parameter $\lambda \geq 0$ in equation (19) controls the smoothness of $\hat{f}(\mathbf{t})$ through weighting the penalty $\|f\|^2_{\mathcal{H}_R}$, where $\|\cdot\|_{\mathcal{H}_R}$ is the norm in $\mathcal{H}_R$. The solution to (19) is a special case of a solution to a more general variational problem that was discovered by Kimeldorf and Wahba [2]. The solution is

$$\hat{f}(\mathbf{t}) = \sum_{i=1}^{n}\mathbf{c}_i R(\mathbf{t}, \mathbf{t}_i), \tag{20}$$

where,

$$\mathbf{c} := (\Sigma + \lambda\mathbf{I})^{-1}(z_1, \cdots, z_n)' \tag{21}$$

$$\Sigma := [R(\mathbf{t}_i, \mathbf{t}_j)]_{ij}, \quad i, j = 1, \cdots, n. \tag{22}$$

The important thing to note about equation (20) is that the solution to (19) at any point $\mathbf{t}$ in the domain is simply a linear combination of the RK evaluated at $\mathbf{t}$ and the $\mathbf{t}_i$, with coefficients depending on the $\mathbf{t}_i$, z_i, and λ. As the notation suggests, the symmetric, positive definite matrix Σ can be thought of as the covariance of the $f(\mathbf{t}_i)$.

An explicit solution to (19) has been given. However, two related issues remain: what are appropriate choices for R and λ? These choices are best made with consideration

---

[2] The methods that follow extend to data from higher dimensional spaces, e.g., 3D images, contiguous 2D images, and sequences of 2D or 3D images acquired over time.

given to the characteristics of the specific problem and to the data. First, consider selection of the RK R. Since R specifies the RKHS, an R that gives rise to a sufficiently rich RKHS such that it contains the true, unknown function (that is to be estimated) is required. Thus, consider selecting a RK from a family of functions such as the Matérn family of radial basis functions (RBF) [18, 19]. Let $\tau := ||\mathbf{s} - \mathbf{t}||$ be the Euclidean distance between points $\mathbf{s}, \mathbf{t} \in \mathbb{R}^d$. Then, the ν-th order[3] Matérn function is:

$$R_\nu(\tau) := e^{-\tau}\pi_\nu(\tau), \quad \nu = 0, 1, 2, \ldots, \tag{23}$$

where $\pi_\nu(\cdot)$ is a ν-th order polynomial of a particular form. RK $R_\nu(\tau)$ can be thought of as an isotropic covariance, which gives rotational invariance of the solution. This is important for imaging applications.

The estimator for WSS is obtained by differentiating the velocity estimator. Let $\mathbf{z} := (z_1, \ldots, z_n)'$ and $\mathbf{A}$ be the influence matrix, which maps to data to the fitted values. Let $\mathbf{A}(\mathbf{t})$ be the influence vector for a generic point $\mathbf{t}$. Note that if $\mathbf{t}$ is a particular $\mathbf{t}_i$ then $\mathbf{A}(\mathbf{t})$ is the ith row of the influence matrix. Using this notation we can express the estimator of f at point $\mathbf{t}$ in matrix form for simplicity

$$\hat{f}(\mathbf{t}) = \mathbf{A}(\mathbf{t})\mathbf{z} \tag{24}$$

$$= \mathbf{r}'(\Sigma + \lambda\mathbf{I})^{-1}\mathbf{z}, \tag{25}$$

where $\mathbf{r} := (R(\mathbf{t}, \mathbf{t}_1), \ldots, R(\mathbf{t}, \mathbf{t}_n))'$. Differentiating $\hat{f}$ with respect to x and y provides the necessary derivatives to estimate WSS. Estimator $\hat{f}(\mathbf{t})$ depends on x and y only through $\mathbf{r}$, which depends on R. Thus,

$$\frac{\partial \hat{f}(\mathbf{t})}{\partial x} = \mathbf{r}_x'(\Sigma + \lambda\mathbf{I})^{-1}\mathbf{z} \tag{26}$$

$$\frac{\partial \hat{f}(\mathbf{t})}{\partial y} = \mathbf{r}_y'(\Sigma + \lambda\mathbf{I})^{-1}\mathbf{z}, \tag{27}$$

where

$$\mathbf{r}_x := \left(\frac{\partial R_\nu(\mathbf{t}, \mathbf{t}_1)}{\partial x}, \ldots, \frac{\partial R_\nu(\mathbf{t}, \mathbf{t}_n)}{\partial x}\right)' \tag{28}$$

$$\mathbf{r}_y := \left(\frac{\partial R_\nu(\mathbf{t}, \mathbf{t}_1)}{\partial y}, \ldots, \frac{\partial R_\nu(\mathbf{t}, \mathbf{t}_n)}{\partial y}\right)'. \tag{29}$$

Substitution gives

$$\hat{W}(\mathbf{t}) = \mu\left(n_x \mathbf{r}_x'(\Sigma + \lambda\mathbf{I})^{-1}\mathbf{z} + n_y \mathbf{r}_y'(\Sigma + \lambda\mathbf{I})^{-1}\mathbf{z}\right) \tag{30}$$

$$= \mu\left(n_x \mathbf{r}_x' + n_y \mathbf{r}_y'\right)(\Sigma + \lambda\mathbf{I})^{-1}\mathbf{z}. \tag{31}$$

From equation (31), we see that to estimate WSS, we need only differentiate the reproducing kernel. In general, to estimate any bounded linear functional we apply the functional to the reproducing kernel in (20).

---

[3] What is referred to here as a ν-th order Matérn RBF is sometimes called order $m = \nu + 1/2$.

4.3. Bayesian confidence intervals for shear

In this section we consider inference for WSS. One disadvantage of our nonparametric estimator for W is that it is biased due to the penalty on the norm of f. This complicates inference for estimates of shear. By interpreting our estimators as Bayes estimators with respect to a (conjugate) normal prior on f, we can construct a Bayesian confidence interval. The original Bayesian confidence interval [20] was constructed for the smoothing spline. Here, we extend Wahba's interval to WSS.

A Bayesian confidence interval for $W(\mathbf{t})$ is constructed from the mean squared error (MSE) of its estimator $\hat{W}(\mathbf{t})$. Recall that

$$\mathrm{MSE}[\hat{W}(\mathbf{t})] = \mathrm{Var}[\hat{W}(\mathbf{t})] + \mathrm{Bias}^2[\hat{W}(\mathbf{t})]. \tag{32}$$

A $(1-\alpha)100\%$ approximate confidence interval has the form

$$\hat{W}(\mathbf{t}) \pm Z_{1-\alpha/2}\sqrt{\mathrm{MSE}[\hat{W}(\mathbf{t})]}, \tag{33}$$

where $Z_{1-\alpha/2}$ denotes the $1-\alpha/2$ quantile of the standard normal distribution. To estimate the MSE, we need to estimate both the variance and the bias. We first consider the variance.

The variance of the estimator is a routine calculation. To simplify the expressions, define

$$\mathbf{h} = \left(n_x \mathbf{r}_x' + n_y \mathbf{r}_y'\right)(\Sigma + \lambda\mathbf{I})^{-1} \tag{34}$$

so that

$$\hat{W}(\mathbf{t}) = \mu\mathbf{h}\mathbf{z}. \tag{35}$$

The variance of the WSS estimator is

$$\begin{aligned} \mathrm{Var}[\hat{W}(\mathbf{t})] &= \mathrm{Var}[\mu\mathbf{h}\mathbf{z}] & (36)\\ &= \mu^2\sigma^2\mathbf{h}\mathbf{h}'. & (37) \end{aligned}$$

The error variance, σ^2, is estimated from the residuals to the fit of the velocity function f. The appropriate degrees of freedom is $n - \mathrm{tr}(\mathbf{A}_{\nu,\lambda})$.

The squared bias is more difficult since we do not know the true velocity function f. It is possible to compute the expected bias from assuming that f is well-approximated in the RKHS and from a Bayesian argument. The true velocity can be represented by its minimum norm interpolant plus error $\rho(\mathbf{t})$, i.e.,

$$f(\mathbf{t}) = \mathbf{r}'\Sigma^{-1}\mathbf{f} + \rho(\mathbf{t}), \tag{38}$$

where

$$\mathbf{f} := (f(\mathbf{t}_1), \ldots, f(\mathbf{t}_n))'. \tag{39}$$

Notice that $\mathbf{r}'\Sigma^{-1}\mathbf{f}$ is $\hat{f}$ when $\lambda = 0$ and the data $\mathbf{z}$ are replaced with the values of the velocity function at the $\mathbf{t}_i$. Assume that f is well-approximated by its minimum norm interpolant at all points. This means that $\rho(\mathbf{t})$ is negligible and we let $\rho(\mathbf{t}) = 0$.

The minimum norm interpolant of WSS is obtained from the minimum norm interpolant of the velocity function by similar calculations to obtain $\hat{W}(\mathbf{t})$ from $\hat{f}(\mathbf{t})$. A minimum norm interpolant of WSS is (ignoring approximation errors)

$$W(\mathbf{t}) = \mu\left(n_x \mathbf{r}'_x + n_y \mathbf{r}'_y\right)\Sigma^{-1}\mathbf{f}. \tag{40}$$

Equation (40) lets us express the bias of $\hat{W}(\mathbf{t})$ as

$$\begin{aligned}
\text{Bias}[\hat{W}(\mathbf{t})] &= \text{E}[\hat{W}(\mathbf{t})] - W(\mathbf{t}) && (41)\\
&= \text{E}[\mu\mathbf{hz}] - \mu\left(n_x \mathbf{r}'_x + n_y \mathbf{r}'_y\right)\Sigma^{-1}\mathbf{f} && (42)\\
&= \mu\left(n_x \mathbf{r}'_x + n_y \mathbf{r}'_y\right)\left((\Sigma + \lambda\mathbf{I})^{-1} - \Sigma^{-1}\right)\mathbf{f}. && (43)
\end{aligned}$$

The only unknown in (43) is $\mathbf{f}$. By placing a prior distribution on $\mathbf{f}$, it is possible to compute the expected square bias.

To estimate the bias, a Bayesian argument that places a prior distribution on $\mathbf{f}$ yields an estimate of the expected square bias. The conjugate prior distribution placed on $\mathbf{f}$ is $\mathbf{f} \sim \mathcal{N}(\mathbf{0}, \sigma^2/\lambda\Sigma)$. This prior corresponds to a posterior estimate of $\hat{f}(\mathbf{t}) = \mathbf{r}'(\Sigma + \lambda\mathbf{I})^{-1}\mathbf{z}$. The expected square bias is

$$\begin{aligned}
\text{E}\left[\text{Bias}^2[\hat{W}(\mathbf{t})]\right] &= \text{E}\left[\mu\left(n_x \mathbf{r}'_x + n_y \mathbf{r}'_y\right)\left(\Sigma + \lambda\mathbf{I})^{-1} - \Sigma^{-1}\right)\mathbf{f}\right]\\
&\quad \cdot \left(\mu\left(n_x \mathbf{r}'_x + n_y \mathbf{r}'_y\right)\left(\Sigma + \lambda\mathbf{I})^{-1} - \Sigma^{-1}\right)\mathbf{f}\right)'\\
&= \frac{\mu^2\sigma^2}{\lambda}\left[\mathbf{h}\Sigma\mathbf{h}' - 2\mathbf{h}\left(n_x \mathbf{r}'_x + n_y \mathbf{r}'_y\right)'\right. && (44)\\
&\quad \left. + \left(n_x \mathbf{r}'_x + n_y \mathbf{r}'_y\right)\Sigma^{-1}\left(n_x \mathbf{r}'_x + n_y \mathbf{r}'_y\right)'\right].
\end{aligned}$$

Collecting the expressions for the variance and the expected square bias, we compute the MSE (or more appropriately named the expected MSE due to the prior)

$$\begin{aligned}
\text{MSE}[\hat{W}(\mathbf{t})] &= \mu^2\sigma^2\mathbf{hh}' + \frac{\mu^2\sigma^2}{\lambda}\left[\mathbf{h}\Sigma\mathbf{h}' - 2\mathbf{h}\left(n_x \mathbf{r}'_x + n_y \mathbf{r}'_y\right)'\right.\\
&\quad \left. + \left(n_x \mathbf{r}'_x + n_y \mathbf{r}'_y\right)\Sigma^{-1}\left(n_x \mathbf{r}'_x + n_y \mathbf{r}'_y\right)'\right].
\end{aligned} \tag{45}$$

Equation (45) can be simplified by substituting the an eigenvalue-eigenvector decomposition for Σ and factoring the eigenvalues (somewhat tedious) to get

$$\text{MSE}[\hat{W}(\mathbf{t})] = \mu^2\sigma^2\mathbf{a}'(\Sigma + \lambda\mathbf{I})^{-1}\Sigma^{-1}\mathbf{a} \tag{46}$$

where

$$\mathbf{a} = n_x \mathbf{r}'_x + n_y \mathbf{r}'_y. \tag{47}$$

By replacing σ^2 in (46) with an estimate, we can have a Bayesian estimator for MSE that can be used to construct confidence intervals for WSS.

4.4. Experimental evaluation and data

The nonparametric methods were applied to two phantom data sets. The goal of the phantom studies was to determine if our method is capable of estimating velocity and shear in situations where we have an approximate knowledge of the truth. With this goal in mind we selected λ and ν to match the expected WSS.

The data consist of PC-MR images of constant flow of a blood substitute, which was matched to the viscosity of human blood. The flow rate was set to 6.9 ml/s for the first two data sets and 6.5 ml/s for the third data set. These are typical flow rates during diastole in humans. With each PC-MR image comes a perfectly co-registered "magnitude image" which has good contrast for resolving the walls of the phantom vessels. The magnitude images were used to identify the vessel wall pixels and segment the interior pixels in the PC-MR images. Vessel wall pixels were identified by the Canny edge detection algorithm [21], which is part of the Matlab (Mathworks, Inc.) image processing toolbox. A periodic cubic smoothing spline was fit to each set of wall pixel locations to compute unit normal vectors along the wall of each of the phantom vessels [22]. The detected vessel wall was eroded by a distance of one pixel to eliminate observations from partially-volumed pixels. The partially-volumed pixels intersect both the vessel wall and the interior of the vessel.

The first data set is from a straight glass tube phantom that has a circular cross-section with a 3.9mm inner diameter radius. For a straight vessel with a circular cross-section, the flow can be expressed analytically as a paraboloid [17]. This allows us to compute that the theoretically-expected WSS is approximately 0.13 N/m^2. The image field of view was 100mm$\times$100mm and the image matrix was 512$\times$512 pixels. The pixel intensities outside of the tube were set to zero and the PC-MR images were reduced to a rectangular region of approximately 50$\times$50 pixels that covered the entire glass tube. The boundary points (i.e., points in the image at the tube wall) were automatically identified.

The second data set is from a modified version of the straight glass tube phantom where the tube was laterally indented on one side to yield a non-convex cross section. For this vessel geometry there is no closed-form parametric model describing fluid flow. The indentation is not so extreme that we expect the WSS to be on the order of 0.13 N/m^2, particularly on the side opposite the indentation where the wall is circular. The image size and pre-processing were identical to that of the first data set.

5. RESULTS

The nonparametric model fits the segmented PC-MR velocity images well for both of the phantom data sets. We first consider the straight glass tube phantom with a circular cross-section. A plot of the raw velocity measurements after segmentation is in Fig. 3 A. The measurements appear slightly more variable in the interior of the glass tube than near the edges. A fit with $\nu = 4$ and $\lambda = 300$ is in Fig. 3 B. As expected, the nonparametric fit appears parabolic. The parameters were tuned to best match the expected WSS (0.13 N/m^2). Fig. 3 C is a diagram of how the WSS estimates (Fig. 3 D) are ordered. Starting at the wall point that is at the largest value along the x-axis, we move counter clockwise. The shear stress estimates in Fig. 3 D are accompanied by 95% Bayesian confidence intervals. The solid horizontal line denotes the expected value of WSS. Based on the circular symmetry of the vessel, we expect WSS to be constant at all wall points. On average the estimates of WSS are very close to the expected value. The estimates of WSS appear to have correlated errors. Of the 35 equally-spaced points along the vessel wall where WSS was estimated, 5 of the intervals do not cover the expected value. This gives roughly 86% coverage of the theoretically expected value.

The straight glass tube phantom with a non-convex cross section presents a case where there is no parametric form describing fluid velocity. Fig. 4 shows an image of the nonparametric fit with $\nu = 4$ and $\lambda = 150$ and WSS estimates with 95% Bayesian confidence intervals. The shear estimates are ordered as in Fig. 3. WSS is roughly constant around the vessel wall except for the region of the indentation where it sharply increases. The horizontal line in Fig. 4 denotes WSS=0.13 N/m^2, which is the expected shear stress for a cylindrical glass tube of 3.9 mm radius. Since there is no closed-form description of velocity, this line serves as an approximate reference. In the regions of the vessel wall away from the indentation, the WSS estimates are very close, on average, to 0.13 N/m^2.

5.1. Discussion

The preliminary results from this study demonstrate the feasibility of our nonparametric estimator of fluid velocity and shear from phase-contrast MR images, which measure fluid velocity. The nonparametric estimator is based on penalized least squares in a reproducing kernel Hilbert space generated by the Matérn family of radial basis functions. The main strength of our method is that it does not require restrictive assumptions about vessel geometry (e.g., convex) or flow (e.g., parabolic). Another advantage of the nonparametric method is that it can be easily extended to higher dimensions for 3D images, 2D images acquired at regularly (or even irregularly) spaced intervals, and multiple images acquired over time. This is important for estimating shear stress over a cardiac cycle since the force driving the blood is not constant. This causes the WSS to vary over the cardiac cycle.

REFERENCES

1. N. Aronszajn, *Trans Amer Math Soc* **68**, 337–404 (1950).
2. G. Kimeldorf, and G. Wahba, *J Math Anal Appl* **33**, 82–95 (1971).
3. P. Craven, and G. Wahba, *Numer Math* **31**, 377–403 (1979).
4. K. J. Friston, P. Jezzard, and R. Turner, *Hum Brain Mapp* **1**, 153–171 (1994).
5. K. J. Friston, A. P. Holmes, J.-B. Poline, P. J. Grasby, S. C. R. Williams, R. S. J. Frackowiak, and R. Turner, *NeuroImage* **2**, 45–53 (1995).
6. K. J. Worsley, and K. J. Friston, *NeuroImage* **2**, 173–181 (1995).
7. E. Bullmore, M. Brammer, S. Williams, S. Rabe-Hesketh, N. Janot, A. David, J. Mellers, R. Howard, and P. Sham, *Magn Reson Med* **35**, 261–277 (1996).
8. K. J. Friston, O. Josephs, E. Zarahn, A. P. Holmes, S. Rouquette, and J.-B. Poline, *NeuroImage* **12**, 196–208 (2000).
9. E. Zarahn, G. K. Aguirre, and M. D'Esposito, *NeuroImage* **5**, 179–197 (1997).
10. M. J. Lowe, and J. A. Sorenson, *Magn Reson Med* **37**, 723–729 (1997).

11. K. J. Worsley, C. H. Liao, J. Aston, V. Petre, G. H. Duncan, F. Morales, and A. C. Evans, *NeuroImage* **15**, 1–15 (2002).
12. C. Chatfield, *The analysis of time series*, Chapman & Hall, 1996.
13. N. Lang, S. C. Strother, J. R. Anderson, F. A. Nielson, A. P. Holmes, T. Kolenda, R. Savoy, and L. K. Hansen, *NeuroImage* **10**, 282–303 (1999).
14. M. H. Friedman, C. B. Bargeron, O. J. Deters, G. M. Hutchins, and F. F. Mark, *Atherosclerosis* **68**, 27–33 (1987).
15. D. N. Ku, D. P. Giddens, C. K. Zarins, and S. Glagov, *Atherosclerosis* **5**, 293–302 (1985).
16. C. K. Zarins, D. P. Giddens, B. K. Bharadvaj, V. S. Sottiurai, R. F. Mabon, and S. Glagov, *Circ Res* **53**, 502–514 (1983).
17. P. S. Barna, *Fluid mechanics for engineers*, Plenum Press, New York, 1969.
18. B. Matérn, *Spatial Variation*, Springer-Verlag, Berlin, 1986, 2nd edn.
19. M. L. Stein, *Interpolation of spatial data: some theory for kriging*, Springer-Verlag, New York, 1999.
20. G. Wahba, *J Roy Stat Soc Ser B* **45**, 133–150 (1983).
21. J. Canny, *IEEE J Pattern Analysis Machine Intelligence* **8**, 679–698 (1986).
22. J. D. Carew, R. Dalal, G. Wahba, and S. B. Fain, Estimating arterial wall shear stress, Tech. Rep. 1088, University of Wisconsin Department of Statistics (2003).

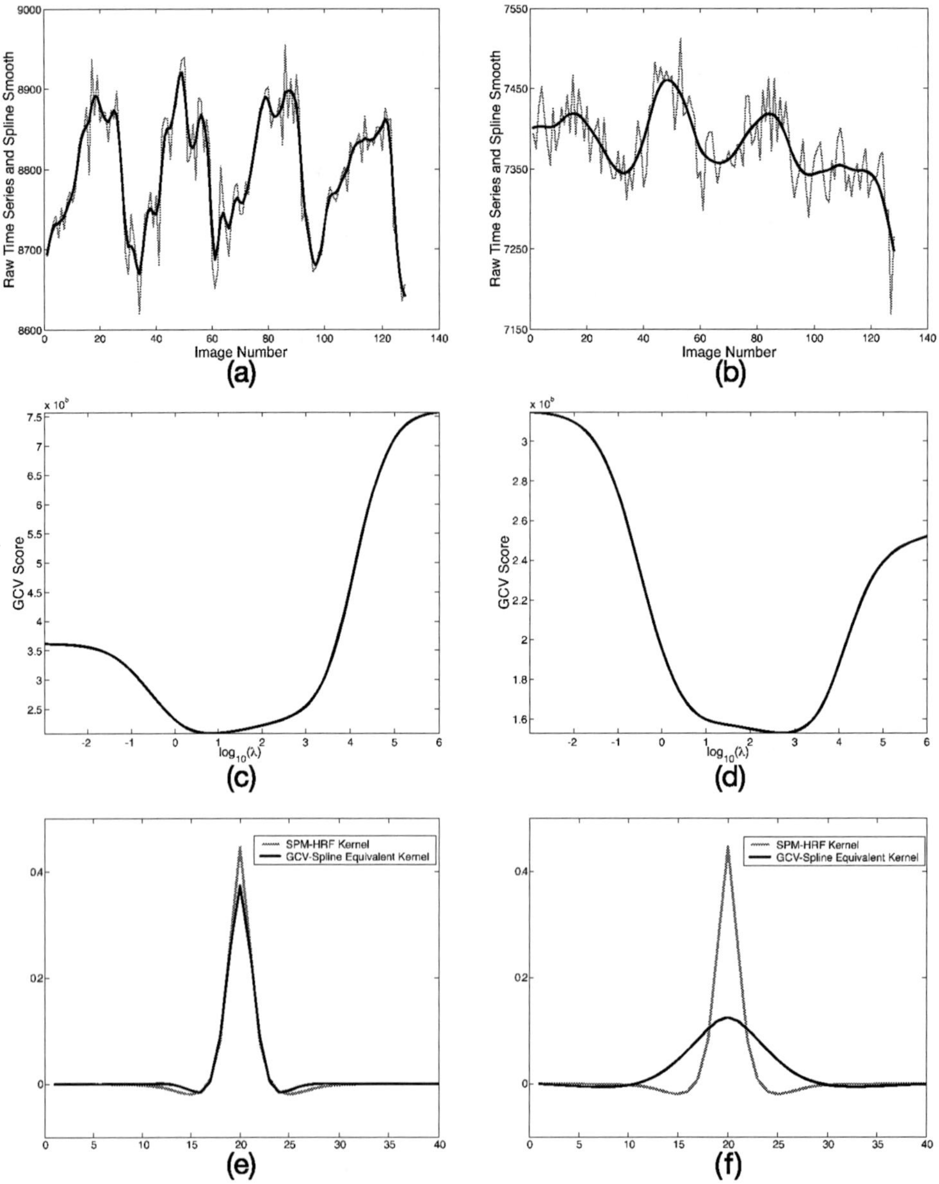

FIGURE 1. Spline smoothing of fMRI time series with the degree of smoothing selected by generalized cross-validation. The time series in (a) required approximately the same amount of smoothing as is provided by the SPM-HRF smoothing kernel. The plot of corresponding GCV score (c) is well-behaved with a minimum at $\lambda = 6.31$. The SPM-HRF kernel is very similar to the equivalent smoothing spline kernel (e). The other time series (b) was smoothed more with the spline method than by the SPM-HRF kernel. The underlying signal is estimated well at the minimum of the GCV score (d) where $\lambda = 501$. The equivalent smoothing spline kernel (f) has a noticeably greater bandwidth than the SPM-HRF smoothing kernel.

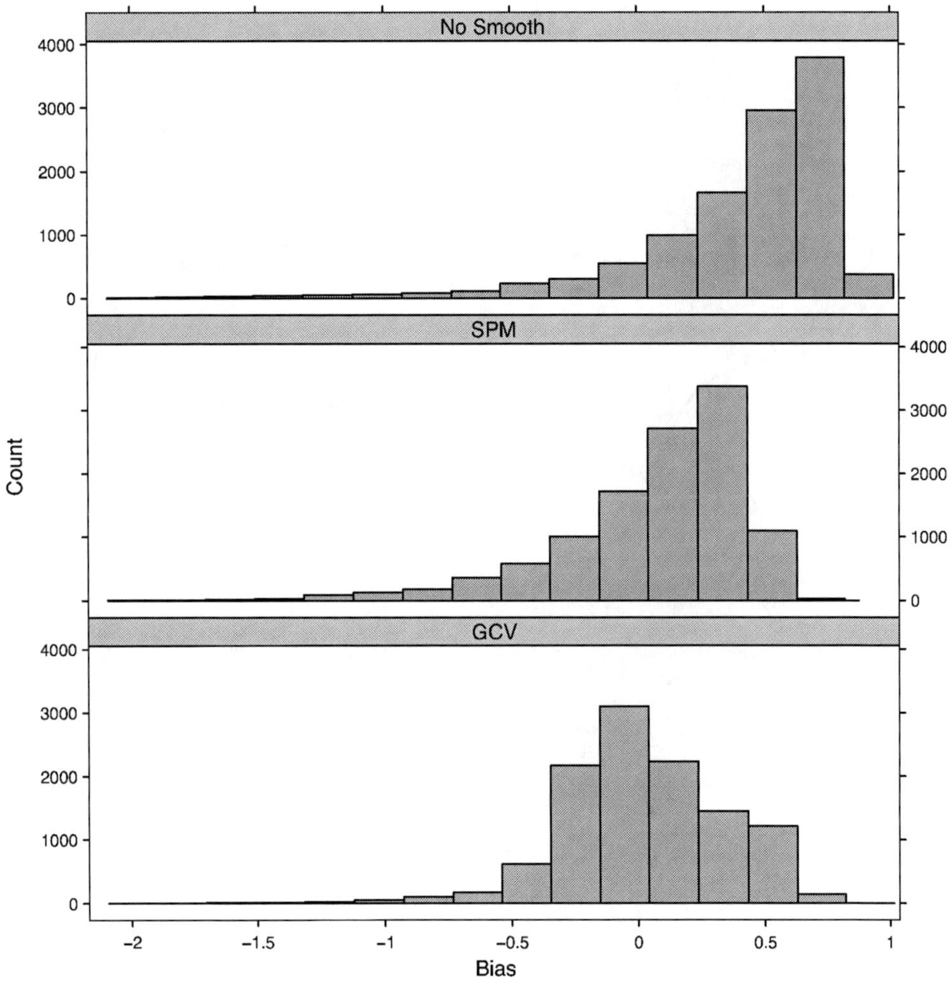

FIGURE 2. Histograms of bias($\hat{\text{var}}(\mathbf{c}'\hat{\beta})$) from the simulated data with no smoothing (top), SPM-HRF smoothing (middle), and GCV-Spline smoothing (bottom). Smoothing with the GCV-Spline method produces $\hat{\text{var}}(\mathbf{c}'\hat{\beta})$ estimates that are, on average, unbiased.

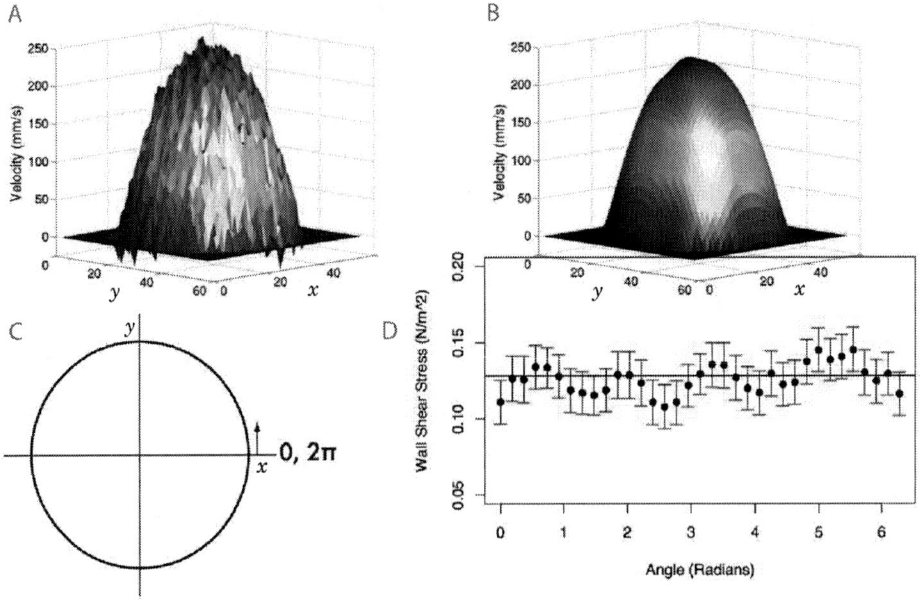

FIGURE 3. Fluid velocity and wall shear stress estimates in a straight glass tube phantom with a circular cross-section. The raw velocity measurements are in A. The nonparametric fit with $\nu = 4$ and $\lambda = 300$ is in B. The fitted velocity function appears parabolic, which is predicted by physical principles. In D are the WSS estimates and 95% Bayesian confidence intervals. The diagram in C shows how the measurements are ordered in D. Starting at the point with the largest value along the x-axis the order of the WSS estimates increase counter clockwise, indexed by the angle.

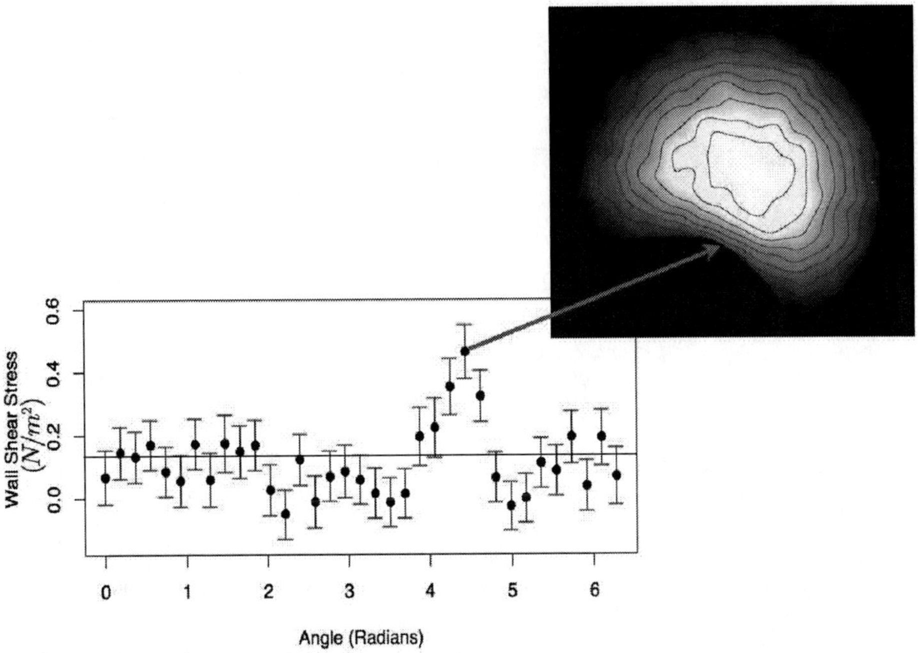

FIGURE 4. Fluid velocity and wall shear stress estimates in a straight glass tube phantom with non-convex cross section. The nonparametric fit uses $\nu = 4$ and $\lambda = 150$. The WSS is nearly constant except for the region near the indentation where the shear stress increases. The increased shear can be seen through the more closely-spaced contour lines on the velocity image near the center of the indentation. The horizontal line shows the expected value of WSS calculated for a vessel with a circular cross-section.

A continuous GRASP to determine the relationship between drugs and adverse reactions[1]

Michael J. Hirsch*, Claudio N. Meneses†, Panos M. Pardalos†, Michelle Ragle† and Mauricio G. C. Resende**

*Raytheon, Inc., Network Centric Systems, P.O. Box 12248, St. Petersburg, FL, 33733, USA.
†Department of Industrial and Systems Engineering, University of Florida, 303 Weil Hall, Gainesville, FL, 32611, USA.
**Algorithms and Optimization Research Department, AT&T Labs Research, 180 Park Avenue, Room C241, Florham Park, NJ 07932 USA.

Abstract. Adverse drug reactions (ADRs) are estimated to be one of the leading causes of death. Many national and international agencies have set up databases of ADR reports for the express purpose of determining the relationship between drugs and adverse reactions that they cause. We formulate the drug-reaction relationship problem as a continuous optimization problem and utilize C-GRASP, a new continuous global optimization heuristic, to approximately determine the relationship between drugs and adverse reactions. Our approach is compared against others in the literature and is shown to find better solutions.

Keywords: Adverse drug reactions/events, global optimization, continuous optimization, heuristic, GRASP, C-GRASP, stochastic local search, nonlinear programming
PACS: 87.55.de, 87.55.kd, 87.55.kh

1. INTRODUCTION

An adverse drug reaction (ADR) is the onset of an untoward medical condition caused by the administration of a therapeutic drug [1]. In the United States, five percent of hospital admissions, 28 percent of emergency room visits, and five percent of hospital deaths can be attributed to ADRs [2]. ADRs are often not recognized during the testing phase of a new drug, as these testing studies are generally small-scale and of short duration. Thus, most ADRs are detected as a result of post-marketing studies, i.e. when the drug is available and prescribed to the general public. The U.S. Food and Drug Administration (FDA), The World Health Organization (WHO), as well as other countries and organizations, have created databases of ADRs reported by practicing physicians and hospitals [3, 4, 5, 6].

The *adverse drug reaction problem* can be set up as follows. Consider n patients, m drugs, and c possible reaction classes. Each of the n patients has taken a known subset of the m drugs. In addition, each patient exhibits symptoms from a subset of the c possible reactions. The goal of the adverse drug reaction problem is to determine, for each drug-reaction pair $\{d,r\}$, a weight or *probability*, that drug d is the cause of the adverse reaction r. Many solutions based on optimization techniques have been proposed for the

---

[1] AT&T Labs Research Technical Report TD-6UPR92.

CP953, *Data Mining, Systems Analysis, and Optimization in Biomedicine*
edited by O. Seref, O. E. Kundakcioglu, P. M. Pardalos

ADR problem [7, 1, 8, 9, 10, 11, 12, 13, 14, 2, 15]. The approach of Mammadov et al. [11, 12, 13] (herein referred to as the *Mammadov Algorithm*) sets up the ADR problem as a nonlinear optimization problem subject to bound constraints.

In this chapter, we propose a modification to the Mammadov et al. formulation. We solve an instance of this continuous optimization problem using real data from the *Australian Adverse Drug Reaction Dataset* with C-GRASP (continuous GRASP), a new continuous global optimization heuristic [16, 17, 18]. To analyze the performance of our algorithm, we also examine the *average precision* metric [11, 12, 13], as well as propose two new metrics.

This chapter is organized as follows. We begin by describing the Australian adverse drug reaction dataset. Then, we introduce both the Mammadov algorithm as well as our modifications of the Mammadov formulation. The C-GRASP heuristic is then discussed. Comparisons are made between our approach and the approach of Mammadov et al. Finally, we provide some concluding remarks and future research directions.

2. AUSTRALIAN ADR DATASET

The aim of the Australian Adverse Drug Reaction Advisory Committee (ADRAC) is to detect signals from adverse drug reactions as early as possible [3]. The ADRAC dataset contains 137,297 records collected from 1971 to 2001. There exist 18 system organ reaction classes, one of those being the cardiovascular class. The cardiovascular reaction class is broken down into four subclasses. We limit our study to data that has at least one reaction from the cardiovascular class. Therefore, for each record of the dataset, there are five possible reactions to consider. Four are from the cardiovascular class and a fifth reaction encompasses all reactions that are not part of the cardiovascular class.

For the data we received[2], each record corresponds to a patient. Each record contains 26 fields. The first ten fields represent up to ten drugs that the patient was prescribed. Each drug is represented as a unique identification number. If a patient took less than ten drugs, then the remaining drug fields are set to 0. Fields 11 through 15 contain a binary vector representing which reactions were observed for the patient. Field 16 contains the year that this ADR was written and fields 17 through 26 contain a binary vector representing which of the drugs in the first ten fields are suspected of causing the observed reactions, in the opinion of the doctor writing the ADR. As an example, consider the following record:

$$\{1984, 4871, 5544, 0, 0, 0, 0, 0, 0, 0, 1, 0, 0, 1, 1, 1972, 0, 1, 0, 0, 0, 0, 0, 0, 0, 0\}.$$

This adverse drug report was written in 1972. The patient was prescribed drugs 1984, 4871, and 5544. Reactions 1, 4, and 5 occurred. The doctor writing the report was of the opinion that drug 4871 caused the observed reactions.

---

[2] The full version of the ADRAC data on CD was sent by Dr Kerri Mackay (TGA) to Musa Mammadov. We received formatted data from Musa Mammadov on April 6, 2006 in private email communication.

3. ADR PROBLEM FORMULATION

We begin this section by presenting the Mammadov algorithm. We describe the problem formulation as well as the solution technique. We then present our modification to this formulation.

We first introduce some essential notation. Let n be the number of patients in the dataset, m be the total number of drugs over all patients, and c be the number of possible reactions. Let X^a be the $n \times m$ binary matrix such that $X^a_{i,j}$ is 1 if drug j was prescribed for patient i, and 0 otherwise, where drug j refers to the j-th drug in the dataset. The matrix X^a corresponds to the patient/drug information if we consider all prescribed drugs for each patient. In a similar way, we can define X^s to represent the patient/drug information if we only consider the suspect drugs for each patient, i.e., $X^s_{i,j}$ is 1 if drug j is a suspect drug for patient i, and 0 otherwise. It is easy to see that $X^s_{i,j} = 1 \Longrightarrow X^a_{i,j} = 1$.

Let Y be the $n \times c$ binary matrix defined such that $Y_{i,k}$ is 1 if reaction k was observed for patient i, and 0 otherwise. Y represents which reactions were observed for each patient. We are tasked with finding the $m \times c$ matrix W, where $W_{j,k}$ is the weight, or probability, that drug j causes reaction k.

The goal of the Mammadov algorithm is to minimize

$$F(W;\, p,\, X) = \frac{1}{n}\sum_{i=1}^{n}\left[\|Y_i\|_1^{-p} \cdot \|\frac{\|Y_i\|_1}{\|X_i W\|_1}(X_i W) - Y_i\|_2^2\right], \tag{1}$$

where $p \in \mathbf{Z}$ and $X \in \{X^a, X^s\}$, such that $0 \le W \le 1$. However, since the size of this optimization problem can be very large, the Mammadov algorithm does not solve this problem exactly. Rather, it solves an approximation to this problem which we next describe.

For drug j, define $X[j]$ to be the set of all patients that were only prescribed drug j, i.e.

$$X[j] = \{i \in \{1, \ldots, n\} \mid X_{i,j} = 1 \wedge X_{i,j'} = 0,\ \forall j' \neq j\}.$$

In a similar way, define $X_{all}[j]$ to be the set of patients that were prescribed drug j. It is clear that $X[j] \subseteq X_{all}[j]$. The solution technique taken in the Mammadov algorithm is to compute $W_{j,k}$, which equals $W^*_{j,k}$ if $a|X[j]| + 100b\frac{|X[j]|}{|X_{all}[j]|} \ge \rho^*$, and $W^{**}_{j,k}$ otherwise.

$$W^*_{j,k} = \left[\sum_{i \in X[j]} \|Y_i\|^{2-p}\right]^{-1} \cdot \left[\sum_{i \in X[j]} \|Y_i\|^{1-p} Y_{i,k}\right], \tag{2}$$

$$W^{**}_{j,k} = \left[\sum_{i \in X_{all}[j]} \|Y_i\|^{2-p}\right]^{-1} \cdot \left[\sum_{i \in X_{all}[j]} \|Y_i\|^{1-p} \frac{rem(Y_{i,k})}{|\Delta''[i]|}\right],$$

Note that $\Delta''[i]$ is the set of all drugs prescribed to patient i whose weights are not calculated using Equation (2), and $rem(Y_{i,k})$ can be seen as the remainder of the reaction available after accounting for the drugs j whose weights are computed using Equation (2). In all experiments to follow, a, b, and ρ^* were set to 1, 1, and 80, respectively.

3.1. Modification to the ADR formulation

As seen above, the Mammadov formulation allows X to be either X^a or X^s. We propose to consider X as a convex combination of X^a and X^s, i.e. $X(\beta) = \beta X^a + (1-\beta)X^s$, where $\beta \in [0,1]$ is unknown. Thus, $X(\beta)$ is a convex combination of all drugs and suspected drugs for each patient. Looked at in another way, this definition of X can be seen as providing a discount measure to the non-suspected drugs, as, for each patient, the suspect drugs are found in both X_i^a and X_i^s. Our formulation,

$$\min F(W;\, p,\, X(\beta)) \ni 0 \le W, \beta \le 1, \tag{3}$$

is a minimization problem with unknowns W and β. Note that the objective function, defined in Equation (1), is a nonlinear function in both W and β.

4. CONTINUOUS GRASP

Feo and Resende [19, 20] describe the metaheuristic GRASP as a multi-start local search procedure, where each GRASP iteration consists of two phases, a construction phase and a local search phase. In the construction phase, interactions between greediness and randomization generate a diverse set of good-quality solutions. The local search phase attempts to improve the solutions found by construction. The best solution over all of the multi-start iterations is retained as the final solution.

Hirsch et al. [16] describe Continuous GRASP (C-GRASP), an adaptation of GRASP to handle continuous optimization problems. C-GRASP works by discretizing the domain into a uniform grid. Both the construction and local improvement phases move along points on the grid. As the algorithm progresses, the grid adaptively becomes more dense. C-GRASP resembles GRASP in that it is a multi-start stochastic search metaheuristic that uses a randomized greedy construction procedure to generate starting solutions for a local improvement algorithm. The main difference is that an iteration of C-GRASP does not consist of a single greedy randomized construction followed by local improvement, but rather a series of construction-local improvement cycles with the output of construction serving as the input of the local improvement, as in GRASP. Unlike GRASP, however, the output of the C-GRASP local improvement procedure serves as the input of the C-GRASP construction procedure.

Hirsch et al. [18] proposed modifications to the original C-GRASP algorithm, resulting in a significant decrease in the number of objective function evaluations required to converge to the global optimum. A detailed description of C-GRASP, as well as applications of C-GRASP to numerous scientific domains is provided in Hirsch [21]. In the remainder of this section, we detail this version of C-GRASP, along with changes specified to handle the formulation (3).

```
procedure C-GRASP(n, ℓ, u, f(·), Maxi, ρ_lo)
1    f* ← ∞;
2    x ← UnifRand(ℓ, u);
3    h ← 0.1;
4    for i = 1,...,Maxi do
5        Impr_C ← false;
6        Impr_L ← false;
7        [x, Impr_C] ← ConstructGreedyRandomized(x, f(·), n, h, ℓ, u, Impr_C);
8        [x, Impr_L] ← LocalImprovement(x, f(·), n, h, ℓ, u, ρ_lo, Impr_L);
9        if f(x) < f* then
10           x* ← x;
11           f* ← f(x);
12       end if
13       if Impr_C = false and Impr_L = false then
14           h ← h/2;    /* make grid more dense */
15       end if
16   end for
17   return(x*);
end C-GRASP;
```

FIGURE 1. Pseudo-code for C-GRASP.

4.1. The heuristic

Pseudo-code for the C-GRASP heuristic is shown in Fig. 1. The procedure takes as input the problem dimension n, lower and upper bound vectors ℓ and u, the objective function $f(\cdot)$, as well as the parameters Maxi, and ρ_{lo}. Parameter Maxi defines the number of iterations to perform in the C-GRASP algorithm while parameter ρ_{lo} defines the portion of the neighborhood of the current solution that is searched during the local improvement phase.

Line 1 of the pseudo-code initializes the objective function value f^* of the best solution found to infinity. While the C-GRASP implementations of [16, 18] are multi-start procedures, because of the size of the ADR problem (see Table 2), we do not implement C-GRASP as a true multi-start algorithm. Rather, we perform a single multi-start iteration. Line 2 initializes the solution x to a random point distributed uniformly over the box in $\mathbf{R}^n$ defined by ℓ and u. The parameter h, that controls the discretization density of the search space, is initialized to 0.1. The construction and local improvement phases are then called sequentially in lines 7 and 8, respectively, Maxi times. The solution returned from the local improvement procedure is compared against the current best solution in line 9. If the returned solution has a smaller objective value than the current best solution, then, in lines 10–11, the current best solution is updated with the returned solution. In line 13, if the variables $\texttt{Impr}_\texttt{C}$ and $\texttt{Impr}_\texttt{L}$ are still set to false, then the grid density is increased by halving h, in line 14. The variable $\texttt{Impr}_\texttt{C}$ is false upon return from the construction procedure if and only if no improvement is made in the construction phase. Section 4.3 shows that the $\texttt{Impr}_\texttt{L}$ variable is `false` on return from the local improvement procedure if and only if the input solution to local improvement is determined to be an *h-local minimum*. We increase the grid density at this stage because repeating the construction procedure with the same grid density will not improve the solution. This approach allows C-GRASP to start with a coarse discretization and adaptively increase the density as needed, thereby intensifying the search in a more dense discretization when a good solution has been found. The best solution found, at the time the stopping criteria are satisfied, is returned.

4.2. Construction procedure

In this section, we describe in detail the construction procedure. The construction algorithm combines greediness and randomization to produce a diverse set of good-quality solutions from which to start the local improvement phase. The construction algorithm is shown in Fig. 2. The input is a solution vector x. To start, line 1 of the algorithm allows `MaxUnFixed` (user-defined parameter) coordinates of x to change in the current construction call (i.e. these coordinates are unfixed). In turn, in line 10 of the pseudo-code, if `ReUse` is `false`, a line search is performed in each unfixed coordinate direction i of x with the other $n-1$ coordinates of x held at their current values. In lines 10 and 11 of the pseudo-code, the value z_i for the i-th coordinate that minimizes the objective function, together with the objective function value g_i, are saved. In line 11, $\check{x}^i$ denotes x with the i-th coordinate set to z_i.

After looping through all unfixed coordinates (lines 7–16), in lines 17–23 a restricted candidate list (RCL) is formed containing the unfixed coordinates i whose g_i values are less than or equal to $\underline{g}+\alpha\cdot(\bar{g}-\underline{g})$, where $\bar{g}$ and $\underline{g}$ are, respectively, the maximum and minimum g_i values over all unfixed coordinates of x, and $\alpha \in [0,1]$ is randomly determined in line 2. In line 24, a coordinate is chosen at random from the RCL, say coordinate $j \in$ RCL. Line 25 checks whether x_j and z_j are equal. If so, line 26 sets `ReUse` to the value `true`. Otherwise, in lines 28–30, `ReUse` is set to `false`, $\texttt{Impr}_\texttt{C}$ is set to `true`, and x_j is set to equal z_j. Finally, in line 30, we fix coordinate j of x, by removing j from the set `UnFixed`. Choosing a coordinate by selecting at random from the RCL ensures both greediness and randomness in the construction phase. The above procedure is continued until all of the n coordinates of x have been fixed. At that stage, x and $\texttt{Impr}_\texttt{C}$ are returned from the construction procedure.

Note that the `ReUse` variable is utilized to speed up computations by avoiding unnecessary line searches. More details can be found in [18]. The parameter α controls the size of the RCL and therefore determines the mix of greediness and randomness in the construction procedure. Different values of α throughout the run allow some construction phases to be more greedy while others to be more random.

```
procedure ConstructGreedyRandomized(x, f(·), n, h, ℓ, u, Impr_C)
    UnFixed ← RandSelect({1,2,...,n}, MaxUnFixed);
    α ← UnifRand(0,1);
    ReUse ← false;
    while UnFixed ≠ ∅ do
        g̲ ← +∞;
        ḡ ← −∞;
        for i = 1,...,n do
            if i ∈ UnFixed then
                if ReUse = false then
                    z_i ← LineSearch(x, h, i, n, f(·), ℓ, u);
                    g_i ← f(x̌^i);
                end if
                if g̲ > g_i then g̲ ← g_i;
                if ḡ < g_i then ḡ ← g_i;
            end if
        end for
        RCL ← ∅;
        Threshold ← g̲ + α·(ḡ − g̲);
        for i = 1,...,n do
            if i ∈ UnFixed and g_i ≤ Threshold then
                RCL ← RCL ∪ {i};
            end if
        end for
        j ← RandomlySelectElement(RCL);
        if x_j = z_j then
            ReUse ← true;
        else
            x_j ← z_j;
            ReUse ← false;
            Impr_C ← true;
        end if
        UnFixed ← UnFixed \ {j};     /* Fix coordinate j. */
    end while
    return(x, Impr_C);
end ConstructGreedyRandomized;
```

FIGURE 2. Pseudo-code for C-GRASP construction phase.

4.3. Local improvement procedure

C-GRASP makes no use of derivatives. Though derivatives can be easily computed for many functions, they are not always available or efficiently computable for all functions. The local improvement phase (with pseudo-code shown in Fig. 3) can be seen

procedure LocalImprovement($x, f(\cdot), n, h, \ell, u, \rho_{lo}$, Impr_L)
1 $x^* \leftarrow x$;
2 $f^* \leftarrow f(x)$;
3 NumGridPoints $\leftarrow \prod_{i=1}^{n} \lceil (u_i - \ell_i)/h \rceil$;
4 MaxPointsToExamine $\leftarrow \lceil \rho_{lo} \cdot$ NumGridPoints$\rceil$;
5 NumPointsExamined $\leftarrow 0$;
6 **while** NumPointsExamined $\leq$ MaxPointsToExamine **do**
7 NumPointsExamined $\leftarrow$ NumPointsExamined $+ 1$;
8 $x \leftarrow$ RandomlySelectElement($B_h(x^*)$);
9 **if** $\ell \leq x \leq u$ **and** $f(x) < f^*$ **then**
10 $x^* \leftarrow x$;
11 $f^* \leftarrow f(x)$;
12 $\text{Impr}_\text{L} \leftarrow$ true;
13 NumPointsExamined $\leftarrow 0$;
14 **end if**
15 **end while**
16 **return**(x^*, Impr_L);
end LocalImprovement;

FIGURE 3. Pseudo-code for C-GRASP local improvement phase.

as *approximating* the role of the gradient of the objective function $f(\cdot)$. From a given input point $x \in \mathbf{R}^n$, the local improvement algorithm generates a neighborhood, and determines at which points in the neighborhood, if any, the objective function improves. If an improving point is found, it is made the current point and the local search continues from the new solution.

Let $\bar{x} \in \mathbf{R}^n$ be the current solution and h be the current grid discretization parameter. Define

$$S_h(\bar{x}) = \{x \in S \mid \ell \leq x \leq u,\ x = \bar{x} + \tau \cdot h,\ \tau \in \mathbf{Z}^n\}$$

to be the set of points in S that are integer steps (of size h) away from $\bar{x}$. Let

$$B_h(\bar{x}) = \{x \in S \mid x = \bar{x} + h \cdot (x' - \bar{x})/\|x' - \bar{x}\|,\ x' \in S_h(\bar{x}) \setminus \{\bar{x}\}\}$$

be the projection of the points in $S_h(\bar{x}) \setminus \{\bar{x}\}$ onto the hyper-sphere centered at $\bar{x}$ of radius h. The *h-neighborhood* of the point $\bar{x}$ is defined as the set of points in $B_h(\bar{x})$.

The local improvement procedure is given a starting solution $x \in S \subseteq \mathbf{R}^n$. The current best local improvement solution x^* is initialized to x in line 1. Lines 3 and 4 determine the number of grid points, based on the current value of the discretization parameter h, and the maximum number of points in $B_h(x^*)$ that are to be examined. This number of grid points is defined by the parameter ρ_{lo} which is the portion of the neighborhood which is to be examined. If all of these points are examined and no improving point is found, the current solution x^* is considered an *h-local minimum*.

Starting at the point x^*, in the loop in lines 6–15, the algorithm randomly selects MaxPointsToExamine points in $B_h(x^*)$, one at a time. In line 9, if the current point

x selected from $B_h(x^*)$ is feasible and is better than x^*, then x^* is set to x, $\texttt{Impr}_\texttt{L}$ is set to true, and the process restarts with x^* as the starting solution. $\texttt{Impr}_\texttt{L}$ is used to determine whether the local improvement procedure improved the best solution. Local improvement is terminated when an *h-local minimum* solution x^* is found. At that point, x^* and $\texttt{Impr}_\texttt{L}$ are returned from the local improvement procedure.

5. METRICS, SCENARIOS, AND RESULTS

In this section, we present metrics used to evaluate the solutions found for the ADR problem. We then discuss some characteristics of the particular data scenarios. Finally, we describe the testing environment and discuss the computational results.

5.1. ADR metrics

Mammadov et al. [11, 12, 13] define only one metric to measure the performance of their algorithm, the average precision metric. In this section, we describe the average precision metric, present an example detailing its computation, and discuss a potential pitfall of this metric. We also suggest two more conventional metrics that have been used extensively in the literature to measure distance between vectors.

Suppose we have a solution W to the ADR problem. We would like some way of measuring the quality of this solution. For each patient i, define the computed reaction of this patient to be $\hat{Y}_i = X_iW$. Recall from Section 3 that the true, or observed, reaction for this patient is given by Y_i. Let $\mathbf{T}_i$ be the set of all permutations of the indices $1,\ldots,c$ such that the permutation applied to $\hat{Y}_i$ produces a non-increasing sequence. Then, for each $\tau \in \mathbf{T}_i$, the precision for τ, $P_\tau(X_i)$, is defined as

$$P_\tau(X_i) = \frac{1}{\|Y_i\|_1} \sum_{k:Y_{i,k}=1} \frac{|R'_{i,k}(\tau)|}{|R_{i,k}(\tau)|},$$

where $R_{i,k}(\tau) = \{\ell \in \{1,\ldots,c\} \mid \ell \text{ is listed before } k \text{ in } \tau\} \cup \{k\}$ and $R'_{i,k}(\tau) = \{\ell \in \{1,\ldots,c\} \mid Y_{i,\ell} = 1 \wedge \ell \text{ is listed before } k \text{ in } \tau\} \cup \{k\}$.

The average precision $P(X_i)$ for patient i is defined to be the average of the best and worst precisions over all $\tau \in \mathbf{T}_i$. The average precision is given as the average over the average precision of each patient, i.e.

$$P_{AP} = \frac{1}{n}\sum_{i=1}^{n} P(X_i).$$

As an example of computing the precision for a patient, suppose that for patient i, the observed reaction is $Y_i = \{1,0,1,1,0\}$ and the computed reaction is found to be $\hat{Y}_i = \{0.9,0.1,0.1,0.9,0.1\}$. The set of all index permutations $\mathbf{T}_i$ that produce a non-increasing sequence for the elements of $\hat{Y}_i$ is listed in Table 1. Looking at the permutation $\tau = \{1,4,5,3,2\}$, we see that $R_{i,1} = R'_{i,1} = \{1\}$, $R_{i,4} = R'_{i,4} = \{1,4\}$, $R_{i,3} = \{1,4,5,3\}$, and $R'_{i,3} = \{1,4,3\}$. Therefore, $P_\tau(X_i) \approx 0.9167$. The best precision for this patient (from permutations that place 3 before 2 and 5) is 1, while the worst precision for this patient (from permutations that place 3 after 2 and 5) is approximately 0.8667. Hence, this patient would have an average precision of $P(X_i) \approx 0.9333$.

TABLE 1. Permutations of $\{1,2,3,4,5\}$ that are elements of $\mathbf{T}_i$.

$\{1,4,2,3,5\}$	$\{4,1,2,3,5\}$
$\{1,4,2,5,3\}$	$\{4,1,2,5,3\}$
$\{1,4,3,2,5\}$	$\{4,1,3,2,5\}$
$\{1,4,3,5,2\}$	$\{4,1,3,5,2\}$
$\{1,4,5,2,3\}$	$\{4,1,5,2,3\}$
$\{1,4,5,3,2\}$	$\{4,1,5,3,2\}$

We now provide an example highlighting a potential shortcoming of the average precision metric. Consider the same patient observed reaction, namely $Y_i = \{1,0,1,1,0\}$, and suppose one approach to solving the ADR problem produces a computed reaction of $\hat{Y}'_i = \{0.1, 0.01, 0.07, 0.04, 0.039\}$, while a second approach produces the computed reaction of $\hat{Y}''_i = \{0.9, 0.01, 0.7, 0.8, 0.039\}$. While it is easy to see that both of these computed reactions will produce an average precision of 1.0 for this patient, it is clear that the second computed reaction $\hat{Y}''_i$ predicts much more closely the true observed reaction than does the first. Hence, for this particular example, it is unclear what the average precision metric is really measuring.

To aid in measuring how well the different solution approaches are performing, in addition to using the average precision metric we also look at the standard least squares residual

$$LS(W) = \frac{1}{n}\sum_{i=1}^{n} \|\hat{Y}_i - Y_i\|_2^2$$

and the average maximum deviation (i.e., average L_∞ norm, with $k \in \{1,\ldots,c\}$).

$$AMD(W) = \frac{1}{n}\sum_{i=1}^{n} \max[|\hat{Y}_i - Y_i|] = \frac{1}{n}\sum_{i=1}^{n} \|\hat{Y}_i - Y_i\|_\infty.$$

Using these conventional metrics will lend stronger support to one algorithm performing better than another.

5.2. Testing scenarios

The ADRAC dataset contains data for the years 1972 to 2001 inclusive. The approach taken by Mammadov et al. [11, 12, 13], which we adhere to, was to consider the years 1996 through 2001. For each of these years, a training dataset was formed using the data from 1972 to the previous year. This is the data that will be used to determine the weight matrix W. Once the weight matrix is determined, it is applied to the current year data to measure how well W predicts the true reactions. As an example, for the year 1998, data from years 1972 to 1997 (the training data) is used in the solution approaches to determine W. Then, W is applied to the year 1998 data (the test data) to measure the performance of W. In addition, it is also useful to measure the performance of W on the training data. Table 2 lists characteristics for each of the training and test data sets.

TABLE 2. ADR data characteristics.

	Training Data			Testing Data
Year	# Patients	# Drugs	# Variables	# Patients
1996	12,600	2,253	11,265	1,049
1997	13,747	2,375	11,875	1,091
1998	15,001	2,497	12,485	1,418
1999	16,684	2,661	13,305	1,746
2000	18,599	2,786	13,930	1,988
2001	20,745	2,916	14,580	1,054

5.3. Testing environment

The computational experiments with C-GRASP were carried out on a Dell PowerEdge 2600 computer with dual 3.2 GHz 1 Mb cache XEON III processors and 6 Gb of memory running Red Hat Linux 3.2.3-53. C-GRASP was implemented in C++ and compiled with GNU `g++` version 3.2.3. The compiler options used were `-O6 -funroll-all-loops -fomit-frame-pointer -march=pentium4`. The algorithm used for random-number generation is an implementation of the Mersenne Twister algorithm described in [22]. The Mammadov algorithm was implemented in Matlab to verify the average precision results, as well as determine values for the least squares and average maximum deviation metrics.

We set the C-GRASP parameters to the values listed in Table 3. While the ideal situation would be to allow all variables to start as *unfixed* in the construction procedure, we limit the number of starting unfixed variables to 10, chosen randomly each time the construction procedure is called. It was estimated that each call to the construction procedure would have taken more than one year of computing time had we allowed all variables to start unfixed, hence the need implement this limit.

TABLE 3. C-GRASP parameter values used in the experiments.

Parameter	Value
`Maxi`	2,000
`MaxUnFixed`	10
ρ_{lo}	0.7

5.4. Computational results

Tables 4–6 display the performance of both the Mammadov algorithm and the enhanced C-GRASP heuristic for the three metrics, when $p = 0$ in the objective functions. It is clear from these tables that, overall, the C-GRASP heuristic performs better for both the training and testing data, for each year, with the three metrics. The Mammadov algorithm, by not considering all of the data when computing W and not forming a convex combination of all drugs and suspect drugs, appears to be limited in its ability to achieve a solution significantly close to the global minimum. C-GRASP does not have this limitation. However, the Mammadov algorithm is much faster than the C-GRASP approach. On average, the Mammadov approach took approximately six minutes. The C-GRASP algorithm on average ran for 90 hours. In our opinion, for problems dealing with the health of individuals, it is much more important to find a correct, close to optimal answer, than to find a quick answer. Of note is that, for each year, each iteration of C-GRASP improved upon the previous iteration. Therefore, letting C-GRASP run for more than 2,000 iterations would probably produce an even better solution. Table 7 displays the β values (discounting values) found by the C-GRASP for each year. Mammadov and Banerjee [9] considered the 'optimal' convex combination of suspect and non-suspect drugs. They found a β value between 0.206 and 0.215. While their approach is not equivalent to our use of β, the similarity of the resulting convex combination factors is worth noting.

6. CONCLUSIONS

From the statistics given in Section 1, determining the relationship between drugs and adverse drug reactions is an extremely important problem in health-care. This chapter has generalized the formulation of Mammadov et al. [11, 12, 13] to include a convex combination of suspected drugs and non-suspected drugs. Depending on the

TABLE 4. ADR average precision results.

Year	Mammadov				C-GRASP	
	All		Suspect			
	Train	Test	Train	Test	Train	Test
1996	0.8145	0.7997	0.8291	0.8025	0.8780	0.8517
1997	0.8153	0.7905	0.8280	0.7979	0.8779	0.8528
1998	0.8158	0.7786	0.8282	0.7802	0.8758	0.8569
1999	0.8152	0.8041	0.8272	0.8091	0.8777	0.8727
2000	0.8169	0.7757	0.8272	0.7788	0.8755	0.8380
2001	0.8135	0.7687	0.8232	0.7685	0.8722	0.8283

data, the number of variables in the optimization formulation can be quite large. We have contrasted the approach taken by Mammadov et al. [11, 12, 13] with our new approach. Using three different metrics, we have shown that the C-GRASP approach produced better solutions than did the Mammadov algorithm.

One area of future research concerns discounting the non-suspected drugs of each patient by a different discount factor. This would have the effect of giving less weight to those patients having the same observed reactions and conflicting suspect drug sets. At the same time, however, this would significantly increase the number of variables in the problem. Another area of future research concerns applying our C-GRASP approach to other adverse drug reaction data. The United States Food and Drug Administration [5], the World Health Organization Uppsala Monitoring Center [6], and the Canadian Adverse Drug Reaction Information System [4] all collect adverse drug reaction reports.

TABLE 5. ADR least squares residual results.

Year	Mammadov				C-GRASP	
	All		Suspect			
	Train	Test	Train	Test	Train	Test
1996	0.0091	0.0330	0.0089	0.0355	0.0090	0.0345
1997	0.0088	0.0321	0.0085	0.0352	0.0087	0.0344
1998	0.0084	0.0264	0.0082	0.0297	0.0085	0.0310
1999	0.0079	0.0245	0.0077	0.0245	0.0080	0.0245
2000	0.0075	0.0239	0.0073	0.0232	0.0076	0.0230
2001	0.0071	0.0331	0.0069	0.0372	0.0074	0.0348

We are currently beginning the task of collecting data from these sites.

ACKNOWLEDGMENTS

This work has been partially supported by NSF, NIH, and CRDF grants.

REFERENCES

1. J. T. Harvey, C. Turville, and S. M. Barty, *International Transactions in Operations Research* **11**, 419–433 (2004).
2. A. M. Wilson, L. Thabane, and A. Holbrook, *British Journal of Clinical Pharmacology* **57**, 127–134 (2003).

TABLE 6. ADR average maximum deviation results.

Year	Mammadov				C-GRASP	
	All		Suspect			
	Train	Test	Train	Test	Train	Test
1996	0.7551	0.7869	0.7306	0.8138	0.6977	0.7836
1997	0.7572	0.7653	0.7321	0.8196	0.7058	0.7719
1998	0.7525	0.7486	0.7321	0.7994	0.7140	0.7802
1999	0.7497	0.7569	0.7296	0.7484	0.7124	0.7305
2000	0.7464	0.7877	0.7267	0.7588	0.7081	0.6950
2001	0.7469	0.7902	0.7284	0.8521	0.7255	0.7544

3. Australian Department of Health and Ageing Therapeutic Goods Administration, Australian Adverse Drug Reactions Bulliten., `http://www.tga.gov.au/adr/aadrb.htm` (Last time accessed: August 2006).
4. Canadian Adverse Drug Reaction Information System, Canadian Adverse Drug Reaction Database., `http://www.cbc.ca/news/adr/database/` (Last time accessed: August 2006).
5. U. S. Food and Drug Administration, Adverse Event Reporting System., `http://www.fda.gov/cder/aers/default.htm` (Last time accessed: August 2006).
6. Uppsala Monitoring Center, Uppsala Monitoring Center - Vigilbase., `http://www.umc-products.com/DynPage.aspx?id=4910` (Last time accessed: August 2006).
7. J. S. Almenoff, W. DuMouchel, L. A. Kindman, X. Yang, and D. Fram, *Pharmacoepidemiology and Drug Safety* **12**, 517–521 (2003).
8. Y. Ji, H. Ying, J. Yen, S. Zhu, R. M. Massanari, and D. C. Barth-Jones, "Team-based multi-agent system for early detection of adverse drug reactions in postmarketing surveillance," in *Annual Meeting of the North American Fuzzy Information Processing Society*, 2005, pp. 644–649.
9. M. Mammadov, and A. Banerjee, "An Optimization Approach Identifying Drugs Responsible for

TABLE 7. Discount factor (β) found by the C-GRASP heuristic.

Year	β
1996	0.27
1997	0.25
1998	0.28
1999	0.25
2000	0.28
2001	0.34

Adverse Drug Reactions," in *Proc. of the Sixteenth Australasian Workshop on Combinatorial Algorithms*, edited by J. Ryan, P. Manyem, K. Sugeng, and M. Miller, 2005, pp. 185–200.
10. M. Mammadov, and A. Banerjee, "An Optimization Approach to the Study of Drug-Drug Interactions," in *Proc. of the Sixteenth Australasian Workshop on Combinatorial Algorithms*, edited by J. Ryan, P. Manyem, K. Sugeng, and M. Miller, 2005, pp. 201–216.
11. M. Mammadov, E. Dekker, and G. Saunders, "An Optimization Approach to the study of drug-reaction relationships on the basis of ADRAC dataset: Neurological class of reactions," in *Proc. of The Sixth International Conference on Optimization: Techniques and Applications (ICOTA6)*, edited by A. Rubinov, and M. Sniedovich, 2004.
12. M. Mammadov, A. Rubinov, and J. Yearwood, "An Optimization Approach to identify the relationship between features and output of a multi-label classifier," in *Data Mining in Medicine*, edited by P. Pardalos, V. Boginski, and A. Vazacopoulos, in print.
13. M. Mammadov, and G. Saunders, "A Comparison of Two Methods to Establish Drug-reaction Relationships in the ADRAC Database," in *Proc. of The Fourth International ICSC Symposium on Engineering of Intelligent Systems (EIS 2004)*, edited by S. S. Raza Abidi, 2004.
14. E. Roux, F. Thiessard, A. Fourrier, B. Begaud, and P. Tubert-Bitter, *IEEE Transactions on Information Technology in Biomedicine* **9**, 419–433 (2005).
15. H. Yu, J. Yang, W. Wang, and J. Han, "Discovering compact and highly discriminative features or feature combinations of drug activities using support vector machines," in *Proc. of the IEEE Bioinformatics Conference*, 2003.
16. M. J. Hirsch, C. N. Meneses, P. M. Pardalos, and M. G. C. Resende, *Optimization Letters* **1**, 201–212 (2007).
17. M. J. Hirsch, P. M. Pardalos, and M. G. C. Resende, *Submitted to Nonlinear Analysis: Real World Applications* (2006).
18. M. J. Hirsch, P. M. Pardalos, and M. G. C. Resende, *Submitted to European Journal of Operational Research* (2006).
19. T. A. Feo, and M. G. C. Resende, *Operations Research Letters* **8**, 67–71 (1989).
20. T. A. Feo, and M. G. C. Resende, *Journal of Global Optimization* **6**, 109–133 (1995).
21. M. J. Hirsch, *GRASP-based heuristics for continuous global optimization problems*, Ph.D. thesis, University of Florida (2006).
22. M. Matsumoto, and T. Nishimura, *ACM Transactions on Modeling and Computer Simulation* **8**, 3–30 (1998).

AntModeler analysis of mechanical stress driven transcription in three cell types

Nan Lin[*], Andy Chen[†], Jennifer R. Mackley[¶], Steven J. Winder[¶] and Hiroki Yokota[†]

[*]*Department of Mathematics, Duke University, Durham, NC 27708, USA*
[†]*Department of Biomedical Engineering, Indiana University – Purdue University Indianapolis Indianapolis, IN 46202 USA*
[¶]*Department of Biomedical Science, University of Sheffield, Western Bank, Sheffield, S10 2TN, UK*

Abstract. Cellular stress activates transcription of various genes that mediate stress-driven proliferation and differentiation in many cells including osteoblasts, endothelial cells, and fibroblasts. In response to mechanical stress, expression of some genes is altered regardless of cell types and that of others in specific cell types. Using the microarray-based expression data for primary fibroblasts isolated from fetal mouse cornea, skin and tendon, we conducted a model-based transcription analysis and predicted transcription-factor binding motifs (TFBMs) responsible for the observed gene alteration. The computational procedure was formulated as a combinatorial optimization problem, and the AntModeler using an ant algorithm was employed to select TFBMs for each of the three fibroblast types. The results indicate that the stress responses are regulated mostly through cell type specific TFBMs together with a limited number of common TFBMs. The predicted role of those TFBMs should be evaluated experimentally.

Keywords: ant algorithm, transcription-factor binding motif (TFBM), mechanical loading
PACS: 87.85.-d , 87.55.de, 87.55.kd

1. INTRODUCTION

Vertebrate organisms are derived from a single embryonic cell through complex processes of proliferation and differentiation. Because of their difference in functionality, cells may react differently to mechanical loading. One example is the fibroblast cells in different organs such as cornea, skin, and tendon: cornea is a covering tissue for pupil, iris, and eye lens, skin is the epithelial tissue covering and protecting underneath tissues, and tendon is part of musculoskeletal tissues originating from a muscle for withstanding tension. Previous studies have shown that some genes in those three cell types react similarly and some others differently in response to mechanical stimulation [1]. The mechanism underlying the difference in their transcriptional regulations, however, has not been fully explored.

The entire genome of many species has been sequenced, and one of the current foci is to identify regulatory elements responsible for tissue specific gene expression [2-4]. It is known that most cells transcribe the genome pervasively, but the mechanism for tissue specific transcription is not well understood [5]. One of the theories is that a differential usage of transcription factor binding motifs (TFBMs) is the cause of these variations. TFBMs are short fragments of DNA sequences for binding of transcription factors, which regulate transcriptional activities with RNA polymerase. Although all cells in any given species possess the identical set of genomic information, differential usage of

CP953, *Data Mining, Systems Analysis, and Optimization in Biomedicine*
edited by O. Seref, O. E. Kundakcioglu, P. M. Pardalos

transcription factors allows cells to function for their specific purpose. We hypothesize that although some of the TFBMs may serve as a general purpose regulator, many other TFBMs are specialized by individual cell types.

The ant algorithm is a search procedure that mimics the foraging behavior of biological ants to efficiently find the "food source." We previously developed search procedures including PROBE [6], singular value decomposition [7], and particle swarm optimization [8]. Uniqueness of the ant algorithm is its analogy to a complex collective behavior in many insect societies, whose mathematical formulation shows existence of stable and unstable equilibrium states [9]. During walking ants deposit pheromone, which attracts other ants to follow their trails. Pheromone is not only deposited but also evaporated over time to update preferred trails. Eventually more ants may follow the trial of the ant that has found the richest food source. Using this principle, the AntModeler was developed and the computational conditions leading to the stable equilibrium state (reaching to the food source) were analyzed [10]. It includes several essential parameters such as a pheromone preference factor and a pheromone evaporation factor. This study utilized this software tool to search for important TFBMs in the upstream DNA sequences of the selected genes which were altered in cornea, skin, and/or tendon in response to mechanical loading.

2. MATERIALS AND METHODS

DNA microarray data: In this study the microarray data for the responses to mechanical stimulation in three different cell types were analyzed [1]. We focused on a group of genes whose mRNA expression levels were significantly altered by mechanical shear stress (Fig. 1). We selected 50 genes for cornea, 46 genes for skin, and 61 genes for tendon using the criterion of the most and least fold changes under mechanical stimulation. Among those genes, 12 were present in all 3 cell types, 1 in cornea and tendon, 8 in cornea and skin, and 11 in skin and tendon. The name of each gene was translated into the numerical gene ID.

In order to predict putative TFBMs for each of the three cell types, we employed the AntModeler whose algorithm was previously reported [11]. In brief, this program utilizes artificial ants to search for putative TFBMs in the 1000-bp upstream region of each gene by mimicking deposition and evaporation of pheromones. First, the mRNA fold change ratio (a logarithmic scale with base 2) in response to mechanical stimulation was modeled using the state vector $\underline{X}$, which represented the stimulatory/inhibitory role of TFBM candidates. We then employed 100 artificial ants and conducted 50 searches using the same computational conditions. In each search, a single ant represented 20 TFBM candidates and 200 computational iterations were included. Two key parameters in the algorithm, i.e., the pheromone preference factor and the pheromone evaporation factor, were set to their default values (500 and 0.5, respectively). We focused on TFBM candidates 4- to 6-bp in length.

Interpretation of the predicted TFBMs with the AntModeler: Using the X-value of the 50 trials, the top 10 TFBM candidates were first identified using DNA sequences of 4- to 6- bp in length. We then evaluated any TFBMs commonly identified for any of the three cell types. Finally, any linkage to biologically known TFBMs and transcription factors was examined using a web-based cisRED software tool (12).

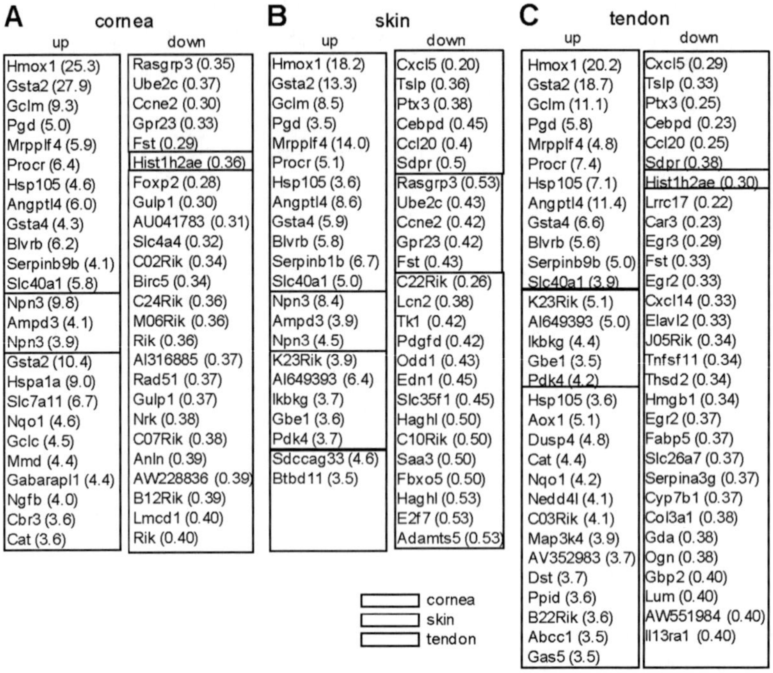

FIGURE 1. Genes analyzed in this study by the AntModeler. The red, yellow, and blue colors represent cornea, skin, and tendon, respectively. Note that coding with multiple colors marks the genes commonly altered among three cell types, and the fold change is indicated in the parenthesis. (A) 50 genes responsive to mechanical loading in cornea. (B) 46 genes responsive to mechanical loading in skin. (C) 61 genes responsive to mechanical loading in tendon.

3. RESULTS

Predicted role of 4-6 bp TFBM candidates: The predicted role of 4-6 bp TFBM candidates is shown for each of the three cell types using the normalized X-value in the range of -1 to 1 (Fig. 2). The positive and negative values represent the stimulatory and inhibitory roles, respectively. The DNA sequences for several major X-value peaks are indicated. Among the different lengths of TFBM candidates, several overlaps were identified across 4- to 6-bp TFBMs. In cornea, for instance, the sequence of ACTC appears in all three lengths. The sequences of AGCG and ATGA were selected more than once in skin, while the common sequences in tendon were AGCG and CGTA.

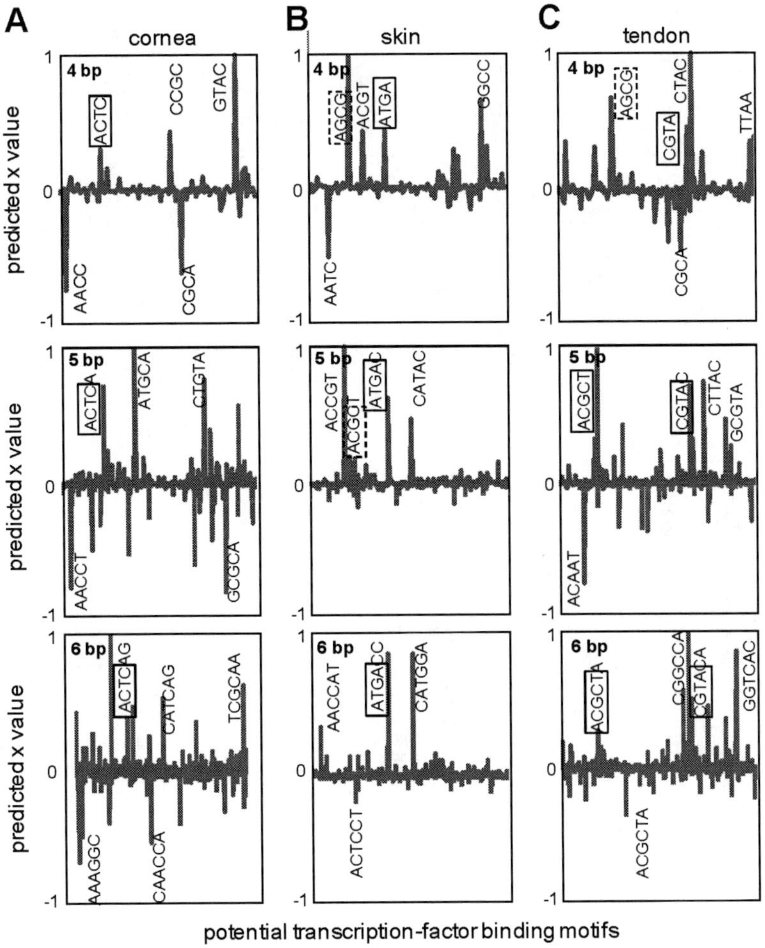

FIGURE 2: **Predicted X values using potential TFBMs of 4 - 6 bp in length.** The top five sequences were indicated, and any continuation of the same sequence was marked with a colored box. (A) Predicted X values in cornea. (B) Predicted X values in skin. (C) Predicted X values in tendon.

Putative TFBMs selected by the AntModeler: The summary of the top 10 putative TFBMs is presented for 4 – 6 bp sequences (Fig. 3). The solid and dotted lines linking within the same cell type indicate the sequence overlaps in the same and complementary strands, respectively. Interestingly, there are the putative TFBMs commonly predicted in skin and tendon. For instance, two 6-bp sequences (CGATGA and ACGCTA) were selected both in skin and tendon. No such 6-bp sequences were found, however, between cornea and skin, or between cornea and tendon.

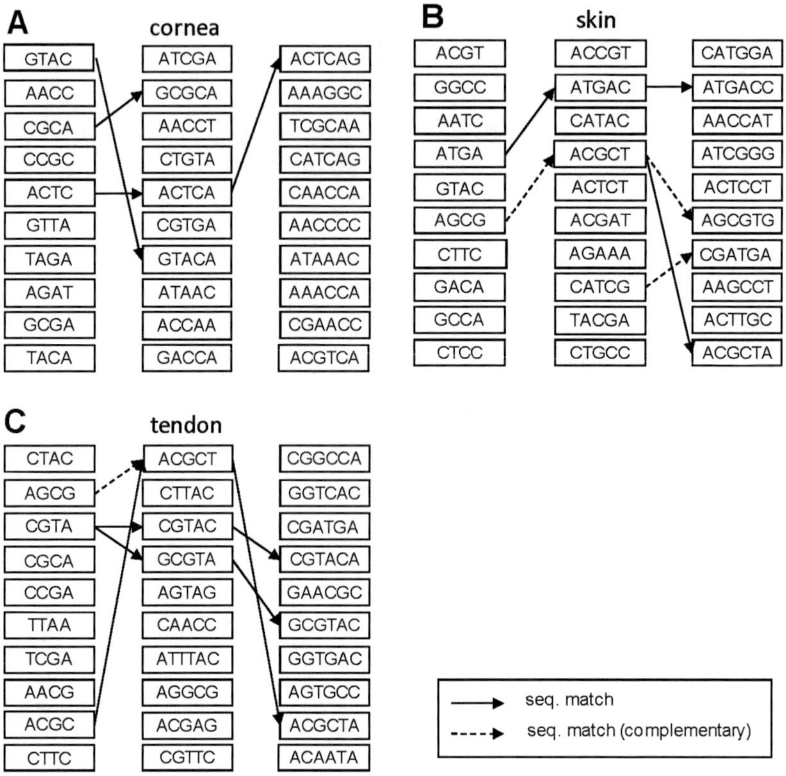

FIGURE 3: **Putative TFBMs in 4 - 6 bp.** The top ten choices are shown. Note that any sequence match among TFBMs is highlighted by the arrows. (A) Putative TFBMs in cornea. (B) Putative TFBMs in skin. (C) Putative TFBMs in tendon.

The linkage of the predicted TFBMs to biologically known motifs and the associated transcription factors was examined using cisRED (Table 1). The table includes the top predictions for each cell type together with the two sequences commonly identified for skin and tendon. Although the top three predictions for cornea did not show any sequence similarity to existing TFBMs, 8 out of 11 predictions for skin and/or tendon gave known transcription factors potentially involved in their mechanotransduction. Those transcription factors included RORα1, ATF3, AR, CAR, AhR, HES1, SREBP1, and CDC5 (skin), YY1, factor δ, NFμE1, HNF3, RSRFC4, and SREBP1 (tendon), and DEAF1, Pax2, CBF2, CREB, and COUP-TF2 (skin and tendon).

Mapping of putative TFBMs in upstream regulatory regions: The locations of the 6-bp putative TFBMs are displayed for 8 selected genes that were upregulated in the three cell types (Fig. 4). First, the putative TFBMs predicted by the AntModeler frequently appeared in the 1000-bp upstream regions: 29 out of the 30 putative TFBMs were located in those regions. Note that if randomly chosen the number of 6-bp sequences assigned in the regions would be only 2. Second, many putative TFBMs appeared multiple times. For example, CATCAG and AAACCA, which were identified for cornea, showed up 6 times as did CATGGA for skin. Third, this graphic representation supports the positive correlation between X-value and the importance of the putative TFBMs. The rectangles that marked the putative TFBMs with the highest five X-values, were found 37 times in total, while the circles that represented the bottom half of the ten selections were placed 22 times. Note that CGATGA that is the common putative TFBM to skin and tendon is located in the upstream region of the gene Mrpplf4.

TABLE 1. Linkage of the predicted binding motifs to transcription factors.

Cell line	Predicted binding motif	Predicted role*	Consensus motif sequence	Associated transcription factors (p-value)
cornea	ACTCAG	+	No match	
	AAAGGC	-	No match	
	TCGCAA	+	No match	
skin	CATGGA	+	CA(A/G/T)GGA	AR (9.2E-4); HNF6 (9.7E-4)
	ATGACC	+	A(A/T)GACC	RORα1 (5.3E-5);ATF3 (4.0E-4) AR (5.2E-4); CAR (5.6E-4)
	AGCGTG	+	AGCGTG	AhR (9.9E-6); HES1 (1.5E-5) SREBP1 (9.4E-4);Cdc5 (9.8E-4)
tendon	CGGCCA	+	CGGCCA	YY1 (1.0E-7); factor δ (2.6E-5) NFμE1 (2.6E-5)
	GGTCAC	+	GG(C/T)CAC	HNF3 (2.8E-4); RSRFC4 (6E-4) SREBP1 (9.2E-4)
	CGTACA	+	No match	
	GCGTAC	+	No match	
skin & tendon	CGATGA	+	CGATGA	DEAF1 (1.8E-4)
			GCCGATGA	Pax2 (7.5E-6); CBF2 (5.5E-4) COUP-TF2 (9.1E-4)
			CCGATGAT	Pax2 (7.5E-6); CREB (2.2E-4) COUP-TF2 (9.1E-4)
	ACGCTA	+	No match	

*Predicted Roles are +: stimulatory and -: inhibitory.

4. DISCUSSION

This study utilized the AntModeler software tool to identify the putative TFBMs in the upstream sequences of the genes responsive to mechanical stimulation in fibroblasts isolated from cornea, skin, and tendon. Through sequence comparisons among varying TFBM candidates of 4, 5, and 6 bp in length, we selected the 6 bp putative TFBMs such as ACTCAG, AAAGGC, and TCGCAA (cornea), CATGGA, ATGACC, and AGCGTG (skin), and CGGCCA, GGTCAC, and CGTACA (tendon). The sequences such as ACTCAG, AAAGGC, and CGATGA were linked to biologically known motifs, while others such as CGGCCA, GGTCAC, and CGATGA did not match to any existing motifs. Those sequences appeared frequently in the upstream region of the upregulated genes such as Hmox1, Gsta2, and Blvrb.

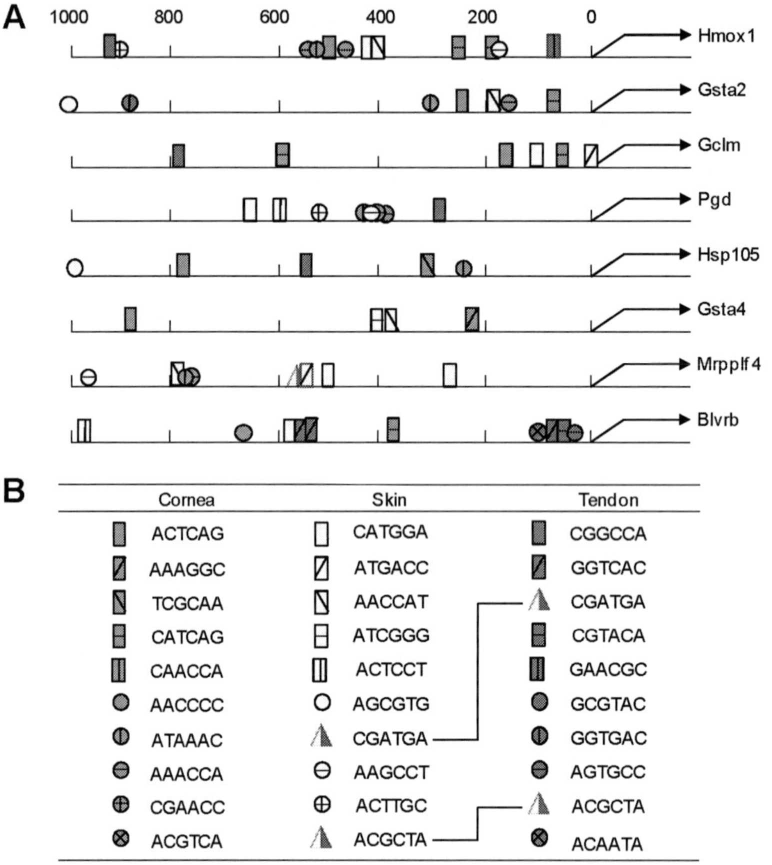

FIGURE 4: Locations of the top ten 6-bp sequences of all three cell types in selected genes. Different sequences in the same cell are marked with different shapes, and each cell's members are color coded the same way as the other three figures.

The present study indicates that most of the putative TFBMs are specific to each of the three cell types, although 4 TFBMs are shared by skin and tendon. The AntModeler predicted only one TFBM shared by cornea and tendon, and by cornea and skin. The result was unanticipated since 20 out of 46 genes were commonly upregulated in more than one cell types in response to mechanical stimulation. However, the predictions suggest that skin and tendon cells are more similar to each other than either of those is to cornea cells. This finding appears to be consistent with the fact that cornea cells are present only in a restricted part of anatomical structure, while skin and tendon cells are ubiquitously distributed. Further studies should be conducted for verification.

The putative TFBMs, shared by skin and tendon, are linked to varying transcription factors. Deaf-1 is involved in tissue development [13], and Pax-2 is required for differentiation of mesenchymal stem cells [14]. The transcription factor COUP-TF2 is a member of the nuclear receptor family [15], and REB is known to play an important role in regulation of various kinases [16]. YY1 is a transcription activator [17], and the androgen receptor (AR) is a ligand-activated transcription [18]. AhR ligands induce cell transformation and steroid-like immunosuppression [19], and RORα is a family of hormone nuclear receptors [20]. The role of those transcription factors in the responses to mechanical stimulation needs to be evaluated using biochemical and molecular assays such as a DNA binding assay and a gel mobility shift assay.

Although the current result provides a set of testable hypotheses, further studies are needed in application of the AntModeler. First, we focused on the 1000-bp upstream sequences, but the recent studies indicate the role of exons and downstream sequences in transcriptional regulation. Second, the analyses were limited to a selected set of genes with significant shear-driven alteration, but the number of genes in the model (46 – 61 genes) is considerably smaller than the genes responsive to mechanical stimulation. Third, the AntModeler includes multiple parameters such as the pheromone preference factor, the pheromone evaporation rate, the number of artificial ants, and the number of iterations. We used the default values based on previous simulational results to achieve a reproducible search of ~ 10 TFBMs, but other computational conditions may result in different predictions. Finally, biological confirmations by experiments are necessary to verify the predictions.

Based on the mathematical formulation derived for Pharaoh's ants [9], we built the simplified model for foraging behavior of artificial ants in the AntModeler:

$$\frac{dx(t)}{dt} = (q + rx(t))(n - x(t)) - s\left\{1 - \left(1 - \frac{x(t)}{n}\right)^2\right\} \tag{1}$$

where t = time, $x(t)$ = number of ants walking to a correct TFBM, q = rate of success by random search, r = rate of success guided by pheromone, n = total number of ants, and s = rate factor of loss due to pheromone evaporation. Using x_1 and x_2 as a pair of potential equilibrium states, $x(t)$ is solved for $x_1 \leq x \leq x_2$ (see Appendix):

$$x(t) = \frac{x_1 x_2 \left(1 - e^{\lambda t}\right)}{x_2 - x_1 e^{\lambda t}} \text{ with } \lambda = (x_1 - x_2)\left(\frac{s}{n^2} - r\right) \tag{2}$$

Equations (1) and (2) suggest that not all ants reach to "food source" depending on the number of ants in a colony as well as the parameters for search activities. In our computational simulation, we examined the number of transient matches to the predicted TFBMs in Fig. 3 as a function of the number of ants and the number of iterations (Fig. 3). The results are preliminary, but they indicate that the number of artificial ants is a critical parameter to achieve stable prediction of TFBMs. Further mathematical modeling and computational simulations are needed to build a linkage between ants in nature and artificial ants.

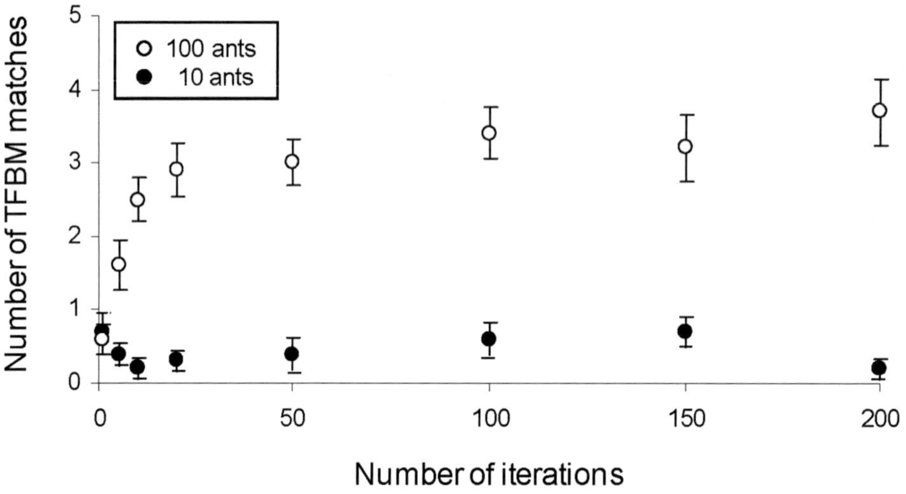

FIGURE 5: Evaluation of the AntModeler in a case with 10 and 100 artificial ants. The number of TFBM matches with the range of computational iterations (0 to 200) is shown using the data for cornea. Note that TFBM candidates of 6 bp in length were analyzed.

In summary, this study provided a set of testable hypotheses regarding TFBMs in three cell types in response to mechanical stimulation. Mechanical stimulation is an important signal in maintaining homeostasis and remodeling various tissue including musculoskeleton and vascular structures. Through application of the AntModeler, we predicted biological known TFBMs as well as potentially novel TFBMs in the responses to mechanical stimulation.

ACKNOWLEDGMENTS

We thank Rob Meagher for technical assistance. This study was in part supported by NIH AR050008.

REFERENCES

1. J.R. Mackley, J. Ando, P. Herzyk and S.J. Winder, *Biochem. J.* **396**, 307-316 (2006).
2. F. Collins, E. Green, A. Guttmacher and M. Guyer, *Nature* **422**, 835-847 (2003).
3. T. Ideker, T Galitski and L Hood, *Annu. Rev. Genomics Hum. Genet* **2**, 343-372 (2001).

4. H. Kitano, *Nat. Biotechnol.* **20**, 337 (2002).
5. The ENCODE Project Consortium, *Nature* **447**, 799-814 (2007).
6. Y. Liu and H. Yokota, *IEEE Computer Society,* 313-318 (2003).
7. Y. Liu, M. Vincenti and H. Yokota, *BMC Bioinformatics* **6**, 276 (2005).
8. Y. Liu and H. Yokota, *IEEE proceedings of Symposium on Computational Intelligence in Bioinformatics and Computational Biology* (2005).
9. M. Beekman, D. Sumpter and F. Ratnieks, *PNAS* **98**, 9703-9706 (2001).
10. A. Chen, H. Yokota, K. Hamamura, G. Wang, W. Xing, S. Mohan and Y. Liu, *Genomics, Proteomics & Bioinformatics,* in press.
11. Y. Liu and H. Yokota, *BMC Genomics* 7 **221,** doi: 10.1186/1471-2164-7-221 (2006).
12. Canada's Michael Smith Genome Sciences Centre and British Columbia Cancer Agency, *Nucleic Acid Res.* **34**, 68-73 (2006).
13. M. Czesak, S. Lemonde, E. Peterson, A. Rogaeva and P. Albert, *Cell J. Neurosci.* **26**, 1864-1871 (2006).
14. S.L. Zhang, Y. Chen, S. Tran, I. Chenier, M. Hebert and J. Ingelfinger, *J. AmSoc. Nephrol.* **18**, 2105-2115 (2007).
15. S. Matsuzaki, M. Canis, J. Pouly, R. Botchorishvili, P. Déchelotte and G. Mage, *Fertil. Steril.* **86**, 548-553 (2006).
16. R.E. Brown and P. Mattjus, *Biochim. Biophys. Acta.* **1771**, 746-760 (2007).
17. Y. Tone, Y. Kojima, K. Furuuchi, M. Brady, Y. Yashiro-Ohtani, M. L. Tykocinski and M. Tone, *J. Immunol.* **179**, 1760-1767 (2007).
18. H. Goyal, T. Braden, C. Williams and J. Williams, *Reproduction* **134**, 199-208 (2007).
19. C.G. Gu, X. Jiang, X.H., Ju, X. Yang and G. Yu, *SAR QSAR Environ. Res.* **18**, 603-619 (2007).
20. A. Rutledge, *Ground Water* **45**, 374-382 (2007).

APPENDIX

An analytical solution of $x(t)$ is expressed:

$$\frac{1}{x_1 - x_2}\log\left|\frac{x - x_1}{x - x_2}\right| = \left(\frac{s}{n^2} - r\right)t + K \tag{A1}$$

where x_1 and x_2 are two equilibrium states at $\frac{dx(t)}{dt} = 0$:

$$x_1, x_2 = n + \frac{q + rn \pm \sqrt{(q + rn)^2 + 4s\left(\frac{s}{n^2} - r\right)}}{2\left(\frac{s}{n^2} - r\right)} \tag{A2}$$

Application of Bayesian networks and data mining to biomedical problems

Alla R. Kammerdiner*, Anatoliy M. Gupal† and Panos M. Pardalos*

*Department of Industrial and Systems Engineering, University of Florida, Gainesville, FL, 32611, USA.
†Glushkov Institute of Cybernetics of National Academy of Sciences of Ukraine, Kyiv, Ukraine

Abstract. During the last several decades, the Bayesian networks have turned into a dynamic area of research. This great interest is owning to the advantages offered by special structure of Bayesian networks, which allows them to be very efficient in modeling domains with inherent uncertainty. Bayesian networks techniques can be successfully applied to mining various types of biomedical data.

This chapter demonstrates how various complex research problems in biology, biomedicine and other fields can be solved by means of the combination of methods from Bayesian networks and data mining.

Keywords: naïve Bayes, Bayesian networks, classification, neuroregeneration model, human genome
PACS: 01.10.Fv

1. INTRODUCTION

The foundation for Bayesian networks comes from statistics. The first formal introduction of the Bayesian networks in the field of artificial intelligence and expert systems can be traced to the works of by Pearl [1] in 1982, and Spiegelhalter and Knill-Jones [2] in 1984. The first practical applications of Bayesian networks, Munin [3] and Pathfinder [4], appeared in 1989 and 1992, respectively. Since then, the scope of research in Bayesian networks has increased dramatically. In particular, various modifications of Bayesian networks has been used for solving many challenging problems in data mining, pattern recognition, image processing and data fusion, engineering, etc.

Mathematically, a *Bayesian networks* on a finite collection of random variables $\mathrm{X} = \{X_1, X_2, \ldots, X_n\}$, can be defined as an ordered pair (G, D), where

- G is a graphical component represented by a directed acyclic graph with nodes corresponding to the random variables $X_1, X_2, \ldots, X_n$, and directed arcs symbolizing conditional dependencies between the variables. The set of all the arcs of G satisfies the following assumption: Each random variable in the graph is conditionally independent of its non-descendants in G, given its parents in G.
- D is a probabilistic component representing the set of parameters that, for each variable X_i, $1 \leq i \leq n$, define the conditional distribution of X_i given its parents in the graph G.

CP953, *Data Mining, Systems Analysis, and Optimization in Biomedicine*
edited by O. Seref, O. E. Kundakcioglu, P. M. Pardalos

A complex concept of a Bayesian network is based on a simple statistical formula called *Bayes rule*:

$$Pr(\mathcal{B} \mid \mathcal{A}) = \frac{Pr(\mathcal{A} \mid \mathcal{B}) \cdot Pr(\mathcal{B})}{Pr(\mathcal{A})}, \quad \forall \text{events } \mathcal{A}, \mathcal{B} \tag{1}$$

which allows for probability updating, i.e. calculating the posterior probability from data given reasonable assumptions about prior probabilities.

For a comprehensive review of the relative definitions, cases and examples of Bayesian networks, as well as the methods and applications of Bayesian networks see, for example, [5, 6].

Bayesian networks offer a very efficient way of modeling situations with inherent uncertainty. In many such situations, the underlying processes are mostly unclear, and the task is to try extracting the information behind such processes from the observed data. Luckily, a rapid technological progress in data collection and storage allows collecting a vast amount of data, and the data mining techniques can greatly facilitate prompt analysis and summarization of massive data. All the above is especially relevant when solving problems in biomedicine. For example, in numerous biomedical problems, one has to perform microarray DNA experiments in order to elucidate the underlying biological processes that are not yet completely understood. It is natural to apply data mining to handle the massive datasets that arise in microarray experiments. In general, it is plausible to try modeling the situation by means of Bayesian networks, when the situation exhibits causality and uncertainty, which is often the case in biological processes.

The chapter is organized as follows. Section 2 discusses the broad scope of the diverse areas of applications of Bayesian network methods, and then presents some illustrative examples of recent applications of Bayesian networks to various problems in expert systems and data mining. Section 3 continues with focuses exclusively on the application of Bayesian network to data mining, and the connection between them. Section 4 presents applications to two important biomedical problems. The chapter is concluded by the summary in Section 5.

2. ON THE BREADTH OF APPLICATIONS OF BAYESIAN NETWORKS

A concept of Bayesian network can be interpreted in different contexts. From a statistical point of view, a Bayesian network can be defined as a compact representation of the joint probability over a given set of variables. From a broader point of view, a Bayesian network is a special type of a graphical model capable of reflecting causality, as well as updating its beliefs in view of received evidence. All these features make a Bayesian network a versatile instrument that can be used for various purposes, including facilitating communication between human and computer, extracting hidden information and patterns from data, simplifying decision making, etc.

Due to their special structure, Bayesian networks have found many applications in various areas such as artificial intelligence and expert systems, machine learning and data mining. Bayesian networks are used for modeling knowledge in text analysis, image processing, speech pattern analysis, data fusion, engineering, biomedicine, gene and protein regulatory networks, and even meteorology. Furthermore, it has been expressed that the inductive inference procedures based on Bayesian networks can be used to introduce inductive reasoning in such previously a strictly deductive science as mathematics [18]. The large scope of different application of Bayesian networks becomes especially impressive taking into account that Bayesian networks is a somewhat new theory that has been around approximately for a mere quarter of a century. Next, several examples of the recent real-life applications of Bayesian networks are considered to illustrate the above point.

Recent research in the field of automatic speech recognition [7] indicates that dynamic Bayesian networks can effectively model hidden features in speech including articulatory and other phonological features. Both hidden Markov models (HMM), which are special case of dynamic Bayesian networks (DBN), and more general dynamic Bayesian networks have been applied for modeling audio-visual speech recognition. In particular, a paper by A.V. Nefian et al. [8] describes an application of the *coupled HMM* and the *factorial HMM* as two suitable statistical models for audio-video integration. The factorial HMM is a generalization of HMM, where the hidden state is represented by a collection of variables also called *factors*. These factors, although independent of each other, all impact the observations, and hence become connected indirectly. The coupled HMM is a DBN represented as two regular HMM whose hidden state nodes have links to the hidden state nodes from the next time slice. The coupled HMM has been also applied to model hand gestures, the interaction between speech and hand gestures, etc. In addition, face detection and recognition problems have been studied with help of Bayesian networks.

Note that different fields of application may call for specialized employment of Bayesian network methods, and conversely, similar approaches can be successfully used in different application areas. For instance, along with the applications to speech recognition above, the coupled hidden Markov models have been employed in modeling multi-channel EGG data [19].

An interesting example of Bayesian network in application to expert systems includes developing strategies for troubleshooting complex electro-mechanical systems presented in [9]. The constructed Bayesian network has the structure of naïve Bayes model. In the decision tree for the troubleshooting model, the utility function is given by the cost of repair. Hence, the goal is to find a strategy minimizing the expected cost of repair.

An interesting recent study [10] describes some applications of Bayesian networks in meteorology from data mining point of view. A large data base of daily observations of precipitation levels and maximum wind speed is collected. The Bayesian network structure is constructed from meteorological data by using various approaches, including batch learning procedure and simulation techniques. In addition, an important data mining application of Bayesian networks is illustrated by giving an example of missing data values estimation from the evidence received.

3. APPLICATIONS OF BAYESIAN NETWORKS TO DATA MINING

Rapid progress in data collection techniques and data storage has enabled an accumulation of huge amounts of experimental, observational and operational data. As the result, massive data sets containing gigantic bulk of information can be found almost everywhere. A well-known example of a massive data set is the data set containing the observed information about human genome. The tremendous need to quickly and correctly analyze or manipulate such enormous data sets facilitated the development of data mining techniques.

Data mining is the research aimed at discovery of various types of knowledge from the large data warehouses. Data mining can also be seen as an integral part of more general process of knowledge discovery in databases. The other two parts of the knowledge discovery in databases are the preprocessing, and postprocessing. As seen above Bayesian networks can also extract knowledge from data, which is called evidence in the Bayesian framework. In fact, the Bayesian network techniques can be applied to solve data mining problems, in particular, classification.

3.1. Application of naïve Bayes to data mining

Many effective techniques in data mining utilize the methods from other multidisciplinary research areas such as database systems, pattern recognition, machine, learning, and statistics. Many of these areas have a close connection to Bayesian networks. In actuality, data mining is utilizing a special case of Bayesian networks, namely, *naïve Bayes*, for performing effective classification. In a data mining context, a classification is a task of assigning objects to their relevant categories. The incentive for performing classification of data is to attain a comprehensive understanding of differences and similarities between the objects in different classes.

In the Bayesian framework, the data mining classification problem translates into finding the class parameter y maximizing the posterior probability of the unknown instance. This statement is also called the *maximum a posteriori principle*, or abbreviated, MAP. Let us show how the naïve Bayes can be represented by the following Bayesian network model:

The graph of the naïve Bayes model consists of one hypothesis variable, say Y, and a finite set of information variables $X_1, X_2, \ldots, X_n$ with the arcs from Y to every X_k, $1 \leq k \leq n$. In other words, the variables form a diverging connection with the information variables $X_1, X_2, \ldots, X_n$ having a common parent Y. The probability distributions are given by the values $Pr(X_k \mid Y)$, for each information variable X_k, $1 \leq k \leq n$.

Let us apply this simple Bayesian network model to construct a so called *naïve Bayes classifier*, which can be used for estimating the posterior probabilities of data having discrete and continuous parameters. Let $x_1, x_2, \ldots, x_n$ denote the known parameter values, and let y to be an unknown parameter. In the context of the naïve Bayes model, $x_1, x_2, \ldots, x_n$ are some realizations of information variables $X_1, X_2, \ldots, X_n$. In other words, we received evidence on $X_1, X_2, \ldots, X_n$, which results in the hypothesis variable Y being in the unknown state y. The posterior probability of y given $x_1, x_2, \ldots, x_n$ can be calculated according to the Bayes rule (1) as follows:

$$Pr(y \mid x_1, x_2, \ldots, x_n) = \frac{Pr(x_1, x_2, \ldots, x_n \mid y) \cdot Pr(y)}{Pr(x_1, x_2, \ldots, x_n)}. \tag{2}$$

Next, from the naïve Bayes model, we can draw additional information about the nature of the relationships among parameters. In particular, since the variables $X_1, X_2, \ldots, X_n$ and Y form a divergent connection with Y as the parent, then from the d-separation rule, $X_1, X_2, \ldots, X_n$ are d-separated, given Y received an evidence. That is, $X_1, X_2, \ldots, X_n$ are conditionally independent of each other, given Y. To put this mathematically,

$$Pr(x_i, x_j \mid y) = Pr(x_i \mid y) \cdot Pr(x_j \mid y), \tag{3}$$

for all i, j, where $1 \leq i, j \leq n$. Hence, the naïve Bayes model implies the following:

$$Pr(x_1, x_2, \ldots, x_n \mid y) = \prod_{i=1}^{n} Pr(x_i \mid y). \tag{4}$$

Then, from (2) it follows that

$$Pr(y \mid x_1, x_2, \ldots, x_n) = \frac{\prod_{i=1}^{n} Pr(x_i \mid y) \cdot Pr(y)}{Pr(x_1, x_2, \ldots, x_n)}. \tag{5}$$

Therefore, the task is to find such value of parameter y that maximizes the expression in the left-hand side of formula (5). Note that the denominator $Pr(x_1, x_2, \ldots, x_n)$ can simply be ignored, since it does not change with the change in the value of parameter y, and so, does not affect the maximum. Hence, it is sufficient to find y that maximizes the product $\prod_{i=1}^{n} Pr(x_i \mid y) \cdot Pr(y)$.

Thus, the above classification method based on the naïve Bayes model, works in the following fashion. To classify an unknown sample $(x_1, x_2, \ldots, x_n, y)$ with unknown y and known values $x_1, x_2, \ldots, x_n$, the naïve Bayes classifier

- computes the product $\prod_{i=1}^{n} Pr(x_i \mid y) \cdot Pr(y)$, for each value of unknown parameter y;
- selects the parameter y_* such that

$$\prod_{i=1}^{n} Pr(x_i \mid y_*) \cdot Pr(y_*) = \max_{y} \prod_{i=1}^{n} Pr(x_i \mid y) \cdot Pr(y). \tag{6}$$

Now consider some common approaches to estimation of probabilities $Pr(y)$ and $Pr(x_i \mid y)$, where $1 \leq i \leq n$. First, note that as a class label parameter y must necessarily follow a discrete distribution. Hence, the probability $P(y)$ can be estimated as the frequency of instances that belong to class y (i.e. instances with $Y = y$) in the training sample D. On the other hand, each of parameters $x_1, x_2, \ldots, x_n$ representing the known information, may follow either continuous or discrete distribution.

For discrete parameters x_i, the conditional probabilities $Pr(x_i \mid y)$ can be estimated similarly to $Pr(y)$ above. In particular, $Pr(x_i \mid y)$ is calculated approximately as the fraction of such instances in class y, where $X_i = x_i$. Unfortunately, the described method of the probability approximation as the corresponding frequency is known to have a potential disadvantage. Indeed, if the training sample D does not contain an observation for some x_i, then the above estimation approach gives $Pr(x_i) = 0$. When the estimated probability equals zero for even a single parameter, the entire posterior probability involving this parameter becomes zero. One of the commonly used ways to resolve this problem is by using the *Laplace correction* to obtain a non-zero estimate:

$$Pr(x_i \mid y) = \frac{n_i + m \cdot p}{n + m}, \tag{7}$$

where n represents the total number of instances of class y in the training dataset D, n_i is the number of instances in D from class y for which $X_i = x_i$, positive real m is a parameter often called the equivalent sample size, and $0 < p < 1$ is some specified prior probability.

For continuous parameters, the common approach to estimation of its conditional probabilities is to make a simplifying assumption about the form of probability distribution. Usually, a continuous parameter is approximated by a normal distribution with some parameters μ and σ. Both the mean of the distribution, $\mu = \mu(y)$, and its standard deviation $\sigma = \sigma(y)$ depend on the class label y, and so, we have:

$$Pr(X_i = x_i \mid Y = y) = \frac{1}{\sqrt{2\pi}\sigma_i(y)} \exp\left\{ -\frac{(x_i - \mu_i(y))^2}{2\sigma_i^2(y)} \right\}, \tag{8}$$

where $\mu_i(y)$ and $\sigma_i(y)$ are the mean and the standard deviation of x_i, respectively.

The classification techniques based on the naïve Bayes model have the following positive characteristics:

- When the amount of missing data is insignificant, Naïve Bayes classifiers can easily handle such data by ignoring the instance during the estimations;
- If a statistically small amount of noisy data is present, naïve Bayes classifiers are robust to the noise, since the noise values is averaged out in the estimations of probability from data;
- For the same reasons, naïve Bayes classifiers are robust to irrelevant attributes.

3.2. Using Bayesian networks for classification

In spite of all the positive characteristics of the naïve Bayes classifier, there is a certain drawback. In general, for many practical data sets, most of the parameters are not totally independent of each other. Hence, the independence assumption, which is essential for the Naïve Bayes classifier to be exact, may not hold. However, this issue can be successfully resolved by designing a suitable Bayesian network for each individual case. Clearly, the simple Bayes classifier can be generalized into a classification method based on designing a specialized Bayesian network, whose structure appropriately models conditional independence assumptions among parameters.

Let x denote the known parameters, and y be unknown class label. The approach to building Bayesian networks classifiers can be summarized as follows:

- Test the parameters to find the conditional independence structure;
- Design a Bayesian network **M** that correctly reflects the conditional independence structure;
- Estimate $Pr(x \mid y)$ from Bayesian network **M** by using a suitable probability updating technique;
- Select parameter y maximizing $Pr(y)Pr(x \mid y)$

Similarly to naïve Bayes classifier, the classification by way of building suitable Bayesian networks are capable of handling the presence of noise in the data as well as the missing values. In addition, the classification process utilizing Bayesian networks is robust to irrelevant parameters. Obviously, Bayesian network classifiers have the same great advantages as the naïve Bayes classifier, while being free of the main disadvantage of the latter. Indeed, in each particular case, the Bayesian network classifier is designed so that the conditional independence assumptions hold automatically for the given parameters. The efficiency of Bayesian network classifiers greatly depends on the topology of the respective network. Therefore, in cases when the independence relationships between parameters can be correctly described by several different Bayesian networks, the model that has the simplest graph structure must always be preferred.

To the defense of the naïve Bayes classifier, it is very simple to implement, and has an easy way of estimating conditional probabilities. That is why, in practice, naïve Bayes classifier is often applied if the independence assumption is not far off. As we can see, there is an understandable tradeoff between the simplicity of Bayesian network reflecting the independence structure, and the precision in the independence assumptions. In practice, for the sake of significantly simplifying the algorithm, one may choose to make some assumptions as long as they are sufficiently close to the true nature of the independence relations among the variables.

Artificial neural networks can serve as an example of the Bayesian network classifier designed for a special case. Artificial neural networks can be viewed as a special case of Bayesian networks, where the nodes are partitioned into n mutually-exclusive *layers*, and the set of arcs represented by the links from the nodes on layer i to the nodes on $i+1$, $1 \leq i \leq n$. Layer 1 is usually called the *input layer*, while layer n is known as the *output layer*.

4. APPLICATIONS TO BIOMEDICAL PROBLEMS

4.1. Modeling neurite regeneration and outgrows with Bayesian networks

Bayesian networks learning procedures can be effectively applied to many biomedical problems involving massive data sets. In particular, Bayesian networks can be exploited to construct comprehensive models of complex biological processes. Take for example the processes of neurite regeneration and outgrows. An advanced study of these processes frequently involves a data collection by performing microarray experiments. Datasets arising in microarray experiments often contain noise and missing data. Bayesian networks are well-suited for handling this stochastic nature of microarray datasets. Since the observations may be entered into Bayesian network as evidence in order to correct the noisy data or fill in the missing values.

Probabilistic graphical models, in particular Bayesian networks, have been found extremely useful for extracting meaningful biological insight from the data sets arising in DNA microarray and protein array experiments. By incorporating simpler submodels, Bayesian networks models give a concise representation of complex cellular networks. Algorithms based on axiomatic principles for inferring such models from data promote a model-based methodology for analysis and exploration, which promising capabilities has been recently demonstrated by numerous applications of this methodology to gene expression data.

Bayesian Networks learning algorithm can be utilized to perform a sophisticated statistical analysis of the data on the multi-layered scale. Another attractive feature of Bayesian networks that makes them so suitable for the purpose of building neuroregeneration model is their ability to accurately model casual relations between variables (i.e. gene and protein expression levels). This ability allows one to model interaction of various components such as genetic interactions, protein-protein interactions, protein pathways, etc.

Moreover, Bayesian networks have been proved to be particularly useful in the analysis of data sets with the number of replicates too small for conventional statistical methods to be reliable, which is a known issue in microarray data analysis. It turns out that even in such lack of experimental replication; Bayesian Networks are capable of extracting knowledge. Additionally, Bayesian networks combined with Bayesian methods can serve as an efficient instrument for avoiding the over fitting of data. It has been shown that it is not necessary to withhold some of the available data for testing and that the Bayesian approach allows us to modify the model so that all available data can be used for training [11].

Finally, Bayesian networks appears to be a very logical choice for constructing, refining and validating the model based on a multi-paradigm approach, because they allow one to utilize knowledge discovered from the analysis of the data, which are obtained on the previous level, when building a more refined model on the next level. For instance, Bayesian learning procedure can use information about the system gathered at the PC-12 cell as a prior knowledge to infer information about the system on the primary neuron level. In the framework of Bayesian networks, different characteristics of the system of neurite regeneration are treated as random variables. Random variables consist of observed attributes (for example, the expression level of a specific gene or protein in a particular experiment) and hidden attributes that are assumed by the model (for instance, the cluster allocation of a specific gene or protein).

4.2. Using Bayesian networks and data mining for genome modeling

Long before its completion in 2000, the Human Genome Project has become one of the most famous examples of a collection of massive databases. Nowadays, genomes of many various organisms have been sequenced. From the broad view of biology, the genome of an organism is all the hereditary material found in the DNA (or RNA, for some viruses). More specifically, a genome is a complete DNA sequence of a collection of chromosomes. The current human genome sequence contains 2.85 billion nucleotides interrupted by only 341 gaps. It covers approximately 99% of the euchromatic genome and is accurate to an error rate of about an event per 100,000 bases. Remarkably, the human genome seems to encode only 20.000 – 25.000 protein-coding genes [12, 13, 14].

Even though DNA is relatively simple and well understood chemically, the human genomeŠs structure is extraordinarily complex and its function is poorly understood. Only less than two percent of its bases encode proteins, and the full interaction of protein-coding sequences still remains to be established. Even less is known about the function of the roughly half of the genome that consists of highly repetitive sequences or of the remaining non-coding, non-repetitive DNA.

Clearly, the task of decoding the information encoded by the human genome, as well as genomes of other sequenced species, represents a great challenge of modern science, especially in the area of data mining. In order to completely uncover the mysteries of genome, it is necessary to learn all the hidden patterns in the massive genomic data that has been accumulated. Therefore, we must utilize many advanced data mining techniques.

A two-chain spiral of DNA encodes hereditary information by a four-letter alphabet composed of the four nucleotides, namely adenine A, cytosine C, guanine G, and thymine T. It is well-known that $C-G$ and $A-T$ are the complementary pairs of bases connecting two chains of DNA. By representing a sequence of nucleotides in a human chromosome by a simple Bayesian network, specifically, the Markov chain, we establish the fundamental principles in encoding complementary bases A and T, C and G by one chain in DNA chromosomes of the human genome and genomes of other investigated organisms. This indicates an opportunity to create a model of the relationships between the letter pairs, the three adjacent bases in a triplet, etc. On the basis of obtained statistical data, it is easy to come to a conclusion that there exist exact rules for forming DNA structure valid for all species.

As reported in [15, 16], it has recently been discovered that for each chain of DNA chromosomes the following complementary relationships hold:

$$\hat{P}_A \cong \hat{P}_T, \tag{9}$$

$$\hat{P}_C \cong \hat{P}_G, \tag{10}$$

where $\hat{P}_A, \hat{P}_C, \hat{P}_G, \hat{P}_T$ denote the frequencies of the letters A, C, G, and T, respectively. For a given base $i = A, C, G, T$, the frequency of i is calculated as follows:

$$\hat{P}_i = \frac{n_i}{n}, \tag{11}$$

where n_i is the total number of base A in the chromosome, and n is the chromosomal length. These frequencies give us the estimates for the probabilities P_i of the respective nucleotides i, $i = A, C, G, T$; and a nucleotide sequence of a given chromosome can be modeled by a finite Markov chain $\{X_i\}_{i=1}^{n}$ of length n, with the prior probabilities known as the *initial probabilities* P_A, P_C, P_G, P_T, such that $P_A + P_C + P_G + P_T = 1$, and the conditional probabilities called the *transition probabilities* $P_{ij} = Pr(X_{k+1} = j \mid X_k = i)$, where $1 \leq k \leq n$, and $i, j = A, C, G, T$. Obviously, $P_{iA} + P_{iC} + P_{iG} + P_{iT} = 1$.

The frequencies $\hat{P}_{ij}$ estimating the transition probabilities P_{ij} can be computed as

$$\hat{P}_{ij} = \frac{n_{ij}}{n_{iA} + n_{iC} + n_{iG} + n_{iT}}, \tag{12}$$

where n_{ik} is the number of nucleotide pairs ik in the data sample. There are two problems in the above approach to estimating the probabilities of transition. First, the denominator in (12) is not being fixed. Second, the estimates in (12) are biased. However, the statistical analysis [17] shows that for $i, j = A, C, G, T$, the differences below are asymptotically normal:

$$\hat{P}_{ij} - P_{ij} \sim \mathcal{N}(0, \sigma_{ij}^2), \quad \text{as} \quad n \to \infty, \tag{13}$$

Where the standard deviation is given by

$$\sigma_{ij} = \frac{P_{ij}(1 - P_{ij})}{nP_i}. \tag{14}$$

In addition, the differences $\hat{P}_{ij} - P_{ij}$ are asymptotically independent for any given $i = A, C, G, T$. Also note that due to the extremely large chromosomal length n, the associated standard deviations in (14) are very small. This shows that Markov chain gives a simple and effective model of genome. In order to completely describe a very lengthy nucleotide sequence using Markov chain, for any given chromosome, it is sufficient to obtain estimates of only 15 parameters. In fact, the Markov chain model of a genome only requires estimating three out of four initial probabilities, and twelve out of sixteen transitional probabilities.

The results of frequency computations for every chromosome of the human genome are presented in Table 1. These values can be used as the estimates for the initial probabilities when constructing a corresponding Markov chain. Table 2 gives numerical results of transitional probability estimation by the frequencies of letter pairs. From Table 2 we see that for chromosome 1, all estimates of the transitional probabilities are strictly positive. Since the chain has finite number of states, this implies that the Markov chain, which approximates chromosome 1, is ergodic. In other words, for any i, the estimates of n-step transitional probabilities $\hat{P}^{n}_{ij}$ converge to limiting probability π_j, as $n \to \infty$, independently on i; and $\pi_A + \pi_C + \pi_G + \pi_T = 1$. The above convergence is geometric.

Thus, the analysis of the collected human genome data shows that the process of chromosome encoding can be modeled by an example of Bayesian network, namely, a Markov chain with four states corresponding to each nucleotide. In addition, since the length of a human chromosome is extremely large, the sequence can be given an approximate representation in form of homogeneous *ergodic* Markov chain.

TABLE 1. Frequencies of four nucleotides for all chromosomes of human genome

Chromosome	Adenine A	Cytosine C	Guanine G	Thymine T
1	0.291	0.209	0.209	0.292
2	0.299	0.201	0.201	0.299
3	0.301	0.198	0.198	0.302
4	0.309	0.191	0.191	0.309
5	0.302	0.197	0.198	0.303
6	0.302	0.198	0.198	0.302
7	0.296	0.204	0.204	0.297
8	0.299	0.201	0.201	0.299
9	0.293	0.207	0.207	0.293
10	0.292	0.208	0.208	0.292
11	0.292	0.208	0.208	0.292
12	0.296	0.204	0.204	0.296
13	0.307	0.193	0.193	0.308
14	0.294	0.204	0.205	0.297
15	0.289	0.211	0.211	0.289
16	0.275	0.223	0.224	0.277
17	0.272	0.228	0.227	0.273
18	0.301	0.199	0.199	0.301
19	0.258	0.242	0.242	0.259
20	0.278	0.220	0.221	0.280
21	0.297	0.204	0.205	0.294
22	0.261	0.240	0.240	0.260
X	0.302	0.197	0.197	0.303
Y	0.299	0.199	0.200	0.301

TABLE 2. Frequencies of different pairs of nucleotides for chromosome 1 of human genome used for estimating the transitional probabilities

$\hat{P}_{ij}$	A	C	G	T
A	0.327	0.172	0.244	0.257
C	0.349	0.261	0.048	0.342
G	0.287	0.211	0.260	0.242
T	0.218	0.205	0.250	0.327

5. SUMMARY

This chapter focuses on the relationship between Bayesian networks and data mining, and the applications of these dynamic research areas to various problems in biomedicine. A general classification procedure based on Bayesian networks is described. The procedure is a generalization of the classification technique known as Naïve Bayes. The application of the Bayesian network methods to biomedical problems is illustrated by two examples. In particular, we outline the ideas behind the study of the causal relations between various components of neurite regeneration model. In addition, we give an example of estimation of probabilities in Markov chain model derived from the human genome data. Just as Naïve Bayes, Markov chain is a special case of an elementary Bayesian network.

With the rapid development of new technologies in collecting biological data such as microarray technologies, data mining has become essential for biomedical research. Recently, the Bayesian networks techniques have been successfully applied to establish the hidden patterns in the datasets arising in microarray experiments. There is an enormous potential in combining these two fields, in particular, to tackle complex research problems in biomedicine and biology. The symbiosis of the Bayesian networks approach and efficient data mining techniques can offer a resourceful instrument for learning the underlying biological processes from massive data.

REFERENCES

1. J. Pearl, "Reverend Bayes on Inference Engines: A Distributed Hierarchical Approach," in *National Conference on Artificial Intelligence*, AAAI Press, Menlo Park, CA, 1982, pp. 133 – 136.
2. D. J. Spiegelhalter, and R. P. Knill-Jones, *Journal of the Royal statistical Society* **147**, 35 – 77 (1984).
3. S. Andreassen, "Knowledge representation by extended linear models," in *Deep Models for Medical Knowledge Engineering*, 1989, pp. 129–145.
4. D. Heckerman, E. Horvitz, and B. Nathwani, *Methods of Information in Medicine* **31**, 90 – 105 (1992).
5. A. R. Kammerdiner, "Bayesian Networks," in *Encyclopedia of Optimization, 2nd ed.*, edited by C A. Floudas and P. M. Pardalos, Springer, Germany (to appear)
6. F. V. Jensen, *An Introduction to Bayesian Networks*, UCL Press, London, 1996.
7. K. Livescu, J. Glass, and J. Bilmes, "Hidden Feature Models for Speech Recognition Using Dynamic Bayesian Networks," in *8th European Conference on Speech Communication and Technology (EUROSPEECH)*, Geneva, Switzerland, September, 2003.
8. A. V. Nefian, L. Liang, X. Pi, X. Liu, K. Murphy, *Journal On Applied Signal Processing* **11**, 1 – 15 (2002).
9. J. Vomlel, "Two applications of Bayesian networks," in *Proceedings of conference Znalosti*, Ostrava, Czech Republic, 2003, pp. 73 – 82.
10. R. Cano, C. Sordo, and J. M. Gutierrez, "Applications of Bayesian Networks in Meteorology," in *Advances in Bayesian networks*, edited by Gamez et al., 2004, pp. 309–327.
11. D. Heckerman, "A Tutorial on Learning With Bayesian Networks," *Microsoft Research technical report 95-06*, 1995.
12. "The International Human Genome Sequencing Consortium. Initial sequencing and analysis of the human genome," *Nature* **409**, 860 – 921 (2001).
13. F. S. Collins, et al., *Nature* **422**, 835 – 847 (2003).
14. "The International Human Genome Sequencing Consortium. Finishing the euchromatic sequence of the human genome," *Nature* **431**, 931 – 945 (2004).
15. I. V. Sergienko, A. M. Gupal, A. A. Vagis, *Journal of Automation and Information Sciences* **37** (4), 153 – 157 (2005).
16. I. V. Sergienko, and A. M. Gupal, *Cytology and Genetics* **38**(4), 76 – 81 (2004).
17. A. M. Gupal, and A. A. Vagis, *Journal of Automation and Information Sciences* **33**, 51–58 (2001).
18. A. M. Gupal, I. V. Sergienko, *Kibernetika i sistemny analiz* **1**, 33–39 (2003).
19. S. Zhong, and J. Ghosh, "HMMs and coupled HMMs for multi-channel EEG classification," in *Proc. IEEE Int. Joint Conf. on Neural Networks*, 2002, pp. 1154 – 1159.

A web server for mining Comparative Genomic Hybridization (CGH) data [1]

Jun Liu, Sanjay Ranka and Tamer Kahveci

Computer and Information Science and Engineering, University of Florida Gainesville FL, 32611 USA

Abstract. Advances in cytogenetics and molecular biology has established that chromosomal alterations are critical in the pathogenesis of human cancer. Recurrent chromosomal alterations provide cytological and molecular markers for the diagnosis and prognosis of disease. They also facilitate the identification of genes that are important in carcinogenesis, which in the future may help in the development of targeted therapy.

A large amount of publicly available cancer genetic data is now available and it is growing. There is a need for public domain tools that allow users to analyze their data and visualize the results. This chapter describes a web based software tool that will allow researchers to analyze and visualize Comparative Genomic Hybridization (CGH) datasets. It employs novel data mining methodologies for clustering and classification of CGH datasets as well as algorithms for identifying important markers (small set of genomic intervals with aberrations) that are potentially cancer signatures. The developed software will help in understanding the relationships between genomic aberrations and cancer types.

Keywords: software package, data mining, CGH
PACS: 07.05.Wr

1. INTRODUCTION

Chromosomal abnormalities have been implicated in cancer development since the turn of the last century, and with advances in cytogenetics and molecular biology, the chromosomal alterations are recognized as critical in the pathogenesis of human cancer. Numerical and structural chromosomal imbalances are one of the most prominent and pathogenetically relevant features of neoplastic cells [1]. Over the last decades, thousands of (molecular-) cytogenetic studies of human neoplasias have led to important insights into the genetic mechanisms of tumor development and description of targets for pharmacologic intervention. The repetitive chromosomal aberration patterns reflect the supposed cooperation of a multitude of tumor relevant genes [2] in most malignant diseases. A systematic analysis of these patterns for oncogenomic pathway description requires the large-scale compilation of (molecular-) cytogenetic tumor data as well as the development of tools for transforming those data into a format suitable for data mining purposes. One of the early techniques used to identify cytogenetic abnormalities in tumor cells is called Metaphase analysis [3, 4, 5]. The Mitelman Database of Chromo-

[1] This work is supported in part by the National Science Foundation under Grant ITR 0325459 and DBI - 0606607. Any opinions, findings, and conclusions or recommendations expressed in this material are those of the author(s) and do not necessarily reflect the views of the National Science Foundation.

CP953, *Data Mining, Systems Analysis, and Optimization in Biomedicine*
edited by O. Seref, O. E. Kundakcioglu, P. M. Pardalos

some Aberrations in Cancer (`http://cgap.nci.nih.gov/Chromosomes/Mitelman`) has become an invaluable resource for cytogenetic aberration data in human malignancies. This database currently contains data from more than 46,000 cancer and leukemia samples analyzed by Metaphase banding. However, utilization of the data collection for data mining purposes so far has been limited by intrinsic problems of the Metaphase banding technique as well as the specific data format.

Over the last decade, the Comparative Genomic Hybridization (CGH) [6] and array- or matrix-CGH techniques [7, 8, 9] have addressed technical problems associated with Metaphase analysis of tumor cells and are now used in many published observations. CGH is a molecular-cytogenetic analysis method for detecting regions with genomic imbalances (gains or losses of DNA segments). Applying microarray technology to CGH measures thousands of copy number information distributed throughout the genome simultaneously [10]. Raw data from array CGH experiments is expressed as the ratio of normalized fluorescence of tumor and reference DNA. Normalized CGH ratio data surpassing predefined thresholds is considered indicative for genomic gains or losses, respectively. In contrast to the array CGH, the chromosomal CGH results are annotated in a reverse *in-situ* karyotype format [11] describing imbalanced genomic regions with reference to their chromosomal location. CGH data of an individual tumor can be considered as an ordered list of status values, where each value corresponds to a *genomic interval* (e.g., a single chromosomal band). The terms *feature* and *dimension* have also been used in the literature to represent the genomic interval. The status can be expressed as a real number (positive, negative, or zero for gain, loss, or no aberration respectively). We use this strategy and represent gain, loss, and no change with +1, -1, and 0 respectively. Chromosomal and array CGH recently account for a significant percentage of the published analyses in cancer cytogenetics [12, 13, 14, 15, 16, 17, 18].

The *Progenetix* database [19] (`http://www.progenetix.net`) is one of the main resources for CGH data. It consists of more than 12,000 cases from 19 different histopathological cancer types. The cancer type with the smallest number of samples, consists of 110 non-neoplastic cases, while the one with largest number of samples, Adenocarcinoma, NOS (ICD-O 8140/3), contains 1054 cases. The Progenetix project stands out as the only collection of molecular-cytogenetic aberration data not limited to results from a limited number of researchers and by encompassing virtually all tumor entities. Each sample in this dataset has 862 genomic intervals. Thus, each sample is represented using a sequence of 862 status values. For convenience, 1, -1, and 0 are used to denote gain, loss, and no change status respectively. The interval specific aberration information from the Progenetix dataset has shown its usefulness for the delineation of genomic aberration patterns with clear prognostic relevance in neuroblastoma [17] and for producing tumor type specific imbalance maps [20, 21].

A key feature of the CGH data is that consecutive values are highly correlated (see Fig. 1 for an example CGH dataset). In other words gain (or loss) in a genomic interval can be observed in a contiguous block of genomic intervals containing that genomic interval. We name such blocks, segments. As a result, CGH data contains significant amount of noise, especially around markers that are signatures of the underlying cancer type. Data mining methods that analyze CGH data needs to filter such noise to have an accurate picture of the underlying cancer.

Despite the abundance and growing importance of CGH data, there is no comprehensive tool exists for segment based analysis of large and heterogeneous set of CGH data. In this chapter, we describe a web based tool for large volume data analysis of CGH datasets. The tool provides various clustering algorithms, distance measures and identifies markers that can be interactively used by researchers. It presents the results are provided in both textual and graphical format. The tool does not require downloading or installing of software. It can be used through a web browser. A preliminary version of this tool is available at `http://128.227.162.207:8007/CGH/Default.html`.

The rest of the chapter is organized as follows. Section 2 describes the key algorithms that are used for analyzing CGH datasets. Section 3 describes the software and benchmarking issues. Section 4 demonstrates the use of the developed software on a sample dataset. Section 5 concludes the chapter.

2. ALGORITHMS

Different strategies for structural analysis of (molecular-) cytogenetic data sets have been successfully applied. These analysis attempts were aimed at the description of pseudo-temporal relationships of cytogenetic events [22, 23] or at the correlation of disease subsets with clinical parameters [24, 25]. Pollack et al, found that DNA copy alterations have significant impact in the progression of the breast cancer [26]. CLAC built a hierarchical tree along the arms of the chromosome to estimate the false discovery rate and to select interesting aberrations [27]. CGHmix used a mixture model to classify copy number changes [28]. BioHMM employed a hidden Markov model on CGH data to identify k hidden states, where k is a given number which is less than the number of genomic intervals [29]. These states define a segmentation of the CGH data. Fridyland et al developed an unsupervised hidden Markov model to find DNA copy numbers [30]. Hodgson et al used Gaussian distribution to model the DNA copy numbers [31]. Eilers and Menezes used penalized quantile regression to smooth CGH data [32]. Olshen et al assumed that the aberrations are discrete events and contiguous blocks of aberrations may be correlated [33]. They developed a binary segmentation to identify these blocks. Venkatraman and Olshen later developed an algorithm to compute the P-value of these segments [34]. Autio et al developed a tool, named CGH-Plotter, that visualizes and identifies amplifications in CGH data [35].

A key feature of the CGH data is that consecutive values are highly correlated. Fig. 1 shows an example dataset. The figure demonstrates that often each gain is followed by a contiguous block of gains (i.e., segment). Similarly, each loss is often followed by a contiguous block of losses. This is because, an aberration can spread from 5 to 15 genomic intervals. Usually, only a single genomic interval in such a contiguous block is a cancer marker while the rest of the aberrations are noise. In addition to the noise within segments, CGH data also contains noisy aberrations that is not necessarily a part of a segment. Such noisy aberrations are often characterized as isolated aberrations that appear in only a small percentage of the cancer patients. Our preliminary research has exploited this property to develop novel and accurate methods for mining information from CGH data. In the following, we briefly describe these methods that form the basis of the CGH analysis tool.

2.1. Distance based clustering

Clustering of the CGH data allows modeling the underlying dataset using a small number of clusters, where each cluster consists of a cohesive set of samples. This leads to a simpler description of the data set and can be effective in getting a better understanding of the underlying data. The problem of clustering is defined as follows.

> Given a set S of n samples $s_1, s_2, \cdots, s_n$, partition S into k subsets $C_1, C_2, \cdots, C_k$, such that the samples assigned to each subset are more similar to each other than the samples assigned to different subsets.

We have developed implementations based on the the following well known clustering methods: K-means clustering [36], complete link bottom-up clustering [37], and top-down clustering [38].

The crucial step in the abovementioned distance based clustering algorithms is the computation of distance (or similarity) between two or more samples. Let $X = x_1, x_2, \cdots, x_m$ and $Y = y_1, y_2, \cdots, y_m$ be two CGH samples. Here, x_i and y_i denote the status of the ith genomic interval of X and Y respectively. An important challenge is to define a measure of similarity between X and Y that utilizes the fact that consecutive values are highly correlated in CGH data. We have used the following measures of distance/similarity in our implementations:

1. *Raw distance measure.* Our first measure assumes that the genomic intervals are independent of each other. This assumption is often made in existing literature to simplify the problem of computing distances [39]. If both samples have gain (or loss) at the same genomic interval then we consider them similar at that position. Otherwise, that genomic interval contributes to the distance between them. Also, we assume that all genomic intervals have the same importance. Thus, each genomic interval contributes the same amount to the total distance. Formally, the distance is computed as $\sum_{j=1}^{m} \text{diff}(x_j, y_j)$, where $\text{diff}(x_j, y_j) = 1$ if $x_j \neq y_j$ or $x_j = 0$, and $\text{diff}(x_j, y_j) = 0$ otherwise. The similarity is obtained by subtracting the distance from m, the number of genomic interval of the CGH samples. We call this distance *Raw* since it is computed on raw CGH data. Raw distance between two samples is small only if the samples have same status for most of the intervals.
2. *Segment-based similarity.* This measure considers the fact that consecutive genomic intervals are usually correlated. A contiguous block of gains (or losses) can be caused by a point-like aberration at a single genomic interval. We use the term *segment* to represent a contiguous block of aberrations of the same type. We call two segments from two samples *overlapping* if they have at least one common genomic interval of the same status. Next, we develop a segment-based similarity measure called *Sim*.
   Given two CGH samples X and Y, Sim constructs maximal segments by combining as many contiguous aberrations of the same type as possible. Formally, the genomic intervals $x_i, x_{i+1}, \cdots, x_j$, for $1 \leq i \leq j \leq m$, define a segment if genomic intervals x_i through x_j are in the same chromosome, the values from x_i to x_j are all gains or all losses, and x_{i-1} and x_{j+1} are different than x_i. Thus, each sample translates into a sequence of segments. After this transformation, Sim assumes that the segments

are independent of each other and gives the same importance to all the segments regardless of the number of genomic intervals in them. Sim computes the similarity between two CGH samples as the number of overlapping segment pairs. Thus, unlike the Raw distance measure, Sim considers an overlap of arbitrary number of genomic intervals as a single match. This is justified because each overlap may indicate a common point-like aberration in both samples which then led to the corresponding overlapping segments.

3. *Segment-based cosine similarity.* Cosine similarity of two vectors measures the cosine of the angle between them and the most commonly used method to compute the similarity between two directional samples in vector-space model [40]. It can be adapted to compute the similariy between segment based representation of two CGH samples. Details are presented in [41]). There are two versions of this measure, one that treats gaps as segments *CosineGaps* and the other one that does not consider gaps (*CosineGaps*) in the calculation.

Once the dataset is partitioned into clusters of samples, the first task is to measure the quality of the resulting clusters. This is, however, nontrivial as clustering is an unsupervised process. One would expect that CGH samples belonging to the same cancer type should have similar imbalances and thus they should belong to the same cluster. The cancer type for each sample in the Progentix database is known. For such datasets, the clustering accuracy can be computed using an external measure. One of the commonly used external measures, termed *Normalized Mutual Information* (NMI) is defined as follows. Let n, m and k denote the total number of samples, the number of different cancer types and the number of clusters respectively. Let $a_1, a_2, \cdots, a_m$ denote the number of samples that belong to each cancer type. Similarly, let $b_1, b_2, \cdots,$ b_k be the number of samples that belong to each cluster. Let $c_{i,j}$, $\forall i, j$, $1 \leq i \leq m$ and $1 \leq j \leq k$, denote the number of samples in jth cluster that belong to the ith cancer type. The *Normalized Mutual Information* (NMI) of the clustering is computed using the following formula:

$$NMI = \frac{\sum_{i=1}^{m}\sum_{j=1}^{k} c_{i,j}\, log(\frac{n \cdot c_{i,j}}{a_i b_j})}{\sqrt{(\sum_i a_i log\frac{a_i}{n})(\sum_j b_j log\frac{b_j}{n})}}.$$

NMI function takes a value in [0, 1] interval. Higher values of this function represent better clustering quality and the measure is independent of the underlying distance/similarity measure. Thus, the results of different distance measures can be compared using NMI. This external measure have been used for evaluation of clusters for various kinds of datasets before [42]. Other distance measures exist that do not require class labeling of samples [43]. A comparison of the different clustering algorithms and similarity measures for CGH datasets is provided in [41]).

If the cancer types of the samples in the dataset are not known priori, it becomes impossible to compute the NMI. In this case an internal measure can be used to evaluate the clustering accuracy. Unlike the external measure, the value of internal measure depends on the distance/similarity measure. Thus, the internal measure of different clusterings obtained by different similarity measures are not comparable. Instead, we use this measure to compare the clusters obtained by applying different clustering methods with same similarity function. In this study, we implement two internal measures. One is the internal measure based on compactness (cohesion) [41], the other is the internal measure based on separation.

Let k denote the total number of clusters. Let $b_1, b_2, \cdots, b_k$ be the number of samples that belong to each cluster. We use s_i and C_r to represent ith sample and the rth cluster respectively. Let $S(s_i, s_j)$ be the function that evaluates the similarity between the ith and jth sample. The internal measure based on compactness is computed as:

$$IC = \sum_{r=1}^{k} \frac{\sum_{i<j, s_i, s_j \in C_r} S(s_i, s_j)}{b_r}.$$

The internal measure based on separation is computed as:

$$IS = \frac{\sum_{r=1}^{k} \sum_{q=1, q\neq r}^{k} \sum_{s_i \in C_r, s_j \in C_q} S(s_i, s_j)}{\sum_{r=1}^{k} \sum_{q=1, q\neq r}^{k} b_r \cdot b_k}$$

Since both internal measures are computed with pairwise similarity, higher values of IC and lower values of IS represent better clustering quality respectively.

2.2. Marker detection

Most cancers result from genomic instability and display various genomic aberrations. A recurrent alteration is a set of aberrations common to sufficiently many CGH samples. The recurrent alterations can be used to characterize the aberration pattern of samples. We call these genomic intervals *markers*. Each marker can be represented by two numbers $<p, q>$, where p and q denote the position and the type of aberration respectively. Let S be a set of N CGH samples $\{s_1, s_2, \cdots, s_N\}$. Let x_d^j denote the status value for sample j at genomic interval d, $\forall d, 1 \leq d \leq D$, where D is the number of intervals. Let $s_j[u,v]$ be the segment of s_j that starts at the uth interval and ends at the vth interval. We use the term *segment* to represent a maximal contiguous block of aberrations of the same type.

Let $\{m_t = <p_t, q_t> | 1 \leq t \leq R\}$, be a set of markers that are ordered along the genomic intervals, i.e. $p_1 < p_2 < \cdots < p_R$. We define the *support* of s_j to m_t as $\sigma(s_j, m_t)$. Here, $\sigma(s_j, m_t) = 1$ only if both of the following conditions are satisfied:

1. *Support:* There exists a segment $s_j[u,v]$ overlapping with m_t, i.e. $u \leq p_t \leq v$ and the type of $s_j[u,v]$ is the same as that of m_t, i.e. $x_u^j = q_t$.
2. *Uniqueness:* There exists no other marker $m_{t'}$, $t' < t$, in the same chromosome band such that $u \leq p_{t'} \leq v$, $x_u^j = q_{t'}$ and $\sigma(s_j, m_{t'}) = 1$.

Otherwise, $\sigma(s_j, m_t) = 0$. We say s_j *supports* m_t or m_t *covers* s_j if $\sigma(s_j, m_t) = 1$. We define the support value of marker m_t as the sum of its support from all the samples.

Marker detection problem can be formally defined as follows. Given a set of CGH samples $S = \{s_1, s_2, \cdots, s_N\}$ and a positive integer R, the goal is to find the set of R markers $M = \{m_1, m_2, \cdots, m_R\}$, $p_1 < p_2 < \cdots < p_R$, such that the sum of support of markers in M, i.e., $\sum_{t=1}^{R} Support(m_t)$, is maximized. Next, we develop a dynamic programming algorithm to solve this problem optimally.

Let $V(d,r) = \sum_{t=1}^{r} Support(m_t)$, for $1 \leq p_t \leq d \leq D, \forall t \in [1,r]$ denote the largest possible support that r markers can get from the genomic intervals in the range $[1..d]$. Here, $[1..d]$ denote the integers $1, 2, \cdots, d$. $V(1,1) =$ support of the single marker at the first genomic interval. The value of $V(d,r)$ in general, where $1 \leq r \leq R, r \leq d \leq D$, can be computed as the maximum of two possible cases.

- *Case 1:* There exists a marker m_r at the dth genomic interval. In this case, $V(d,r)$ can be computed as the maximum sum of support of m_r and the optimal value of locating $r-1$ markers. Formally, it can be written as $V(d,r) = \max_{b-1 \leq d' \leq d-1}\{V(d',r-1) + Support(m_r)\}$, where b denotes the least genomic interval that is in the same chromosome as the dth genomic interval. Note that $V(d',r-1)$ may correspond to different set of $r-1$ markers for different values of d'. The type of marker m_r can be either gain or loss. We choose the marker type as the one that leads to a larger $Support(m_r)$ value.
- *Case 2:* There is no marker at the dth genomic interval. $V(d,r) = V(d-1,r)$ in this case.

Thus, $V(d,r)$ can be computed using the following recursive equation,

$$V(d,r) = \max \begin{cases} \left.\begin{array}{l} V(d-1,r-1) + Support(m_r) \\ V(d-2,r-1) + Support(m_r) \\ \cdots \\ V(b-1,r-1) + Support(m_r) \end{array}\right\} & \text{if } m_r \text{ appears at interval } d \\ V(d-1,r) & \text{otherwise} \end{cases}$$

The marker set M that leads to $V(D,R)$ corresponds to R markers with the largest sum of supports. The overall time complexity of computing $V(D,R)$ is $O(DNR)$ (proof omitted).

Several clustering based algorithms can be developed using markers. We have developed two clustering strategies using these markers, each with a different goal. The first one, prototype-based approach, maximizes the support for the markers. The second one, similarity-based approach, develops a new similarity measure called RSim that computes the number of common markers. In the following, we briefly describe the Prototype based clustering algorithm to enunciate how these markers can be used for clustering.

Our prototype-based clustering algorithm partitions the dataset into subsets such that each subset has a prototype common to the samples in that subset. The user specifies the number of clusters, k, and the number of markers per cluster, R. It starts with a random partitioning of the samples. It then iteratively maximizes an objective function, which we call *cohesion* function, in two steps. We discuss the cohesion function later. The two steps are described as follows.

- **Refinement step:** For each cluster, this step identifies the optimal set of R markers. These markers serve as the prototype of this cluster.

- **Reassignment step:** This step assigns each sample to the cluster whose prototype is supported the most by that sample.

Essentially, both steps optimize a cohesion function alternatively until the value of this function does not change, i.e. converges to a stable value. We formally prove this below.

Given a set of CGH samples $S = \{s_1, s_2, \ldots, s_N\}$. Let K denote the number of clusters. Let c_i denote the ith cluster. A clustering of samples can be represented by a K-partition, $C = c_1 \cup c_2 \cup \cdots c_K$, that divides samples into K disjoint clusters. Let R be the number of markers in each cluster. Let $M_i = \{m_{i,t} | 1 \leq t \leq R\}$ denote the set of markers (i.e., prototype) for cluster i, where $1 \leq i \leq K$, and $m_{i,t}$ denote the tth marker in the ith cluster. Each marker $m_{i,t}$ is a tuple $<p_{i,t}, q_{i,t}>$, where $p_{i,t}$ and $q_{i,t}$ denote the position and type of this marker respectively. Let M denote the set of prototypes $M = \{M_1, M_2, \ldots, M_K\}$. We define a cohesion function as follows:

$$cohesion(C,M) = \sum_{i=1}^{K} \sum_{s_j \in c_i} \sum_{t=1}^{R} \sigma(s_j, m_{i,t}).$$

The cohesion function computes the intra-cluster similarity between samples and cluster prototypes.

Detailed experimental results for different number of clustering approaches are presented in [44]. In the following, we briefly describe the application of our techniques to Head-Neck Squamous Cell Carcinoma (HNSCC) with 339 samples with and without clustering.

Fig. 1 plots the dataset we used to obtain these markers. Each row in this figure corresponds to a sample and each column corresponds to a genomic interval. The figure supports the hypothesis that consecutive genomic intervals are correlated as each gain or loss is often followed by a contiguous block of gain or loss. The figure also shows that the markers we found have high support in the dataset. Interestingly, the markers found represent a superset of the markers found by simple manual analysis.

Fig. 2 presents the clustering of the samples into 4 clusters. This figure demonstrates that the dataset can be effectively represented by four clusters and that the markers found within each cluster can be substantially different. This is an interesting observation as this can not be seen without clustering (see Fig. 1).

2.3. Marker selection for classification

Markers can be used for the development of a classification tool that can serve as an effective filtering process for cancer diagnosis and treatment. The markers described in the previous section are useful in effectively modeling the different clusters and provide a description of the underlying imbalances. Another important use of these markers is for classification - for classification the key is to find markers that can discriminate between different cancer types.

The first step in classification is selection of markers that will be used in building a classifier for supervised classification of multi-class CGH data, i.e. data with multiple

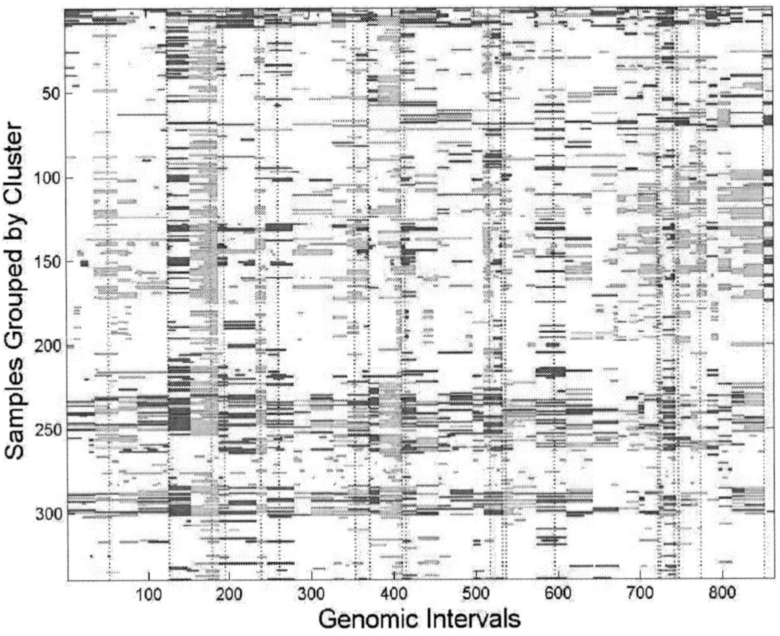

FIGURE 1. Application of the marker finding algorithm to Head-Neck Squamous Cell Carcinoma with 339 samples without clustering. In the figure, X-axis denotes the index of 862 genomic intervals. Y-axis denotes the index of cases. The green and red dots in each row denote gains and losses of the corresponding patient. The black vertical dashed lines show the positions of the markers we found. Notice the strong correlations in consecutive intervals.

cancer types. A number of methods can then be applied using these markers for classification. These include decision tree based methods, neural network based methods, coverage based methods, Bayesian methods and support vector based methods [43]. For some classifier (e.g. decision tree classifiers) that are chosen for building the classifier are the most discriminating. In the following, we briefly describe the different methods that have been used for marker (or feature) selection.

Feature selection [45] methods have been widely studied in gene selection of microarray data. Features using filter based methods are selected based on discriminant criteria [46, 47], mutual information [48] or statistical tests (t-test, F-test) [49, 50]. More sophisticated methods that minimize redundancy, have also been developed [51, 48]. Recently, wrapper based methods using SVM-RFE (support vector machine recursive feature elimination) [52] techniques have been found to be useful in analyzing noisy proteomics and microarray data [53], computing feature ranking [54] and recursive feature elimination for multi-class gene expression data [55].

Fu et al. [56] proposed a probability model of selection to evaluate the importance of genes in microarray data for multi-class data. This method adopts a one-versus-others

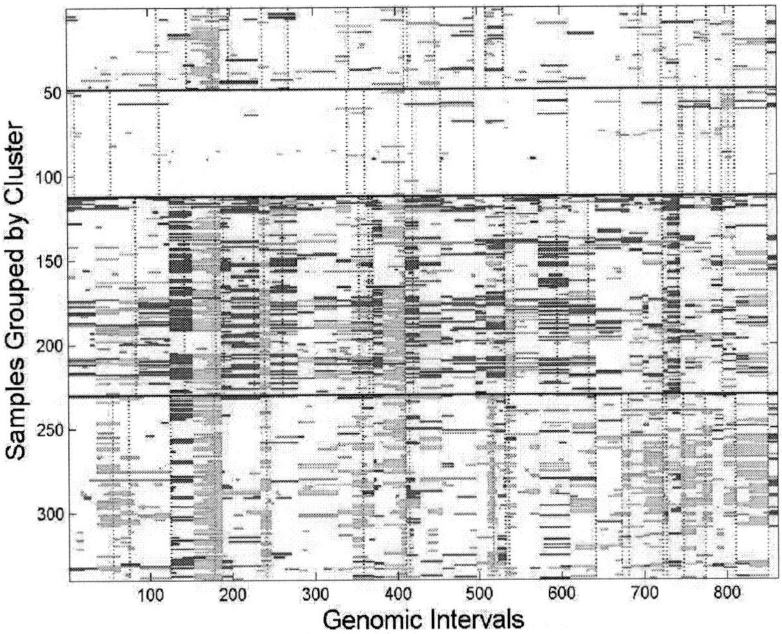

FIGURE 2. Application of the marker finding algorithm to Head-Neck Squamous Cell Carcinoma with 339 samples with clustering (number of clusters = 4). In the figure, X-axis denotes the index of 862 genomic intervals. Y-axis denotes the index of cases. The green and red dots in each row denote gains and losses of the corresponding patient. The black vertical dashed lines show the positions of the markers we found. The samples in the same cluster are placed in consecutive rows. The clusters are separated using blue horizontal lines. The samples that belong to the same cluster show similar patterns. There are significant differences between samples that belong to different clusters.

strategy to convert the multi-class problem into a series of binary class problems. It ranks features and selects them locally for each binary class problem. It combines these rankings and uses 10-fold cross validation to determine the important features. Under the assumption that vital genes will to be selected more consistently than others across multiple folds, it determines a final set of features whose repeatability is greater than the threshold associated with the minimum predictive error. We have adapted this method to develop marker selection algorithm for CGH data. Given the local nature of each of the classifiers that are used in this method, we use the term *Local selection* to represent this selection algorithm.

We have also developed a new SVM-based feature selection method for multi-class data, *Global Selection*, that adopts a recursive feature elimination similar to SVM-RFE but utilizes the ranking of a marker across all the classifiers for eliminating a marker. The preliminary results show that better classification can be achieved for many datasets as opposed to using all the 862 intervals for classification (results not shown).

3. SOFTWARE ENVIRONMENT

We have developed a web based data mining tool for mining CGH data using the above algorithms for clustering, marker detection and marker selection. It has the following features:

- It allows data import from tab-delimited text files in Progenetix (`http://www.progenetix.de`) format
- It can perform data clustering using multiple algorithms and distance measures, perform detection of important markers and perform selection of discriminative markers that help build a reliable classifier
- It provides results in textual and Graphical formats
- It provides multiple metrics for evaluation of results

The application is developed using Microsoft Internet Information Server (`http://www.microsoft.com`). The web user interface is developed using Microsoft ASP.NET and C# (`http://www.microsoft.com`). When a user submit a request, the front-end C# program calls the executable files to perform the computation. The results are written to HTML files stored at the server side. The front-end program polls these results and return them to users.

The underlying algorithms are developed using MATLAB (`http://www.mathworks.com`). A MATLAB compiler is used (`http://www.mathworks.com/products/compiler/`) to generate executable files.

A preliminary version of this tool is available at (`http://128.227.162.207:8007/CGH/Default.html`). New features and algorithms are constantly being added to this tool.

In the following we briefly describe the Input data format and data generation process.

3.1. Input data format

The programs accepts tab-delimited text files with both CGH data and genomic interval information. We follow the format of Progenetix, the largest source for published CGH data (with more than 12,000 cases) (`http://www.progenetix.de`). The Progenetix format is a chromosomal band specific matrix suitable for mining experiments. It currently supports a resolution of 862 genomic intervals from 23 chromosomes. Data files consist of a header row followed by rows of data. Progenetix also provides online tools that can convert other formats of CGH data to the matrix format.

We also expect to develop the following functionality in the near future:

- Format Conversion: CGH data is available in multiple formats. For example, the new CGH data is collected in 200K format. We will develop transformation methods that CGH dataset in multiple formats can be converted into the same format (generally to the format with smaller number of intervals).
- Relationship to other types of data: We expect that CGH data will be used in conjuction with other data such as Gene expression array data, SNP data, and proteomics

data. We will provide appropriate mapping routines that allow a researcher to correlate the different regions in these data sets with the corresponding intervals in the CGH data.

3.2. Benchmarking

Progenetix database [19] contains over 12,000 samples that span 19 different histopathological cancer types. Thus, it contains samples with varying properties. For testing new methods, they need to be evaluated over multiple datasets with different properties such as 1) Number of samples contained in the dataset 2) Number of cancer types contained in the dataset 3) The similarity level between samples from different cancer types, which indicate the difficulty of classification. However, there are no standard benchmarks for smoothed CGH data that take the three properties into account. We have developed methods to select subsets of the Progenetix database or other databases in a principled manner to create datasets with desired properties.

The dataset generator will accept the following three parameters as input:

- Approximate number of samples (denoted as N)
- Number of cancer types (denoted as C)
- Similarity range (denoted as $[\delta_{min},\delta_{max}]$) between samples belonging to different cancer types

Here is an outline of the dataset generation process:

1. For each cancer type, partition all the samples belonging to this cancer type into several disjoint groups using clustering. Each cluster corresponds to the different aberration patterns for a given cancer type.
2. Compute the pairwise similarity between pairs of groups obtained in the first step.
3. Construct a complete weighted graph where each vertex denotes a group of samples and the weight of an edge equals to the similarity between two groups that are connected by this edge.

This graph can then be used to find a set of samples of a given size N (by choosing a subset of groups that sum to N), given number of cancer types, and based of level of similarity between groups (by only considering groups that have a similarity within the range of $[\delta_{min}, \delta_{max}]$).

Fig. 3 shows an example of how such a dataset may work. Consider a dataset containing 1,000 CGH samples - 400 samples belonging to cancer type c_1 and the other 600 samples belonging to cancer type c_2. Assume that each cancer type is clustered into 2 clusters. This results in 4 groups of CGH samples, which are denoted as $g_i, 1 \leq i \leq 4$. Let the size of g_1, g_2, g_3 and g_4 be 150, 250, 450, and 150 respectively. The pairwise similarity between any two groups is shown in the figure. Using this, one can construct a complete weighted graph where each vertex denotes a group and the weight of each edge equals to the similarity between two groups that are connected by this edge. Suppose that a dataset needs to be generated with $N = 400$, $C = 2$, $\delta_{min} = 0.025$ and $\delta_{max} = 0.035$.

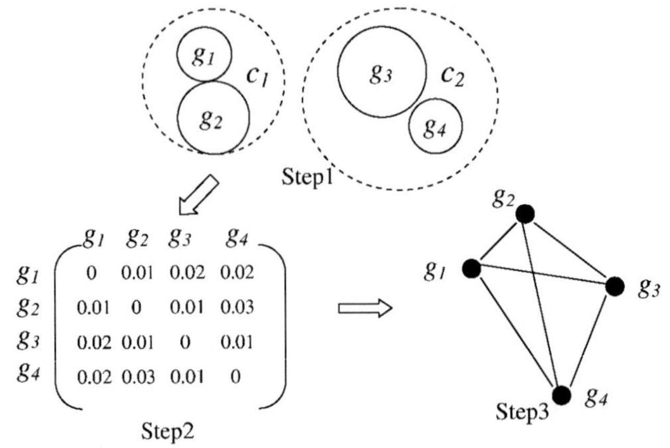

FIGURE 3. A working example of the algorithm to select benchmark datasets. c_i and g_j denote the ith cancer type and the jth group of samples, respectively. In the first step, the samples are partitioned in each cancer type into two disjoint groups. In the second step, pairwise similarity metrics are computed. In the third step, a complete weighted graph is generated.

The graph can be parsed to find out that g_2 and g_4 satisfy the three conditions and a new dataset can be generated by combining the samples in g_2 and g_4.

The advantage of the above dataset generator is that a large number of datasets can be created with variable number of samples and cancer types as well as variable level of similarities between the chosen cancer types. This allows for testing the accuracy and performance of a new method across a variety of potential scenarios - each of them a subset containing samples from a real dataset.

4. EXAMPLE: DISTANCE-BASED CLUSTERING OF SAMPLE DATASET

In this section, we briefly describe the clustering of a small dataset using our web based tool to demonstrate its functionality.

Our web tool provides five distance based clustering algorithms: topDown, bottomUp, k-means, topDown+k-means and bottomUp+k-means. The first three are well known clustering algorithms in the literature. The last two algorithms are the combination of topDown or bottomUp with k-means. They work as follows. They first find clusters using the top down (or bottom up) clustering algorithm. They, then, feed these clusters into k-means algorithm as the initial cluster. They aim to avoid the poor results obtained by the k-means algorithm due to the random initial clusters. The distance-based clustering tool provides four distance measures, Raw, Sim, cosineGaps and cosineNoGaps as described in Section 2.

Users can choose any one of the 20 combinations of the five algorithms and four distance measures as shown in Fig. 4. The default algorithm and distance measure is topDown and Sim respectively because this combination produces the best clusters ac-

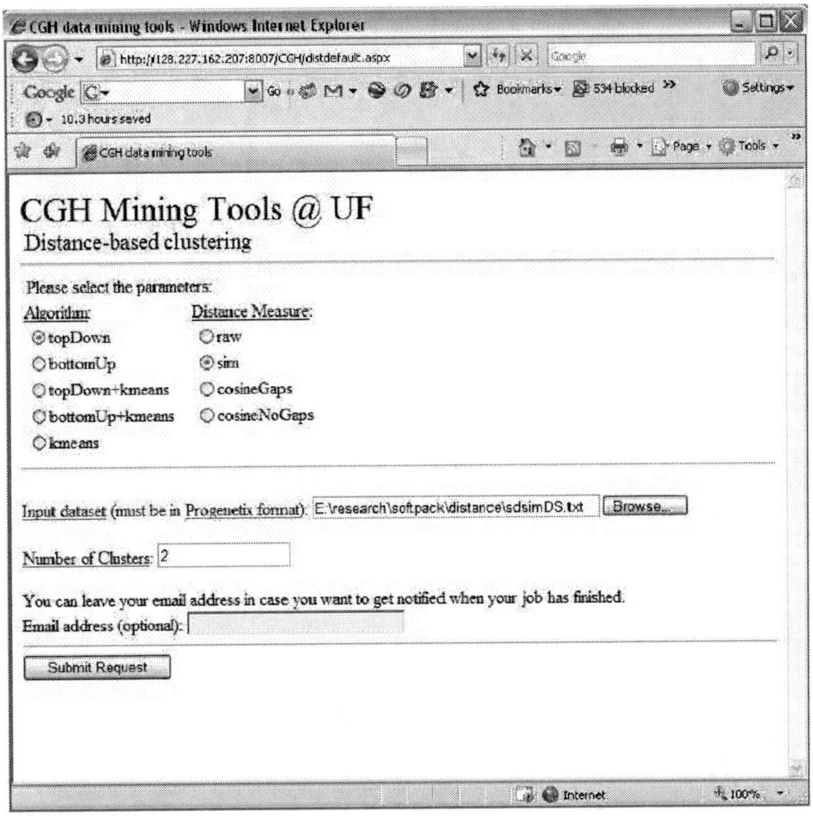

FIGURE 4. The snapshot of distance-based clustering. This page allows users to choose algorithm and distance measure.

cording to our experimental results. The interface allows the user to upload a database file containing all the samples. It also allows the the user to specify the number of clusters. Clustering is usually a computational intensive task. Depending on the database, this process can take several minutes. The web server allows users to provide their email addresses so that they can be notified when the results are ready. The server stores the results in a temporary html file and emails the link for this file. The user can browse this file later. If the user chooses to keep the browser open, the server automatically refreshes it with the page that contains the results.

Fig. 5 provides the snapshot of the resulting page after applying the clustering algorithm to the sample dataset. In this example the dataset contains 391 samples from two cancer types, namely Retinoblastoma, NOS (ICD-O code = 9510/3) and Neuroblastoma, NOS (ICD-O code = 9500/3). The numbers of samples in the two cancer types are 120 and 271 respectively. The server identifies the clusters using the parameters provided by the user. It also provides clustering by applying the centroid shrinking technique [57] to

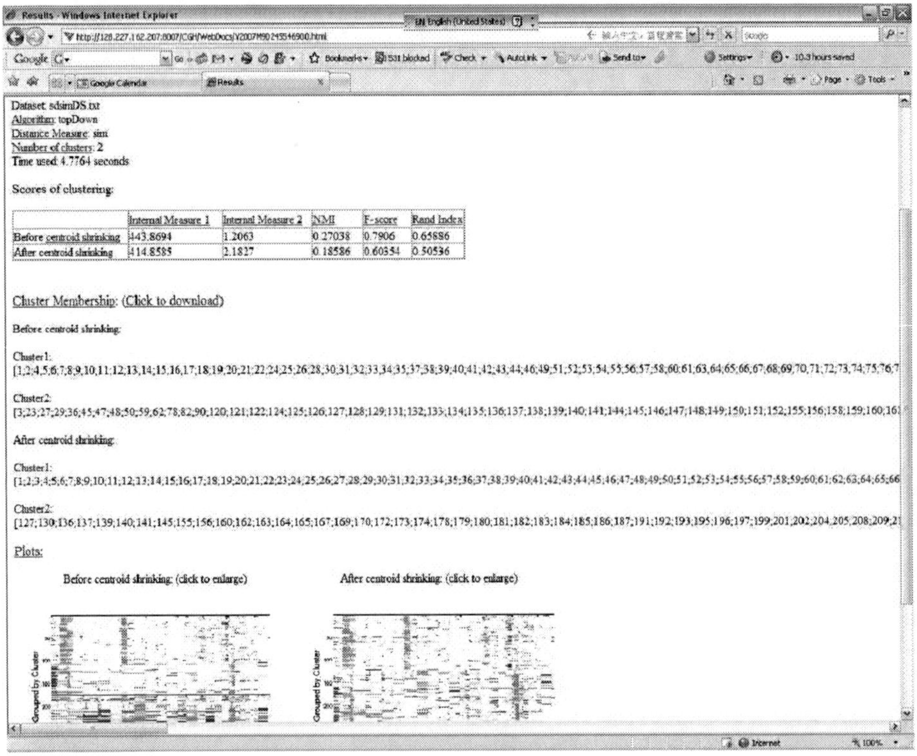

FIGURE 5. The snapshot of distance-based clustering. This page presents the clustering results of the sample dataset.

refine the clusters. The results page consists of three parts.

1. *Quality of clustering.* We provide a table that reports the quality of the clusters according to a number of internal and external measures [43].
2. *Cluster membership.* We list the index of samples that belong to each cluster. We also provide a link to a downloadable text file that contains this information.
3. *Cluster Plot.* We plot the clusters as JPEG files and embed them in the results page as thumbnails. Users can click the thumbnails to view the full-size pictures. In the plot shown in Fig. 6, the X-axis denotes the index of genomic intervals, the Y-axis denotes the samples that grouped by clusters. Different clusters are separated by the horizontal lines.The genomic intervals with gain and loss imbalances are plotted in green and red, respectively. Those with no imbalance are not plotted.

In both the query and result pages, all field names are clickable. Clicking on a field name brings the help page that contains a description of that field.

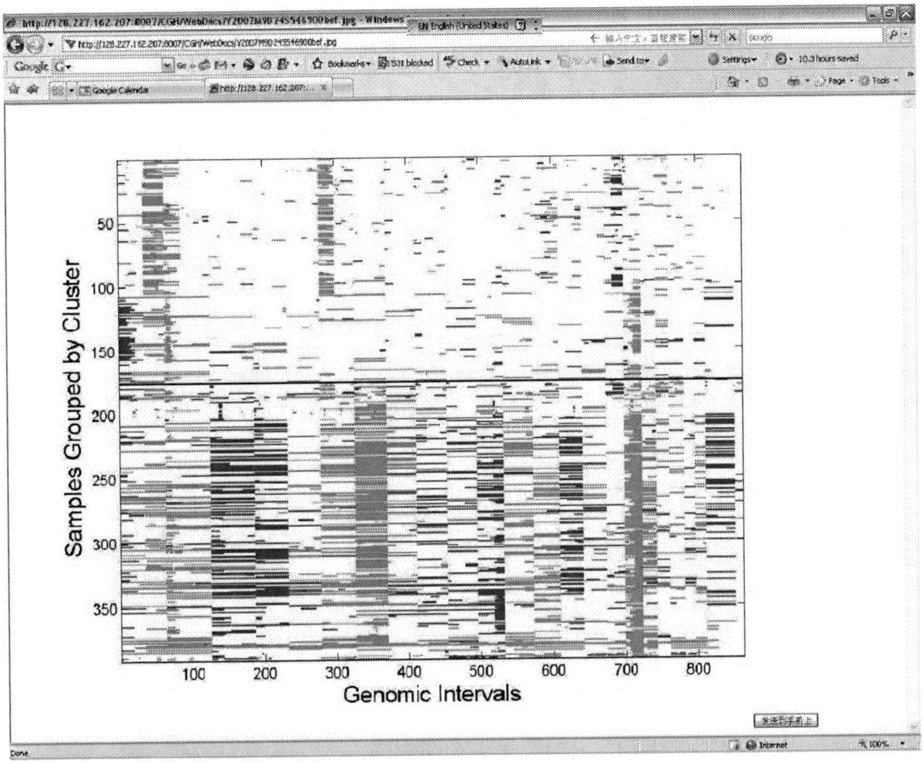

FIGURE 6. The snapshot of distance-based clustering. This page shows the plot of clusters.

5. CONCLUSIONS

It employs novel data mining methodologies for clustering and classification of CGH datasets as well as algorithms for identifying important markers that are potentially cancer signatures. It also provides a visualization of the dataset and the results. The developed software will help in understanding the relationships between genomic aberrations and cancer types. n the future, we will add additional functionality to support different CGH formats and provide relationships to other datasets such as Gene expression array data, SNP data, and proteomics data.

ACKNOWLEDGMENTS

The authors would like to thank Dr. Michael Baudis for providing the CGH datasets and several discussions on analyzing CGH data. The authors would also like to thank Jaaved Mohammed and James Carter for their help in implementing and testing the clustering algorithms, and Raja Appuswamy for helping develop an initial web-based interface.

REFERENCES

1. F. Mitelman, J. Mark, G. Levan, and A. Levan, *Science* **176**, 1340–1341 (1972), ISSN 0036-8075.
2. B. Vogelstein, and K. Kinzler, *Trends Genetics* **9**, 138–141 (1993).
3. J. Tijo, and A. Levan, *Hereditas* **42**, 1–16 (1956).
4. K. Patau, *Am J Hum Genet* **12**, 250–276 (1960).
5. P. Crossen, *Clin Genet* **3**, 169–179 (1972).
6. A. Kallioniemi, O. Kallioniemi, D. Sudar, D. Rutovitz, J. Gray, F. Waldman, and D. Pinkel, *Science* **258**, 818–821 (1992).
7. S. Solinas-Toldo, S. Lampel, S. Stilgenbauer, J. Nickolenko, A. Benner, H. Dohner, T. Cremer, and P. Lichter, *Genes Chromosomes Cancer* **20**, 399–407 (1997).
8. D. Pinkel, R. Segraves, D. Sudar, S. Clark, I. Poole, D. Kowbel, C. Collins, W.-L. Kuo, C. Chen, Y. Zhai, S. H. Dairkee, B. marie Ljung, and J. W. G. D. G. Albertson, *Nature Genetics* **20**, 207–211 (1998).
9. J. R. Pollack, C. M. Perou, A. A. Alizadeh, M. B. Eisen, A. Pergamenschikov, C. F. Williams, S. S. Jeffrey, D. Botstein, and P. O. Brown, *Nature Genetics* **23**, 41–46 (1999).
10. D. Pinkel, and D. G. Albertson, *Nat Genet* **37 Suppl** (2005), ISSN 1061-4036.
11. F. Mitelman, editor, *International System for Cytogenetic Nomenclature*, Karger, Basel, 1995.
12. J. Gray, C. Collins, I. Henderson, J. Isola, A. Kallioniemi, O. Kallioniemi, H. Nakamura, D. Pinkel, T. Stokke, and M. Tanner, *Cold Spring Harb Symp Quant Biol* **59**, 645–652 (1994).
13. M. Bentz, C. Werner, H. Dohner, S. Joos, T. Barth, R. Siebert, M. Schroder, S. Stilgenbauer, K. Fischer, P. Moller, and P. Lichter, *Blood* **88**, 1437–1444 (1996).
14. R. Desper, F. Jiang, O.-P. Kallioniemi, H. Moch, C. H. Papadimitriou, and A. A. Schaffer, *Journal of Computational Biology* **6**, 37–52 (1999).
15. M. Hoglund, A. Frigyesi, T. Sall, D. Gisselsson, and F. Mitelman, *Genes Chromosomes Cancer* **42**, 327–341 (2005).
16. T. Mattfeldt, H. Wolter, R. Kemmerling, G. Gottfried, and H. A. Kestler, *Analytical Cellular Pathology* **23**, 29–37 (2001), ISSN 0921-8912.
17. J. Vandesompele, M. Baudis, K. De Preter, N. Van Roy, P. Ambros, N. Bown, C. Brinkschmidt, H. Christiansen, V. Combaret, M. Lastowska, J. Nicholson, A. O'Meara, D. Plantaz, R. Stallings, B. Brichard, C. Van den Broecke, S. De Bie, A. De Paepe, G. Laureys, and F. Speleman, *J Clin Oncol* **23**, 2280–2299 (2005).
18. S. Joos, C. Menz, G. Wrobel, R. Siebert, S. Gesk, S. Ohl, G. Mechtersheimer, L. Trumper, P. Moller, P. Lichter, and T. Barth, *Blood* **99**, 1381–1387 (2002).
19. M. Baudis, and M. L. Cleary, *Bioinformatics* **17**, 1228–1229 (2001).
20. M. Mao, R. Hamoudi, P. Zhao, and M. Baudis, *Cancer Genetics Cytogenetics* **160**, 141–151 (2005).
21. M. Mao, R. Hamoudi, I. Talbot, and M. Baudis, *Cancer Genetics Cytogenetics* **167**, 1–14 (2006).
22. R. Desper, F. Jiang, O.-P. Kallioniemi, H. Moch, C. H. Papadimitriou, and A. A. Schaffer, *J Comput Biol* **6**, 37–51 (1999).
23. M. Höglund, A. Frigyesi, T. Säll, D. Gisselsson, and F. Mitelman, *Genes Chromosomes Cancer* **42**, 327–341 (2005).
24. T. Mattfeldt, H. Wolter, R. Kemmerling, G. Gottfried, and H. A. Kestler, *Anal Cell Pathol* **23**, 29–37 (2001).
25. J. Vandesompele, M. Baudis, K. De Preter, N. Van Roy, P. Ambros, N. Bown, C. Brinkschmidt, H. Christiansen, V. Combaret, M. Lastowska, J. Nicholson, A. O'Meara, D. Plantaz, R. Stallings, B. Brichard, C. Van den Broecke, S. De Bie, A. De Paepe, G. Laureys, and F. Speleman, *J Clin Oncol* **23**, 2280–2299 (2005).
26. J. R. Pollack, T. Sorlie, C. M. Perou, C. A. Rees, S. S. Jeffrey, P. E. Lonning, R. Tibshirani, D. Botstein, A.-L. Borresen-Dale, and P. O. Brown, *Proceedings of the National Academy of Sciences* **99**, 12963–12968 (2002).
27. P. Wang, Y. Kim, J. Pollack, B. Narasimhan, and R. Tibshirani, *Biostat* **6**, 45–58 (2005).
28. P. Broet, and S. Richardson, *Bioinformatics* **22**, 911–918 (2006).
29. J. C. Marioni, N. P. Thorne, and S. Tavare, *Bioinformatics* **22**, 1144–1146 (2006).
30. J. Fridlyand, A. M. Snijders, D. Pinkel, D. G. Albertson, and A. N. Jain, *J. Multivar. Anal.* **90** (2004).
31. G. Hodgson, and J. H. H. et al, *Nature Genetics* **29**, 459–464 (2001).
32. P. H. C. Eilers, and R. X. de Menezes, *Bioinformatics* **21**, 1146–1153 (2005).

33. A. B. Olshen, E. S. Venkatraman, R. Lucito, and M. Wigler, *Biostat* **5**, 557–572 (2004).
34. E. S. Venkatraman, and A. B. Olshen, *Bioinformatics* **23**, 657–663 (2007).
35. R. Autio, S. Hautaniemi, P. Kauraniemi, O. Yli-Harja, J. Astola, M. Wolf, and A. Kallioniemi, *Bioinformatics* **19**, 1714–1715 (2003).
36. J. B. MacQueen, "Some Methods for classification and Analysis of Multivariate Observations," in *Proceedings of 5-th Berkeley Symposium on Mathematical Statistics and Probability*, 1967.
37. B. King, *Journal of the American Statistical Association* **69**, 86–101 (1967).
38. M. Steinbach, G. Karypis, , and V. Kumar, "A comparison of document clustering techniques," in *KDD Workshop on Text Mining*, 2000.
39. F. Picard, S. Robin, M. Lavielle, C. Vaisse, and J.-J. Daudin, *BMC Bioinformatics* **6**, 27 (2005).
40. G. Salton, *Automatic text processing: the transformation, analysis, and retrieval of information by computer*, Addison-Wesley Longman Publishing Co., Inc., Boston, MA, USA, 1989, ISBN 0-201-12227-8.
41. J. Liu, J. Mohammed, J. Carter, S. Ranka, T. Kahveci, and M. Baudis, *Bioinformatics* **22**, 1971–1978 (2006).
42. A. Strehl, and J. Ghosh, "Cluster Ensembles – A Knowledge Reuse Framework for Combining Partitionings," in *Proceedings of AAAI 2002, Edmonton, Canada*, AAAI, 2002, pp. 93–98.
43. P.-N. Tan, M. Steinbach, and V. Kumar, *Introduction to Data Mining, (First Edition)*, Addison-Wesley Longman Publishing Co., Inc., Boston, MA, USA, 2005, ISBN 0321321367.
44. J. Liu, S. Ranka, and T. Kahveci, *Bioinformatics* **23**, 450–457 (2007).
45. R. Kohavi, and G. H. John, *Artificial Intelligence* **97**, 273–324 (1997).
46. T. Golub et al., *Science* **286**, 531–537 (1999).
47. P. Pavlidis, J. Weston, J. Cai, and W. N. Grundy, "Gene functional classification from heterogeneous data," in *RECOMB*, 2001, pp. 249–255.
48. C. Ding, and H. Peng, "Minimum Redundancy Feature Selection from Microarray Gene Expression Data," in *CSB '03: Proceedings of the IEEE Computer Society Conference on Bioinformatics*, IEEE Computer Society, Washington, DC, USA, 2003, p. 523, ISBN 0-7695-2000-6.
49. C. H. Q. Ding, "Analysis of gene expression profiles: class discovery and leaf ordering," in *RECOMB '02: Proceedings of the sixth annual international conference on Computational biology*, ACM Press, New York, NY, USA, 2002, pp. 127–136, ISBN 1-58113-498-3.
50. F. Model, P. Adorján, A. Olek, and C. Piepenbrock, *Bioinformatics* **17 Suppl 1** (2001), ISSN 1367-4803.
51. L. Yu, and H. Liu, "Redundancy based feature selection for microarray data," in *KDD '04: Proceedings of the tenth ACM SIGKDD international conference on Knowledge discovery and data mining*, ACM Press, New York, NY, USA, 2004, pp. 737–742, ISBN 1-58113-888-1.
52. I. Guyon, J. Weston, S. Barnhill, and V. Vapnik, *Machine Learning* **46**, 389–422 (2002).
53. X. Zhang, X. Lu, Q. Shi, X.-q. Xu, H.-c. Leung, L. Harris, J. Iglehart, A. Miron, J. Liu, and W. Wong, *BMC Bioinformatics* **7**, 197 (2006), ISSN 1471-2105.
54. K. B. Duan, J. C. Rajapakse, H. Wang, and F. Azuaje, *IEEE Trans Nanobioscience* **4**, 228–234 (2005), ISSN 1536-1241.
55. S. Ramaswamy et al., *Proc Natl Acad Sci U S A* **98**, 15149–15154 (2001), ISSN 0027-8424.
56. L. Fu, and C. Fu-Liu, *BMC Bioinformatics* **6** (2005).
57. R. Tibshirani, T. Hastie, B. Narasimhan, and G. Chu, *PNAS* **99**, 6567–6572 (2002).

Monkey search: a novel metaheuristic search for global optimization

Antonio Mucherino and Onur Seref

Center for Applied Optimization, University of Florida, Florida, USA

Abstract. We propose a novel metaheuristic search for global optimization inspired by the behavior of a monkey climbing trees looking for food. The tree branches are represented as perturbations between two neighboring feasible solutions of the considered global optimization problem. The monkey mark and update these branches leading to good solutions as it climbs up and down the tree. A wide selection of perturbations can be applied based on other metaheuristic methods for global optimization. We show that Monkey Search is competitive compared to the other metaheuristic methods for optimizing Lennard-Jones and Morse clusters, and for simulating protein molecules based on a geometric model for protein folding.

Keywords: Metaheuristics, Global Optimization, Protein Folding
PACS: 87.55.de, 87.55.kd, 87.15.Cc

1. INTRODUCTION

Many meta-heuristic search methods have been developed over the years for solving difficult global or combinational optimization problems. Many of these methods were inspired by optimal behaviors in nature. These methods are designed to find the global optimum solution of a hard optimization problem. The main challenge these methods face is the existence of multiple local minima of these optimization problems, which the methods that are not well designed usually converge to and stop. The success of these search methods rely on their handling such local minima and converging towards a solution which is close to the global solution.

In the field of meta-heuristic optimization, many methods have been proposed, which are inspired by natural phenomena or animal behavior. For instance, the Simulated Annealing (SA) algorithm [10] simulates the physical annealing process by decreasing the temperature of a given system to reach a stable low-energy configuration, which can correspond to the global optimum of an optimization problem. The Genetic Algorithms (GA) [8] mimic the evolution of a population of solutions. As the Darwinian Theory suggests, good solutions with better objective function values are allowed to survive and create new children by the crossover mechanism, which may even undergo mutations to avoid local minima. The solutions with worse objective function values are destroyed by natural selection. Ant Colony (AC) algorithms [4] simulate the behavior of a colony of ants that searches for food. Since each ant leaves a chemical called pheromone to mark its path, a colony in which each ant seems to work randomly begins to follow an optimal path as soon as there is enough pheromone on it. The Harmony Search (HS) [7] simulates musical improvisation, in which the optimal harmony is sought by improvising on the notes of previously played harmonies . The problem with some

CP953, *Data Mining, Systems Analysis, and Optimization in Biomedicine*
edited by O. Seref, O. E. Kundakcioglu, P. M. Pardalos

global optimization problems is that the gradient of the objective function is hard to calculate. The Differential Evolution (DE) algorithm [19] extend the concept of gradient by defining the direction which leads one feasible solution to another. The DE algorithm works with a population of solutions and has similarities with GAs.

In this study, we propose a new method based on an agent discovering and searching trees of solutions for solving global optimization problems. The general behavior of the search agent resembles a monkey looking for food by climbing up trees, from which the method acquires its name. The basic assumption for such behavior is that the monkey explores trees and learns which branches lead to better food resources. In our search method, food is represented by desirable solutions and the branches of the tree of solutions are represented as perturbations between two neighboring feasible solutions in the considered optimization problem. These perturbations can be totally random; however they can also be customized and based on other strategies previously developed for global optimization. This structure of our search is not meant to be a method for global optimization to be compared to the other aforementioned methods, it rather exploits the perturbation mechanisms of these methods in defining its branches.

The movement of the monkey is up and down the branches of a tree. Unknown branches are discovered using perturbations and whenever a better solution is found, the monkey climbs down the unique path that connects this solution to the root, while marking the path with this better solution. After reaching the root, the monkey climbs back up the tree using the previously discovered branches. The path the monkey follows on the way up the tree is probabilistically determined at each branch, based on the previous marks, which is eventually intended to lead the monkey to branches with better solutions. The monkey climbs until it reaches an undiscovered frontier, and continues its search from there on with new unknown branches.

In this study, we test the Monkey Search method on the global optimization problem of finding stable conformations of clusters of atoms regulated by the *Lennard Jones* potential energy [11] and by the *Morse* potential energy [13]. These difficult to optimize energy functions have been already considered for testing methods for optimization [5, 6, 12, 16, 17, 20, 21]. We also consider an optimization problem arising from a model developed for studying the protein folding problem, which has a growing need of fast and efficient optimization methods. This model is called the "tube model" due to its protein representation, and it simulates protein conformations based on geometric features of proteins [1, 3, 9].

The chapter is organized as follows. Section 2 provides a detailed description of the Monkey Search method. In Section 3 computational results and comparisons between Monkey Search and other metaheuristics are presented on Lennard-Jones and Morse clusters, and also on the tube model. Finally in Section 4, conclusions are drawn.

2. MONKEY SEARCH METHOD

Monkey Search (MS) is an agent based search algorithm, in which the searching agent discovers branches of a tree of solutions [18]. This behavior is not an exact replica of the behavior of monkeys in real life, however the general behavior of the search agent resembles a monkey wandering trees in search of food.

The monkey in our algorithm randomly climbs and discovers new branches of a tree. Each time a better solution is found, the monkey climbs down the unique path that connects the solution to the root, and marks this path with this solution value. The monkey then climbs up the tree again, choosing the previously marked branches randomly with higher probability towards the branches with better values, until it reaches an undiscovered frontier of the tree. The behavior of preferring branches with better values is intended for the monkey to converge to a specific area of the tree in which the global solution may be discovered. The monkey is not limited with a single tree; the search domain for the objective function may rather be considered as a jungle.

The branches of trees contain feasible solutions. Starting from any solution, which is at the root of a branch, a new neighbor solution is given at the tip of the same branch. The functional distance between the two solutions is determined by the random perturbation. The trees have a binary tree structure, which means that there are two new branches extending from each branch, each with a neighbor solution. Each time the monkey finds a solution with a better objective function value, it stops climbing and stores this solution as the current best solution. The monkey also stops when it reaches the top of the tree, which is the depth of the tree. Whenever either stopping condition is realized, the monkey starts climbing down branch by branch and looking for some other tree branch which could lead to better solutions. On its way back, the monkey updates and marks all the branches it passes for later exploration. The tree is considered to contain a preset number of paths from the root that reach the entire height of the tree. Each time the monkey reaches the allowed height of the tree, one of such paths is assumed to be explored. When all such paths are explored, the monkey starts climbing a new tree starting from the best solution, or a combination of better solutions.

At each step of the algorithm, new feasible solutions are created starting from the current one, and using a randomized perturbation function, which generates a new solution when applied to the current one. We aim to exploit other metaheuristic searches in defining these perturbation functions. For instance, one of the perturbations we apply is the crossover between solutions from genetic algorithms. The other perturbations we apply are inspired by other metaheuristics mentioned in the Introduction. However, perturbation functions are not limited by other heuristics, and can be derived from the optimization problem itself, leading to a wide domain of applications for Monkey Search.

At the end of each branch while climbing up, the monkey chooses between the left or the right branch. Therefore, at each step, we need to apply two different perturbations to the current solution, and the perturbation to apply can be chosen among a set of perturbations adaptively to provide more flexibility. The choice of perturbations may start uniformly, and converge towards those with higher success percentage. For avoiding that some kind of perturbation becomes predominant, the choice of perturbations is randomly reset back to uniform.

The immediate information available to the monkey at any point is the current solution X_{cur}, and the best solution X_{best} on the current tree. Both these solutions can be used in the perturbations for generating new solutions. The perturbations we apply in Monkey Search in this study are as follows:

- Random changes to X_{cur}, as in the Simulated Annealing methods;

- Crossover operator applied for generating a child solution from the parents X_{cur} and X_{best}, as in Genetic Algorithms;
- The mean solution built from X_{cur} and X_{best}, inspired by Ant Colonization;
- Directions that lead X_{cur} to X_{best}, as in Directional Evolution;
- Creating harmonies (solutions) from X_{cur} and X_{best} and introducing random notes, as in Harmony Search;

Note that our perturbations do not consider a population of solutions like in in Genetic Algorithms. They consider X_{cur} and X_{best} only, which assumes the existence of a predominant individual, X_{best}, which may be the case in some species such as in a pride of lions.

For avoiding local optima, we keep a predetermined number of n_m best solutions updated by each successive tree. Since we use the solutions in this set frequently, we call this set the *memory* of our search algorithm. In order to force the search in different spaces of the search domain, we start with n_w trees with randomly generated solutions as root. Usually $n_m < n_w$, and we keep the best n_m of them in the memory. After the first n_w trees, the root solution of a new tree is randomly chosen from the memory. We update the memory whenever a new solution is better than the ones that are already in the set, by including this solution to the memory and removing the closer solution in memory that is worse than the new one. In order to avoid local optima, if the new solution is too similar to an existing solution in the memory, it is ignored.

Monkey Search stops when the difference between the objective function values of all solutions in the memory is below a threshold. It is worth to note that Monkey Search is based on the following parameters which influence the convergence of our algorithm: the height h_t of the trees, the number n_t of times the monkey reaches the top of the tree, the size of the memory n_m, and the starting number n_w of random trees.

3. APPLICATIONS AND COMPUTATIONAL RESULTS

In this section, we present the application of Monkey Search on two sets of biomedical problems. The first set of problems involve Lennard-Jones and Morse potential energy functions for cluster conformations, both of which are well-known hard global optimization problems. The second application is the tube model based solely on the number of amino acids in a protein that determines a folding pattern from its mechanical properties. Monkey Search is applied to different problem sizes of both problem sets and compared to other metaheuristics.

3.1. Lennard-Jones and Morse clusters

Finding the ground state conformations of clusters of atoms represents one of the hardest global optimization problems. Many researches have been studying the so-called Lennard Jones (LJ) clusters [12, 16, 20] and Morse clusters [5, 6], previously used for testing global optimization methods [6].

The LJ energy is very simple to explain but only sufficiently accurate for the noble gases, and it also provides approximations for the structures of nickel and gold clusters. The LJ pairwise potential is given conventionally in reduced terms

$$v(r) = \frac{1}{r^{12}} - \frac{2}{r^6},$$

where r is the inter-particle Euclidean distance between any two atoms of the cluster. All the atoms are considered to be identical. The problem is to find the global minimum of the potential energy defined by

$$V_{LJ} = \sum_{i=1}^{n-1} \sum_{j=i+1}^{n} v(r_{ij}),$$

where r_{ij} is the inter-particle distance between atoms i and j. Note that the V_{LJ} values do not change by translating or rotating the clusters. This simple function is not easy to optimize, because it has high number of local minima. For instance, a cluster containing $n = 13$ atoms has about 1000 local minima, and the number of these local minima grows faster as n increases [20]. It has been proven in [17] that, when the size of the cluster is less than 1600, the stable structures have an icosahedral structure, even if there are exceptions, in which the clusters have lower energy when they have a decahedral or octahedral structure.

The Morse potential is a simple model, and, compared to LJ potential, it has a parameter ρ, which determines the width of the potential well (bottom point of the potential curve) and allows to model a wide variety of materials [5]. This potential depends on the interatomic distances r:

$$v_m(r) = [\exp\{\rho(1-r)\} - 1]^2 - 1,$$

and the potential energy is given by

$$V_M = \sum_{i=1}^{n-1} \sum_{j=i+1}^{n} v_m(r_{ij}),$$

where r_{ij} is the distance between the atoms i and j.

The LJ energy has the same curvature at the bottom of the well as the Morse energy when $\rho = 6$, and the Morse clusters show a wide variety of structural behavior as a function of ρ. When Morse clusters have less than 8 atoms, all the global minima of these clusters are independent of ρ. However, when the clusters have 8 atoms or more, motifs may change from icosahedral to decahedral to close-packed as the value of ρ is increased. There are, however, exceptional dimensions for which the transition from the icosahedral to the close-packed structure is direct.

Since the optimization process of both LJ and Morse clusters may be very hard, many proposed methods [5, 16, 20] for finding the stable cluster conformations subject to such energies are based on the structural motifs cited above. In this way, the degrees of freedom of each cluster can be reduced, and the optimization process can be easier. However, such methods are doomed to failure if the solution does not belong to the

TABLE 1. Percent success and mean computational time obtained by carrying out experiments with different values for the parameters n_t and h_t

n_t / h_t	30		50		70		100	
30	4%	30s	26%	60s	38%	90s	70%	2m
50	14%	50s	40%	90s	74%	2m	86%	4m
70	38%	1m	60%	2m	82%	4m	100%	6m
100	42%	1m	72%	3m	94%	5m	100%	8m

chosen motif, and they cannot be applied for solving any problem but the one they have been designed for.

In this study, we do not consider any information available on the low-energy cluster conformations, such as the structural motifs they may have. We check the efficiency of our algorithm in solving these problems without reducing the degrees of freedom of the atoms in the cluster. A two-phase method for optimizing LJ and Morse clusters has been proposed in [12], in which the knowledge on the particular motifs the cluster may assume is not exploited, even if the search is guided towards the global optimum by using information derived from the global shape of the expected solutions. We compared our Monkey Search to Simulating Annealing, Harmony Search, and the two-phase method, proving that Monkey Search performs competitively. Monkey Search algorithm is implemented using C++ programming language running on Windows XP operating system. A 4GHz duo-processor AMD Athlon is used for computational experiments. An experiment is considered to be successful if the obtained LJ or Morse cluster has an energy corresponding to those available on the Cambridge Cluster Database [2].

In order to investigate the efficiency of Monkey Search by using different values for the parameters h_t and n_t, computational experiments are carried out, in which different values for h_t and n_t are used for optimizing an LJ cluster formed by 15 atoms. In all of the computational experiments reported below, the memory size n_m is fixed and set to 10, the number of trees the monkey starts climbing from a random solution is set to $2n$, and the maximum number n_{max} of trees the monkey is allowed to climb is set to 3000, but this limit has never been reached in our experiments.

In Table 1, percent success over 50 runs and mean computational times are reported for each couple (n_t,h_t). As we can see from this table, by increasing the parameter values, Monkey Search can obtain better results, but naturally, with the increased computational burden. Note that our method's performance improves faster as h_t is increased. Since we obtained a 100% percent success when $n_t = 70$ and $h_t = 100$, we fixed these two parameters in the following computational experiments.

In Table 2, the results related to the optimization of LJ clusters are presented and compared to the results obtained by the two-phase method [12]. In Table 3, the results related to the Morse clusters with different choices of the parameter ρ are presented. As we can see from the Tables 2 and 3, the percent success of Monkey Search decreases by increasing the number n of atoms in the cluster and by increasing the ρ value in the case of Morse clusters. As pointed out above, optimizing a LJ cluster is as hard as optimizing a Morse cluster with $\rho = 6$, and this is reflected on the obtained percent

TABLE 2. Percent success and mean computational time related to LJ clusters having different dimensions compared to the percent success of the two-phase method

Lennard Jones			
n	% success	time	% success in [12]
10	100%	1m	52%
12	100%	4m	94%
15	100%	6m	22%
18	58%	11m	19%
20	28%	13m	36%
22	10%	16m	24%
23	6%	17m	29%
25	10%	20m	35%

TABLE 3. Percent success and mean computational time related to Morse clusters having different dimensions and ρ values

Morse					
n	% success				time
	$\rho = 3$	$\rho = 6$	$\rho = 10$	$\rho = 14$	
10	100%	100%	100%	86%	4m
12	100%	100%	96%	58%	5m
15	100%	88%	42%	14%	10m
18	88%	24%	24%	2%	12m
20	34%	30%	4%	0%	15m
22	14%	22%	2%	0%	18m
23	10%	0%	0%	0%	20m
25	10%	0%	0%	0%	22m

success. Monkey Search results are found to be very competitive compared to the two-phase method. Moreover, as the results in Table 1 suggest, the percent success obtained by Monkey Search can be improved by modifying the parameters n_t and h_t.

We also compared our Monkey Search to other metaheuristic algorithms, such as Simulated Annealing and Harmony Search for LJ clusters. Both of these metaheuristics were not able to find the global optimum. In Table 4, the best results for some choices of the dimension n are presented for 50 runs. Harmony Search generally stops far from the optimal solutions, and Simulated Annealing can get closer to the optimal solution but it is not able to find the precise location of the optimal clusters, whereas Monkey Search always found the optimal conformations. In the Harmony Search experiments, the harmony memory size *hms* has been set to 10, 25, 50 and 100, for each of the 50

TABLE 4. The LJ energy related to the best results obtained by Harmony Search, Simulated Annealing, and Monkey Search, after 50 runs for each dimension n

n	HS	SA	MS
10	-26.957428	-28.422322	-28.422532
12	-32.293843	-37.967146	-37.967600
15	-40.959475	-52.321407	-52.322627
18	-41.079693	-66.280907	-66.530949
20	-49.213888	-77.170622	-77.177043
22	-51.988805	-86.194146	-86.809782
23	-50.888812	-91.363302	-92.844472
25	-56.919838	-100.675583	-102.372663

runs. The presented results in which $n = 10$ to 15 have been obtained with $hms = 25$, whereas $hms = 10$ for each result for dimension $n = 18$ to 25. The starting temperature in the Simulated Annealing experiments has been set to 10, $100 \cdot n$ steps are performed at constant temperature, and then the temperature parameter is decreased by 1%.

3.2. The tube model

The tube model is a geometric model for protein folding proposed in [1, 9] and recently modified in [3]. Protein folding is one of the major challenges in biology and medicine: nobody knows how a chain of amino acids folds forming the typical compact and functional structures the proteins have inside the cells. The most promising approach in this area is the *ab-initio* one, in which the energy in the proteins is modeled and is minimized in order to find the energetically stable conformations. The considered tube model can be classified as ab-initio, but it is very different from the others because it is mainly based on geometric features of proteins. One cannot predict the protein conformation by using this model, but just simulating conformations which may be similar to those proteins actually have.

The model is named as the tube model because a protein conformation is modeled as a tube with a certain diameter such that the tube does not cross itself. The diameter of the tube has particular values when the protein secondary structure such as α-helices and β-sheet are considered. The model leads to the formulation of a global optimization problem, in which some requirements on the protein conformations are imposed. The optimization problem can be formulated in different ways, it depends on the requirements one wants to consider and the relationships among them. More details on the tube model and on the different formulations of the optimization problem can be found in [14].

The global optimization problem we consider in this work can be described as follows:

$$\min_X f(X)$$

where $X = \{x_1, x_2, \dots, x_n\}$ is a protein conformation represented through the spatial coordinate of each C_α Carbon atom (one for each amino acid), and f is the objective function which takes into account three geometric requirements. Its mathematical formulation is:

$$\begin{aligned} f(X) = \ & \gamma_1 \sum_{i=1}^{n-3} \sum_{j=i+3}^{n} d(x_i, x_j) + \\ & \gamma_2 \sum_{i=1}^{n-2} \sum_{j=i+2}^{n} exp_+(th - d(x_i, x_j)) + \\ & \gamma_3 \sum_{i=1}^{n-2} (d(x_i, x_{i+2}) - c)^2 \end{aligned}$$

where γ_1, γ_2, γ_3, th and c are five positive and real constants. The function exp_+ corresponds to the exponential function when its argument is non-negative, and it is zero otherwise; d represents the Euclidean distance.

The considered optimization problem simulates protein conformations which are compact and contain α-helices. The first term of $f(X)$ minimizes all the relative Euclidean distances; the second one avoids the case that the conformation falls on itself, by keeping a threshold value of $th = 4.30$; finally, the last term forces each x_i and x_{i+2} to have a relative distance equal to $c \in [5.0, 6.0]$ units, which is a typical distance in α-helices. Even though only a C_α atom is considered for each amino acid, we represent the protein conformation by the backbone dihedral angles representation, in order to avoid unnatural conformations. Hence, the objective function $f(X)$ is evaluated after a change of representation. The constants γ_1, γ_2 and γ_3 are set, respectively, to 1.0, 10.0 and 1.0.

We are interested in simulating a large number of protein conformations in a short amount of time. Therefore we rely on simple requirements for conformations and check if some protein conformation which actually exist in Nature is found. For this reason, Monkey Search must perform very fast. We set the maximum number n_{max} of trees the monkey is allowed to climb to 40 only. Since the number of variables which are involved is greater than those in the LJ and Morse clusters, and since n_{max} is limited to 40, we used larger values for the parameters h_t and n_t. The memory dimension n_m is fixed at 10, the initial random number of trees n_w is $2n$, and the computational hardware and software setup is the same we used for carrying out the LJ and Morse experiments.

As the reference protein, we have chosen the one which is referenced in the Protein Data Bank (PDB) [15] by the code `2cro`. This protein contains 65 amino acids and it is one of the smallest all-alpha proteins currently known. Table 5 shows the results related to experiments, in which different values for the parameters h_t and n_t have been used. For each pair of h_t and n_t values, 50 runs have been carried out and the best result has been reported. From this table, the best result has been obtained with $h_t = 200$ and $n_t = 125$. Fig. 1 shows the related protein conformation and the `2cro` protein. The conformation found is very compact, it contains α-helices and it has the same globular shape of our reference protein. This is surprising because such a conformation has been simulated by using a model which is very simple with respect those usually used in the field of protein

TABLE 5. Simulation of a protein conformation having 65 amino acids by using different values for h_t and n_t. For each pair of h_t and n_t values, the best result over 50 runs is reported

h_t	n_t	$f(X)$	time	h_t	n_t	$f(X)$	time
100	10	656185	40s	200	50	652685	3m
100	25	585785	50s	200	100	594501	5m
100	50	586891	1m	200	125	**549413**	7m
100	75	577496	2m	200	150	597495	9m
100	85	586284	3m	200	175	552085	10m
100	100	644809	5m	200	200	598234	13m

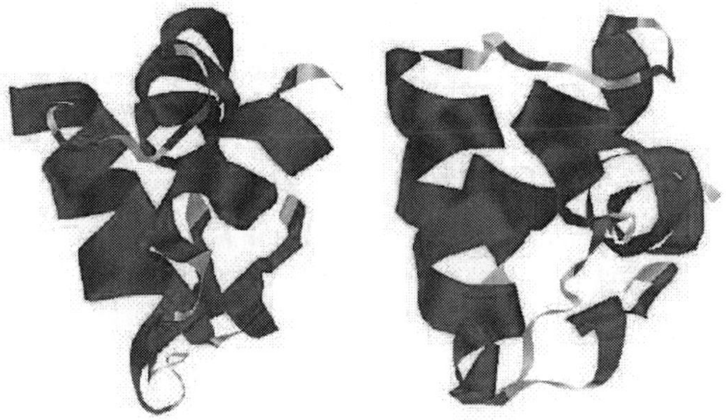

FIGURE 1. On the left, the best protein conformation obtained by Monkey Search; on the right, the actual conformation of the `2cro` protein

folding. Even if the other simulated conformations have a larger objective function value, they preserve the globular shape of `2cro` as well, but they are less compact and contains less α-helices. Further details on the simulated protein conformations are reported in [14].

As for the LJ and Morse clusters, we compared the performance of Monkey Search on this problem to Simulated Annealing and Harmony Search. Simulated Annealing performed $100 \cdot n$ (with $n = 65$) steps at constant temperature, where the temperature was initially set to 10 and decreased by 1% at once. After 50 runs, Simulated Annealing found a conformation having a function value equal to 848922 as the best solution, with a mean computational time of 6 minutes. In the experiments related to Harmony Search, the harmony memory size *hms* has been set to 10, 25, 50 and 100, and 50 runs have been carried out for each of them. The best solution Harmony Search found had a function value equal to 729381 with a mean computational time of 4 minutes. As we can see, Monkey Search performs much better than Simulated Annealing and Harmony Search on this problem. Moreover, Harmony Search works better than Simulated Annealing on the tube model, whereas Simulated Annealing can provide better LJ and Morse clusters.

In fact, each metaheuristic method can be more efficient for solving some optimization problem and less efficient for solving another. The fact that the perturbations in Monkey Search are inspired by Simulated Annealing, Harmony Search and also by other metaheuristic searches may be the reason why Monkey Search performs better than the other two metaheuristics.

4. CONCLUSIONS

A novel metaheuristic method for global optimization has been proposed which resembles the behavior of a monkey climbing trees in its search for food. Monkey Search discovers a tree of solutions where branches contain neighbor solutions on their ends. The monkey marks these branches on its way climbing down whenever a better solution is found, such that these marks can be used to choose branches to climb, on the way up. This marking scheme is intended for the monkey to focus on a part of tree where it can find better solutions. The flexibility of Monkey Search comes from the definition of the perturbations which define the neighbor solutions. Such perturbations exploit ideas and strategies from other metaheuristics and optimization methods and combines them all into an unique method.

In this study, we show that Monkey Search performs significantly better than some other metaheuristic methods such as Simulated Annealing and Harmony Search for solving a set of global optimization problems emerging from molecular structures. Regarding the LJ and Morse clusters, Monkey Search is able to find better solutions with a higher percent success, and it is also very competitive compared to the two-phase method, which partially exploits information on the known solutions. Regarding the tube model for protein folding, after 50 trials, MS is able to provide better solutions compared to Simulated Annealing and Harmony Search, providing protein conformations satisfying the considered geometric requirements.

Further experiments on the optimal monkey behavior and the use of other global optimization problems are considered next as future research to explore the potential of Monkey Search in a more general sense. Since the structure of Monkey Search allows other search methods to be embedded in it, Monkey Search provides endless possibilities, which would potentially produce competitive results compared to these individual metaheuristics.

REFERENCES

1. J. R. Banavar, A. Maritan, C. Micheletti, and A. Trovato, *Geometry and Physics of Proteins* **47**, 315-322 (2002).
2. Cambridge Cluster Database, `http://www-wales.ch.cam.ac.uk/CCD.html`.
3. G. Ceci, A. Mucherino, M. D'Apuzzo, D. di Serafino, S. Costantini, A. Facchiano, and G. Colonna, "Computational Methods for Protein Fold Prediction: an Ab-Initio Topological Approach, Data Mining in Biomedicine", *Optimization and Its Applications*, edited by Panos Pardalos et al., vol. 7, Springer, 2007.
4. M. Dorigo, and G. Di Caro, "Ant Colony Optimization: A New Meta-Heuristic", in *Proceedings of the Congress on Evolutionary Computation*, 1999, pp. 1470–1477.
5. J.P.K. Doye, and D.J. Wales, *J. Chem. Soc. Faraday Trans.* **93**, 4233–4244 (1997).

6. J.P.K. Doye, R.H. Leary, M. Locatelli, and F. Schoen, *INFORMS Journal On Computing* **16**, 371–379 (2004).
7. Z. W. Geem, J. H. Kim, G. V. Loganathan, *SIMULATIONS* **76**, 60–68 (2001).
8. D. E. Goldberg, *Genetic Algorithms in Search, Optimization & Machine Learning*, Reading, MA: Addison-Wesley, 1989.
9. T. X. Hoang, A. Trovato, F. Seno, J. R. Banavar, and A. Maritan, *PNAS* **101**, 7960–7964 (2004).
10. S. Kirkpatrick, C. D. Jr. Gelatt, and M. P. Vecchi, *Science* **220**, 671–680 (1983).
11. J. E. Lennard-Jones, "Cohesion", in *Proceedings of the Physical Society*, **43**, 1931, pp. 461–482.
12. M. Locatelli, and F. Schoen, *Computational Optimization and Applications* **21**, 55–70 (2002).
13. P. M. Morse, *Physical Review* **34**, 57–64 (1929).
14. A. Mucherino, O. Seref, P.M. Pardalos, working paper.
15. Protein Data Bank, `http://www.rcsb.org/pdb/`.
16. J.A. Northby, *J. Chem. Phys.* **87**, 6166–6177 (1987).
17. B. Raoult, J. Farges, M.F. De Feraudy, and G. Torchet, *Philos. Mag.* **60**, 881–906 (1989).
18. O. Seref, and E. Akcali, "Monkey Search: A New Meta-Heuristic Approach", in INFORMS Annual Meeting, San Jose, CA, 2002.
19. R. Storn, and K. Price, *Journal of Global Optimization* **11**, 341–359 (1997).
20. Y. Xiang, H. Jiang, W. Cai, and X. Shao, *J. Phys. Chem.* **108**, 3586–3592 (2004).
21. T. Zhou, W.-J. Bai, L. Cheng, and B.-H. Wang, *Physical Review E* **72**, 016702 (2005).

Approximation algorithms of non-unique probes selection for biological target identification[1]

My T. Thai*, Ping Deng†, Weili Wu† and Taieb Znati**

*Department of Computer and Information Science and Engineering, University of Florida, Gainesville, FL, 32611.
†Department of Computer Science, University of Texas at Dallas, Richardson, TX 75083.
**Department of Computer Science, University of Pittsburgh, Pittsburgh, PA, 15215.

Abstract. Non-unique probes are used to identify the targets, i.e., viruses, present in a given sample. Since the number of selected non-unique probes is equal to the number of hybridization experiments, it is important to find a minimum set of non-unique probes, which is NP-complete. Using d-disjunct matrix, we present two $(1+(d+1)\log n)$-approximation algorithms to identify at most d targets. Based on our selected non-unique probes, we also present the decoding algorithms with linear time complexity. In addition, our solutions are fault tolerant. The proposed algorithms can identify at most d targets in the presence of experimental errors.

Keywords: non-unique probe, non-adaptive group testing, pooling designs, d-disjunct matrix
PACS: 02.30.Mv, 87.55.kd

1. INTRODUCTION

Given a set of n targets and a sample containing at most d, $d \geq 1$ of these targets, an *identification problem* asks to identify the presence (or absence) of all targets in the sample using the minimum number of tests. The identification problem arises from many applications in different domains, such as biology, medicine, environmental science, industrial qualification control, and pattern recognition [2, 4, 6, 7, 8]. In biology, a sample is a biological object, such as blood and targets are viruses or bacteria, for example.

A *probe* is a short oligonucleotide of size 8-25, used for identifying targets in a biological sample through hybridizations using DNA microarrays. A probe is called a *unique* probe if it hybridizes to only one specific target; otherwise, called a *non-unique* probe. Since unique probes have a strong separability of targets, identifying the presence of targets in a sample by using unique probes is straightforward.

However, finding unique probes for every target is a difficult task because of the strong similarity of closely related targets. Fortunately, we can use non-unique probes for the identification problem. Since each non-unique probe can hybridize with more than one target, the identification problem with non-unique probes becomes much more difficult. Specifically, we need to properly select a set of non-unique probes so that we can decode the hybridization results. Note that the number of selected probes is exactly the number

[1] Weili Wu: Support in part by National Science Foundation under grant ACI-0305567. Taieb Znati: Support in part by National Science Foundation under grant CCF-0548895.

CP953, *Data Mining, Systems Analysis, and Optimization in Biomedicine*
edited by O. Seref, O. E. Kundakcioglu, P. M. Pardalos

of hybridization experiments needed. By selecting a minimal set of probes, we can reduce the number of hybridization experiments, thereby reducing the experimental cost and effort.

The identification problem using non-unique probes can be solved in three steps [3] as follows:

1. Find a large set of non-unique probes as candidates and let a binary matrix M represent the probe-target hybridizations [9, 10]. An example of matrix M is given in Table 1. Let a_{ij} denote an entry at cell $M[i, j]$. $a_{ij} = 1$ iff probe p_i hybridizes to target t_j; otherwise, $a_{ij} = 0$.
2. Select a minimum subset of probes obtained in Step 1 so that these probes can identify up to d targets. In other words, find a minimum submatrix H of M with the same number of columns. Note that submatrix H must be able to identify up to d targets. An example of submatrix H is given in Table 2.
3. Decode the presence or absence of targets in a given sample from hybridization results, called *test outcomes* V, where V is a column vector. If $V_i = 1$, then probe p_i hybridizes to at least one target in the sample; otherwise, $V_i = 0$. In Table 2, an example of V is given when two targets t_1 and t_3 present in a given sample.

TABLE 1. Probe-target Incidence Matrix M.

	t_1	t_2	t_3	t_4
p_1	1	0	0	1
p_2	1	1	0	1
p_3	0	0	1	1
p_4	1	1	0	0
p_5	0	1	1	1
p_6	0	1	1	0
p_7	0	1	0	1
p_8	1	0	1	0
p_9	1	1	1	0
p_{10}	1	0	1	1

TABLE 2. Selected submatrix H and outcome vector V where t_1 and t_3 present in a given sample

	$\mathbf{t_1}$	t_2	$\mathbf{t_3}$	t_4	Test Outcomes V
p_1	1	0	0	1	1
p_3	0	0	1	1	1
p_4	1	1	0	0	1
p_6	0	1	1	0	1
p_7	0	1	0	1	0
p_8	1	0	1	0	1

In this chapter, we study the steps 2 and 3 of the identification problem with non-unique probes, that is, finding a minimum submatrix H with an efficient decoding algorithm. Unsurprisingly, these two problems are highly related. The design of submatrix H must satisfy two conditions: (1) All unions of up to d columns in H must be different. Here, the union of a set of columns is equal to the boolean sums of these columns. (2)

The time complexity of a decoding algorithm yielding from the design of H must be efficient.

However, constructing such a submatrix H is NP-hard [1]. In [3, 4], the authors proposed two heuristics to construct a submatrix H using the greedy and Linear Programming based techniques. However, the time complexity of the decoding algorithms using their submatrix H is high, which is $O(n^d)$. In this chapter, we present a $[1+(d+1)\log n]$-approximation algorithm to construct a submatrix H where the time complexity of the decoding algorithm is $O(n)$.

More challenging, the presence of errors due to the noise of hybridizations makes the problem become even harder. With the experimental errors, the test outcomes may consist of *false negatives* or *false positives*. In the former, a test i yields a negative outcome, i.e. $V_i = 0$ when a probe p_i *does* hybridize to at least one target in the sample. Likewise, in the latter, a test i yields a positive outcome, i.e. $V_i = 1$ when a probe p_i does *not* hybridize to any targets in sample. In this case, we present an algorithm to construct a minimum submatrix H to correctly identify up to d targets with the presence of at most k errors in experiments. Unfortunately, the decoding algorithm in the case of error tolerance become much more complicated. In this chapter, we also present a solution and its hardness to this problem.

The rest of this chapter is organized as follows. Section 2 describes the preliminaries and problem formulations as well as all the definitions used throughout the chapter. The related work on the non-unique probe selection is presented in Section 3. The solutions to both minimum submatrix H and decoding algorithms are discussed in Section 4. Section 4 also presents the solutions in the presence of experimental errors. Finally, Section 5 concludes our chapter.

2. PROBLEM FORMULATIONS AND PRELIMINARIES

Let an $m \times n$ binary matrix M represent the probe-target hybridizations. The rows in M represent the set of non-unique probes p_i, $\forall i = 1, \ldots, m$ whereas the columns in M represent the set of n targets t_j, $\forall j = 1, \ldots, n$. Let a_{ij} denote an entry at the cell $M[i, j]$. $a_{ij} = 1$ iff p_i hybridizes to t_j; otherwise, $a_{ij} = 0$. In this chapter, we will use probe p_i and row p_i (target t_j and column t_j) interchangeably.

Consider a binary matrix M, we have the following definitions:

Definition 1 Isolated Column: *A column is called an isolated column iff there is a row containing only one 1-entry at the intersection with that column.*

In our context, a column (target) t_j is isolated if there exist a probe p_i hybridizes to only one target t_j, that is, p_i is a unique probe. Since we study the non-unique probe selection problem, we can assume that the given matrix M does not contain any isolated column. Likewise, the constructed matrix H does not contain any isolated column either.

Definition 2 Union of d columns: *The union of d columns is the boolean sum of these d columns.*

Definition 3 Hamming distance: *The Hamming distance of two column vectors is defined as the number of different components between them.*

Definition 4 k**-error-correcting:** *A matrix M is said to be k-error-correcting if the Hamming distance of any two unions of d columns must be at least* $2k+1$

Definition 5 d**-separable:** *A matrix H is called d-separable if all unions of d columns are different.*

Definition 6 $\bar{d}$**-separable:** *A matrix H is called $\bar{d}$-separable if all unions of at most d columns are different.*

Definition 7 d**-disjunct:** *A matrix H is called d-disjunct if the union of d columns does not contain any other column.*

Definition 8 (d,k)**-disjunct:** *H is called (d,k)-disjunct if for any column t_j, t_j must have at least $k+1$ 1-entries not contained in the union of other d columns.*

Since all unions of up to d columns in submatrix H must be different, H should be either disjunct or separable. Note that d-disjunct matrices have a decoding complexity of $O(hn)$ where n is the number of targets and h is the number of selected probes. This time complexity is much lower than that of $\bar{d}$-separable matrices, which is $O(n^d)$. Therefore, for the low computational complexity of decoding algorithms, we choose to construct a disjunct submatrix H (instead of a separable submatrix H).

Let $S(d,n)$ be a sample space such that for a given sample $s \in S$, s will have *exactly* d targets from a set of n targets. Likewise, let $S(\bar{d},n)$ be a sample space such that for a given sample $s \in S$, s will contain *at most d* targets. In this chapter, we consider the following four problems in the above two $S(d,n)$ and $S(\bar{d},n)$ sample spaces with and without error tolerance.

2.1. Error free

Problem 1: MIN-d-DS (Minimum d-Disjunct Submatrix): Given an $m \times n$ binary matrix M, find a minimum d-disjunct $h \times n$ submatrix H where $h \leq m$.

Problem 2: Decoding Algorithm: Given a d-disjunct matrix H, a sample s, and the test outcomes vector V, find an algorithm to identify all the targets t_j present in the given sample s.

In [5], Du and Hwang showed that for any fixed d, MIN-d-DS is NP-hard [1]. Therefore, we need to study the approximation algorithms to find a near optimal solution for the MIN-d-DS problem.

2.2. Error tolerance

In order to identify all the targets with at most k experiment errors in hybridizations, the matrix H must have the k-error-correcting property. Note that by the definition, it is enough to see that the (d,k)-disjunct matrix is a k-error-correcting $d-$separable matrix.

Therefore, in the case of error tolerance, we need to study the following two problems:

Problem 3: MIN-(d,k)-DS (Minimum (d,k)-Disjunct Submatrix): Given an $m \times n$ binary matrix M, find a minimum (d,k)-disjunct $h \times n$ submatrix H where $h \leq m$.

Problem 4: Decoding Algorithm: Given a (d,k)-disjunct matrix H, a sample s, and the test outcomes vector V, find an algorithm to identify all the targets t_j present in the given sample s in the case of at most k experimental errors.

3. RELATED WORK

Although non-unique probes selection is of great importance in many applications, *little research* has been done on this problem. To the best of our knowledge, there are only two work addressing this problem in literature [3, 4]. In [3], the authors proposed a greedy heuristic to construct a d-separable submatrix by adding probes one by one. Indeed, at first, they considered a 1-separable submatrix, and then randomly adding a number of pairs of targets sets to distinguish all the targets sets. Therefore, the constructed submatrix is not totally d-separable. They also considered this problem with error tolerance.

Later, Klau *et al.* presented an integer programming based heuristic [4]. The master ILP guarantees the pairwise separation of all targets. The slave ILP guarantees the separability between all pairs of small target groups by adding violated group inequalities to the master ILP.

Both heuristics in [3, 4] studied 1-separable and "near" $\bar{d}$-separable submatrix with k errors. Therefore, the corresponding decoding algorithm is complicated and time-consuming. In addition, there is no approximation guarantee for these heuristics.

In this chapter, we present two $(1+(d+1)\log n)$-approximation algorithms using d and (d,k)-disjunct matrices. This approach will have a decoding algorithm with linear time complexity.

4. MAIN RESULTS

In this section, we present solutions to the four mentioned problems for a given sample in both $S(d,n)$ and $S(\bar{d},n)$. Owing to the noise of hybridization, the test outcomes will contain some errors. Therefore, the designed algorithms must be robust. In this section, we also present our solutions in the case of error tolerance.

4.1. Error free

Recall that H is a d-disjunct matrix iff the union of d columns does not contain any other columns. Therefore, we have:

Fact 1: H is a d-disjunct matrix iff for any $(d+1)$ columns $t_0, t_1, \ldots, t_d$, there exists a row p_i such that an entry $H[i,0] = 1$ and $H[i,j] = 0$ for all $j = 1 \ldots d$.

Such a row p_i is said to *cover* a pair $(t_0, < t_1, ..., t_d >)$. Hence, to construct matrix H from M, we need to find a set of probes covering all pairs $(t_j, < t_k, ..., t_{k+d-1} >)$ where $j, k, \ldots, (k+d-1) \in \{1, ..., n\}$ and $j \notin \{k, \ldots, (k+d-1)\}$.

$H \leftarrow 0; P \leftarrow \emptyset$
while all pairs $(t_0, < t_1, ..., t_d >)$ are *not* covered **do**
  Select a probe covering the most uncovered pairs, say p
  $P = P \cup \{p\}$
end while
for $j = 1 ... |P|$ **do**
  for $i = 1 ... m$ **do**
    if $p_i \in P$ **then**
      $H[j,:] = M[i,:]$
    end if
  end for
end for
Return H

Algorithm 1: Constructing d-disjunct Submatrix

Theorem 1 *Algorithm 1 produces an approximation solution within a factor of* $(1 + (d+1)\log n)$.

Proof. Let $|P| = h$ and opt be the size of an optimal solution. Let $p_1, ..., p_h$ be a set of probes which were chosen by Algorithm 1 in this order. Let E be a set of all pairs $e = (t_0, < t_1, ..., t_d >)$. Hence $|E| \leq n^{d+1}$. Let A_i be the set of all new pairs e that p_i can cover and let $a_i = |A_i|$. Note that $E = \bigcup_{i=1}^{k} A_i$.

For each pair $e \in A_i$, we assign to e a cost $c(e) = 1/a_i$. Then we have:

$$h = \sum_{A_i} \sum_{e \in A_i} c(e) = \sum_{e \in E} c(e) \leq \sum_{k=1}^{E} \frac{opt}{|E| - k + 1}$$

$$\leq (1 + \log(|E|))opt \leq (1 + (d+1)\log n)opt$$

□

After finding a d-disjunct submatrix H, the decoding is simple based on the following Lemma:

Lemma 1 *[5] For testing based on a d-disjunct matrix, the number of targets not appearing in any negative results is no more than d*

Therefore, if given sample s has at most d targets, these targets cannot appear in the negative results. Thus in the decoding algorithm, we just need to remove all items appearing in the negative results and the remaining ones should be the targets we want to identify. This algorithm works for both $S(d,n)$ and $S(\bar{d},n)$. Note that the time complexity of this decoding algorithm is $O(hn)$ where h is the number of selected probes in H.

4.2. Error tolerance

In this section, we consider the non-unique probes selection problem where there exists at most k experimental errors. More specifically, we study the problem 3 and 4 as defined in Section 2. We present solutions on finding the (d,k)-disjunct submatrix H and decoding the hybridization results for both $S(d,n)$ and $S(d,n)$ sample spaces.

4.2.1. $S(d,n)$ Sample space

Consider a constructed submatrix H. Since matrix M does not contain any *isolated column*, the submatrix H does not contain any isolated column either. It is well-known that every $(d+k)$-disjunct matrix without isolated column is a (d,k)-disjunct matrix [5]. Therefore,

we can use Algorithm 1 to construct the $(d+k)$-disjunct matrix H, thus H is (d,k)-disjunct, which can solve for the problem with at most k errors in $S(d,n)$ sample space. In this case, the approximation ratio is $[1+(d+k+1)\log n]$.

Note that although we can construct a $(d+k)$-disjunct matrix to obtain the (d,k)-disjunct matrix, the number of probes in that matrix is maybe bigger than that in a (d,k)-disjunct found directly. Therefore, we need to study how to find a (d,k)-disjunct matrix H *directly* from a given matrix M.

Recall that a_{ij} is an entry at cell $M[i,j]$. Let $k' = k+1$. The problem of finding a minimum (d,k)-disjunct submatrix is equivalent to the following integer programming:

$$\begin{array}{ll} \min & \sum_{i=1}^{m} x_i \\ \text{subject to} & \sum_{i=1}^{m} a_{ij} x_i \geq k' \text{ for } j = 1,2,\ldots,n \\ & \sum_{i=1}^{m} (|a_{ij} - (a_{ik} + \ldots + a_{i(k+d-1)})|) x_i \geq k' \text{ for } j,k,\ldots,k+d-1 \in \{1,\ldots,n\} \\ & \qquad \text{and } j \notin \{k,\ldots,k+d-1\} \\ & x_i \in \{0,1\} \text{ for all } i = 1,2,\ldots,m \end{array} \tag{1}$$

where $x_i = 1$ iff probe p_i is selected; otherwise, $x_i = 0$.

Remarks:

- The first constraint is to make sure that each target is covered by at least k' probes
- The second constraint is to make sure that each pair $(t_j, < t_k, \ldots, t_{k+d-1} >)$ can be covered by at least k' probes

Again, let E be a set of all pairs $(t_j, < t_k, \ldots, t_{k+d-1} >)$. Let matrix B be a binary matrix where rows represent m probes p_i and columns represent the pairs. Let b_{ij} be each entry at cell $B[i,j]$. $b_{ij} = 1$ iff probe p_i covers the pair at column j.

In a (d,k)-disjunct matrix, for any pair $(t_j, < t_k, \ldots, t_{k+d-1} >)$, we can find at least $k+1$ rows such that the intersection entries of these rows at column t_j are 1 whereas the intersection entries of these rows at d columns $< t_k, \ldots, t_{k+d-1} >$ are all 0. Therefore, the integer programming (1) should be equivalent to the following:

$$\begin{array}{lll} \min & \sum_{i=1}^{m} x_i & \\ \text{subject to} & \sum_{i=1}^{m} b_{ij} x_i \geq k' \text{ for } j = 1,2,...,|E| & \\ & x_i \in \{0,1\} \text{ for all } i = 1,2,...,m & \end{array} \tag{2}$$

To solve (2), at each iteration, we select a probe covering the most *unsatisfied* pairs. A pair is unsatisfied if it has not been covered by at least k' probes yet. The details of this algorithm are described in Algorithm 2.

$I \leftarrow \{1,2,...,m\}$
$J \leftarrow \{1,2,...,|E|\}$
$H \leftarrow 0; P \leftarrow \emptyset$
while $J \neq \emptyset$ **do**
  Find $i_0 \in I$ such that
  $\sum_{j \in J} b_{i_0 j} = \max_{i \in I} \sum_{j \in J} b_{ij}$
  $P \leftarrow P \cup \{i_0\}$
  $I \leftarrow I - \{i_0\}$
  $J \leftarrow J - \{j \mid \sum_{i \notin I} b_{ij} \geq k'\}$
end while
for $j = 1...|P|$ **do**
  for $i = 1...m$ **do**
    if $p_i \in P$ **then**
      $H[j,:] = M[i,:]$
    end if
  end for
end for
Return H

Algorithm 2: Constructing (d,k)-disjunct Submatrix

Theorem 2 *Algorithm 2 produces an approximation solution within a factor of* $1 + (d+1)\log n$

Proof. Since the proof is similar to the proof of Theorem 1, we omit the proof here. □

We now consider the decoding algorithm.

Lemma 2 *[5] Suppose testing is based on a (d,k)-disjunct matrix. If the number of error tests is no more than k, then the number of negative results containing a target is always smaller than that of the number of negative results containing a non-target.*

Proof. For the convenience of readers, we present the proof of Lemma 2 here. Let i be a target and j be other items in a hybridization (but not a target). Suppose the number of negative hybridization containing i is l. Then these l hybridization must receive error tests. Therefore, there are at most $k-l$ error tests turning negative outcomes to positive outcomes. Moreover, we note that if no error exists, the number of negative results containing j is at least $k+1$ by the definition of (d,k)-disjunctness. Hence, the number of negative results containing j is at least $(k+1)-(k-l) = l+1 > l$

□

From the above lemma, we see that to decode the targets from testing based on (d,k)-disjunct matrix for $S(d,n)$ sample space, we only need to compute the number of negative results containing each item and select d smallest ones. This runs in time $O(hn)$.

4.2.2. $S(\bar{d},n)$ Sample space

In the $S(\bar{d},n)$ sample space, the decoding algorithm is much more complicated. Although Lemma 2 still holds in $S(\bar{d},n)$, we do not know how many smallest one we should select. Fortunately, Du and Hwang [5] have proven the following lemma:

Lemma 3 *[5] There exists a decoding algorithm for a k-error-correcting d-disjunct matrix H, running in time $O((n+h)h^k)$ where h is the number of rows (selected probes) in H.*

where a k-error-correcting d-disjunct matrix is defined as follows:

Definition 9 ***k-error-correcting d-disjunct matrix***: *A matrix H is called k-error-correcting d-disjunct matrix if H is d-disjunct and the Hamming distance between two union of* at most *d columns is at least* $2k+1$.

Thus, to study the Problem 3 in the $S(\bar{d},n)$ space with a possible decoding algorithm, we need to find a minimum k-error-correcting d-disjunct submatrix H. Once we construct a k-error-correcting d-disjunct submatrix H, we can use a decoding algorithm mentioned in [5], of which the time complexity is $O((n+h)h^k)$. This time complexity is quite high.

Interestingly, the following lemma gives an efficient way to construct such a k-error-correcting d-disjunct submatrix H with a linear decoding algorithm.

Lemma 4 *Every $(d,2k)$-disjunct matrix is k-error-correcting d-disjunct matrix*

Proof. Given a matrix H as a $(d,2k)$-disjunct matrix. Since H is $(d,2k)$-disjunct, H is d-disjunct. Therefore, for any two different subsets of at most d columns in H, there must be one not contained by the other. By the definition of $(d,2k)$-disjunct matrix, the union of the former contains at least $2k+1$ 1-entries not appearing in the union of the latter. This implies that the Hamming distance between these two unions is at least $2k+1$. Hence H is also an k-error-correcting matrix.

□

From Lemma 4, instead of directly constructing a k-error-correcting d-disjunct submatrix H, we find a $(d,2k)$-disjunct submatrix H by using Algorithm 2.

Now, we consider the decoding algorithm using a $(d,2k)$-disjunct submatrix H.

Lemma 5 *Suppose testing done on a $(d,2k)$-disjunct matrix H with at most k errors, an item is a target iff it appears in at most k negative results.*

Proof. Since there are at most k errors, a target can appear in at most k negative results (due to errors). However, a non-target item appears in at least $2k+1-k=k+1>k$

negative results. It implies that an item is a target iff it appears in at most k negative results.

□

Based on the above Lemma, the decoding algorithm becomes quite simple. For each item, we just need to count the number of negative results containing it. If this number is less than k, then this item must be a target. Hence, the time complexity of this decoding algorithm is $O(hn)$.

5. CONCLUSIONS

In this chapter, we study the non-unique probes selection, which is used to identify the targets present in a given sample. In biology, a sample is a biological object, such as blood and a target is a virus or bacteria. Since the number of selected non-unique probes is equal to the number of hybridizations, we need to find a minimum number of selected probes in order to reduce the experimental cost and time. In addition, the selected probes must yields an efficient decoding algorithm, which is used to infer the test outcomes. Finally, the designed algorithms must be robust in order to handle the experimental errors due to noises.

We presents several algorithms using a d-disjunct matrix for both cases, error free and error tolerance. For each case, we present efficient algorithms to select a minimum number of probes as well as to decode the results. These algorithms are able to identify all targets in a given sample with at most d targets. Particularly, we present a $(1+(d+1)\log n)$-approximation algorithm where n is a number of targets. Our non-unique probes selection can also identify the presence of at most d targets in a case of at most k experimental errors. In addition, we also present and analyze the time complexity of how to decode the test results obtained from the microarrays experiments. The time complexity of our decoding algorithms is $O(hn)$ where h is the number of selected probes.

Note that d-disjunct matrices have a stronger property than that of $\bar{d}$-separable matrices. Hence, a d-disjunct matrix should have more rows (probes) than that of a d-separable matrix. That is a trade-off for having a faster decoding algorithm. In addition, the IP formulation mentioned above is a general covering problem [11]. Therefore, we can use similar techniques in [11] to show that both Algorithm 1 and 2 can produce an approximation ratio within a factor of $(1+\log\Delta)$ where $\Delta = \max_{p_i} |\{e \mid p_i \text{ covers } e \;\&\; e \in E\}|$.

REFERENCES

1. M.R. Garey, and D.S. Johnson, *Computers and Intractability - A Guide to the Theory of NP-completeness,* W.H. Freeman & Co., 1979.
2. B.M.E. Moret, and H.D. Shapiro, *SIAM Journal on Sceientific and Statistical Computing* **6**, 983–1003 (1985).
3. A. Schliep, D.C. Torney, and S. Rahmann, "Group Testing with DNA Chips: Generating Designs and Decoding Experiments", in *Proceedings of the Computational Systems Bioinformatics (CSB '03)*, 2003.

4. G.W. Klau, S. Rahmann, A. Schliep, M. Vingron, and K. Reinert, *J. of Bioinformatics* **20**, 186–193 (2004).
5. D.Z. Du, and F.K. Hwang, *Pooling Designs: Group Testing in Molecular Biology*, World Scientific, Singapore, 2006.
6. Steinfath, J. O' Brien, H. Seidel, S. Meier-Ewert, H. Lehrach, and U. Radelof, *Bioinformatics* **16**, 890–898 (2000).
7. J. Borneman, M. Chrobak, G. D. Vedova, A. Figueroa, and T. Jiang, *Bioinformatics* **17**, 339–348 (2001).
8. X. Wang, and B. Seed, *Bioinformatics* **19**, 796–802 (2003).
9. M. S. Rahmann, "Rapid large-scale oligonucleotide selection for microarrays", in *Proceedings of the First IEEE Computer Society Bioinformatics conference(CSB)*. Stanford, 2002, pp. 54–63.
10. S. Rahmann, "Fast and sensitive probe selection for DNA chips using jumps in matching statistics", in *Proceedings of the 2nd IEEE Computational Systems Bioinformatics Conference(CSB)*, Stanford, 2003, pp. 57–64.
11. V. V. Vazirani, *Approximation Algorithms*, Springer-Verlag, 2001.

Modeling and *in vitro* and *in vivo* characterization of a tissue engineered pancreatic substitute

C.L. Stabler[*,¥], C. Fraker[*,¥], E. Pedraza[*], I. Constantinidis[£,¶1] and A. Sambanis[€,†]

[*] *Department of Biomedical Engineering, University of Miami, 219A McArthur Annex, Coral Gables, FL 33124-0621*
[¥] *Diabetes Research Institute, University of Miami, 1450 NW 10th Avenue, Miami, FL 33136*
[£] *Division of Endocrinology, Department of Medicine, University of Florida, PO Box 100226, Gainesville, Florida 32610-0226*
[¶] *National High Magnetic Field Laboratory, 1800 E. Paul Dirac Dr., Tallahassee, FL 323103706*
[€] *Department of Biomedical Engineering, Georgia Institute of Technology/Emory University, 313 Ferst Drive, Atlanta, GA 30332-0363*
[†] *School of Chemical & Biomolecular Engineering, Georgia Institute of Technology, 311 Ferst Drive, Atlanta, GA 30332-0100*

Abstract. This study investigated the model-based design, fabrication and *in vitro* and *in vivo* experimental characterization of a pancreatic substitute consisting of mouse insulinoma cells encapsulated in agarose in a disk-shaped construct. Two construct prototypes were examined: (i) a single disk construct comprised of agarose and βTC3 cells; and (ii) a buffered disk construct, consisting of agarose and βTC3 cells, coated with an additional layer of pure agarose. Diffusional studies of glucose and insulin were performed to characterize the transport properties of the material. Three dimensional oxygen diffusion-reaction models were used to predict the appropriate cell loadings for the two construct prototypes under varying external oxygen tensions. *In vitro* and *in vivo* experiments found the overall viable cell number for each construct prototype plateaued to the same value, regardless of the initial cell seeding number, when constructs were placed under identical environmental conditions. Furthermore, mathematical model calculations correlated well with experimental *in vitro* and *in vivo* results of cell viability, indicating oxygen tension to be the dominating factor in establishing total viable cell number in these constructs. These results indicate that modeling is useful for the development of tissue engineered constructs when permissive matrices and continuous cell lines are used. The applicability of this modeling and experimental methodology in the development of agarose-based constructs for use as a bioartificial pancreas is discussed.

Keywords. Tissue construct modeling, pancreatic substitute, agarose, encapsulated cells, βTC3 insulinoma cells
PACS: 87.80.Rb

---

[1] The authors would like to dedicate this paper to the memory of the colleague, friend, and mentor Ioannis (Yanni) Constantinidis, who died suddenly and untimely on April 16, 2007.

CP953, *Data Mining, Systems Analysis, and Optimization in Biomedicine*
edited by O. Seref, O. E. Kundakcioglu, P. M. Pardalos

1. INTRODUCTION

Diabetes mellitus is a serious pathological condition involving damaged pancreatic islets, impaired insulin production, and poor glycemic control. Typically, treatment for insulin-dependent diabetics includes blood glucose monitoring and multiple daily insulin injections. The 1993 Diabetes Control and Complications Trial, which found that tight control of glucose levels delays the onset and reduces the intensity of diabetes-related complications, outlined the need for a more physiological control of blood glucose levels beyond what is achieved with bolus insulin injections [1]. While the implantation of human islets has been successful in recent clinical trials [2, 3], the supply of this cell source, of which significant propagation is unavailable, is insufficient to provide islets for any substantial number of insulin-dependent diabetics. Additionally, the life-long immunosuppression required to sustain treatment is problematic.

A tissue engineered pancreatic substitute, consisting of insulin-secreting cells and biocompatible materials, holds significant promise for treating insulin-dependent diabetes through continual regulation of insulin secretion in response to glucose. Use of transformed β cells, such as the continuous β cell lines developed by Efrat et al [4, 5], relaxes the cell availability problem posed by pancreatic islets, while also utilizing cells that have lower oxygenation requirements for insulin secretion [6]. Encapsulation of the implanted cells can partially alleviate the need for immunosuppression. Development of a pancreatic substitute based on encapsulated insulin-secreting cells requires (i) a thorough characterization of the transport of nutrients and metabolites and their respective consumption and production by the cells, based on the construct architecture; (ii) *in vitro* evaluation of the construct in terms of cell viability and the overall secretion dynamics; and (iii) *in vivo* evaluation of the construct cell viability and functionality.

In this study, we describe the design, fabrication and characterization of a pancreatic substitute consisting of mouse insulinoma cells encapsulated in agarose in a disk-shaped construct. Purified agarose has been shown to be an appropriate hydrogel for cell encapsulation [7-12], while the disk-shaped design affords ease in implantation and complete retrievability. Two construct designs were examined, mathematical models of oxygen transport and consumption in constructs were generated, and the constructs were experimentally evaluated *in vitro* and *in vivo*. Advantages and shortfalls of the implemented design approach, and possible compromises in construct functionality imparted by their architectural features, are discussed.

2. MATERIALS AND METHODS

2.1. Cell and cell culture

βTC3 cells were obtained from the laboratory of Shimon Efrat, Albert Einstein College of Medicine, Bronx, NY. Cells were cultured as monolayers in T-flasks and fed every 2-3 days with fresh medium consisting of Dulbecco's Modified Eagle's

Medium (DMEM) with 25 mM glucose and supplemented with 15% horse serum, 2.5% bovine serum, 1% penicillin-streptomycin, and L-glutamine to a final concentration of 6 mM (Sigma, St. Louis, MO). Upon confluency, βTC3 cells were trypsinized using 0.25% Trypsin with EDTA (Sigma, St. Louis, MO) and either split for propagation or used in construct fabrication (passage numbers 33-48).

2.2. Construct fabrication

Two types of agarose constructs were used in this study (Fig. 1). One was a single 2% agarose disk with entrapped βTC3 cells and final dimensions of 1.8 mm height and 11 mm diameter (Fig. 1A). The other was a buffered agarose disk, consisting of an inner, cell-containing agarose disk coated with an additional layer of pure 2% agarose, to final dimensions of 3 mm height and 16 mm diameter (Fig. 1B). The presence of this buffer layer was found necessary to minimize the contribution of interabdominal fat on the NMR spectrum acquired from implanted constructs. These *in vivo* studies are reported elsewhere[13, 14].

Constructs were fabricated as described previously [14,15]. Briefly, Lexan molds were used to fabricate constructs of freshly trypsinized βTC3 cells mixed with agarose (SeaPlaque, Cambrex, NJ)-PBS solution. The molds were filled to capacity with the agarose/cell mixture (0.2 mL in volume) and allowed to cool at room temperature until gelled (~ 3 mins). For buffered agarose constructs (Fig. 1B), the cell-containing disks were removed from their molds and placed into larger Lexan molds and filled to capacity after the inner disk was centrally aligned. All constructs were allowed to solidify at room temperature for 5 min before being placed in fully supplemented DMEM (*in vitro)* or HBSS (*in vivo)* experiments.

2.3. Diffusivity measurements

Glucose and insulin diffusivities through agarose were measured by mixing glucose or porcine insulin (both from Sigma) with 2% (w/v) agarose and casting 1.5 mm slabs into wells of a six well plate. The concentrations of glucose and insulin in the slabs were 18 mg/mL and 0.1 mg/mL, respectively. Three mililiters of PBS solution were added to each well, marking time zero of the experiment and plates were placed on an orbital shaker at room temperature. For glucose, 5μL samples were collected from each well at intervals of 1-15 min from 0 to 120 min. For insulin, 150 μL samples were collected, while 150 μL of pure PBS was added, in order to maintain a constant volume, from each well at intervals of 1-15 mins from 0 to 240 min. A final sample was collected after 24 hours and compared to the previously collected sample to verify equilibrium. Glutaraldehyde-deactivated cells were incorporated into the agarose construct at a density of $5x10^7$ cell/mL to evaluate the effect of cells on effective diffusivity. Experiments for both glucose and insulin were repeated using these cells as described above. To assess the effects of an agarose/agarose interface on the effective diffusivity, an additional 0.6 mm layer of pure 2% agarose was added to the existing molds. For each experimental type, a minimum of three independent experiments were conducted, and data were fitted to the appropriate one-dimensional diffusion equation to estimate the effective diffusivity.

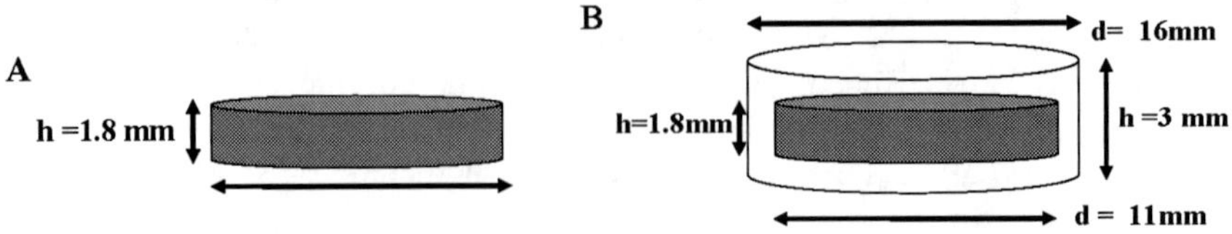

FIGURE 1. Schematic representation and dimensions of single agarose disk construct (A) and buffered agarose disk construct (B). The shaded region represents the inner disk containing βTC3 cells entrapped in 2% agarose, while the white region indicates pure 2% agarose.

2.4. *In vitro* construct characterization

For *in vitro* studies, both constructs types containing different initial cell numbers were examined. For single agarose constructs, initial cell loadings were $14x10^6$, $10x10^6$, and $7x10^6$ cells, while buffered agarose constructs initially contained $7x10^6$ and $3x10^6$ cells. For each experiment, twelve identical constructs were fabricated and split into two groups of 6 constructs each. Each group was placed in a 125 mL spinner flask (Wheaton, Millville, NJ) containing 35 mL of fully supplemented DMEM. The spinner flasks were placed on a spinner plate in a humidified 37°C, 5% CO_2 / 95% air incubator and were agitated at 30 rpm. The time duration of each experiment was 13 days. Complete media changes were performed every other day. Samples for insulin analysis were collected every 24 hrs. Two constructs from one spinner flask were removed on Days 0, 1, 3, 5, 7, and 13 for MTT and histology testing.

2.5. Animal studies

All animal experiments were approved by the Institutional Animal Care and Use Committee (IACUC) at Emory University. Male C57BL/6J mice of at least 20 grams were obtained from Jackson Laboratories (Bar Harbor, Maine) and used for surgical implantations. Only buffered agarose disk constructs were used for *in vivo* transplantation. One agarose construct was implanted within the peritoneal cavity of each mouse, sedated using an IP injection of ketamine/xylazine (150/15 mg/kg body weight), via a 2 cm midline celiotomy. Three cell loadings of $14x10^6$, $8x10^6$, and $4x10^6$ cells were evaluated. A minimum of 20 mice were used for each cell loading, with a minimum of three mice euthanized for construct cell viability testing on Days 1, 2, 3, 4, 5, and 14. Agarose disk constructs were explanted using forceps following euthanasia via CO_2 asphyxiation.

2.6. Analytical techniques

Glucose concentrations in samples were measured using a calibrated Elite Glucometer (Bayer, PA). Insulin concentrations in samples from the diffusion experiments, where insulin was the only protein in solution, were measured with a Coomassie blue-based total protein assay (Pierce, IL). A calibration curve was generated for each assay run using a stock insulin solution diluted to the appropriate range. Insulin from *in vitro* incubation experiments was measured with an ultrasensitive mouse insulin EIA kit (ALPCO, NH) following the manufacturer's protocol. Measurements of viable cell number in agarose construct using the CellTiter 96 MTT (3-(4,5-dimethylthiazol-2-yl)-2,5-diphenyl-tetrazolium bromide) assay (Promega, Madison, WI) have been described previously [15]. Samples designated for histology were fixed in 3% glutaraldehyde for 48 hrs and then prepared for paraffin embedding and sectioning. After sectioning into 5μm samples, the slides were stained with hematoxylin/eosin (H/E).

2.7. Mathematical modeling

All time dependent, three dimensional studies were carried out using finite difference with Comsol Multiphysics Software (Burlington, MA), version 3.3. Time dependent data was collected using a time domain of 60 minutes.

2.8. Statistical analysis

All measurements are presented as the mean ± SD. All statistical analyses were performed using a two-sample t-test. Results were considered significantly different when p-values were less than 0.05.

3. RESULTS

3.1. Effective diffusivity in agarose constructs

Equation (1) describes the diffusion of solutes out of the agarose and into the solution for the experimental conditions outlined in the Methods section, assuming one-dimensional diffusion, uniform concentration of the solute in solution, adequate mixing of the bathing solution these two are the same, keep one or the other, constant effective diffusivity, and a partition coefficient of 1 [16, 17].

$$\frac{C_b}{C_o} = \frac{1}{1+\alpha}\left[1-\sum_{n=1}^{\infty}\frac{2\alpha(1+\alpha)}{1+\alpha+\alpha^2 q_n^2}\exp\left(-\frac{Dq_n^2 t}{L^2}\right)\right] \tag{1}$$

where the parameters are defined as: D effective diffusivity of solute out of the agarose slab; L slab thickness; C_o initial concentration of solute in agarose; C_b concentration of solute in the bulk solution; α ratio of the bulk liquid volume to agarose slab volume; and q_n the nth root of the equation $\tan q_n = \frac{3q_n}{3+q_n\alpha}$. To simulate experimental results, only three terms of the series in equation (1) were required, based on the magnitude of the term $\exp\left(-\frac{Dq_n^2 t}{L^2}\right)$.

Equation (1) was then expressed in the form of the total amount of solute M_t in solution at time t as a fraction of M_{inf}, the corresponding quantity after infinite time, and the effective diffusivity D for each solute was determined by fitting equation this equation to experimental data, with D as the only adjustable parameter. Effective diffusivities were measured to be 4.26×10^{-4} cm^2/min and 3.22×10^{-4} cm^2/min for the cell-free and cell-containing constructs, respectively, or 103% and 78% of glucose diffusivity in water at 25°C [18]. Insulin effective diffusivities for the cell-free and cell-containing constructs were measured to be 1.15×10^{-4} cm^2/min and 9.19×10^{-5} cm^2/min, respectively, or 96% and 77% of the diffusivity in water at 25°C [19]. In all cases, results from the buffered agarose/agarose constructs were statistically identical to those from the corresponding single layer constructs.

3.2. Diffusion-reaction modeling of construct with cells

To provide an estimate for the appropriate cell number that may be supported in the agarose constructs, diffusion/reaction models were developed to simulate the axial concentration profile of dissolved oxygen, since dissolved oxygen is typically the limiting nutrient for extravascular constructs [20-23]. To simulate the oxygen profile within the construct, the primary assumptions were made: (i) cellular oxygen consumption following Monod's model; and (ii) homogeneous distribution of cells in the agarose matrix, and (iii) no convective effects through the agarose constructs.

Based on these assumptions, the resulting equation for oxygen concentration within the cylindrical construct is:

$$\frac{\partial c}{\partial t} = D_{eff}\left(\frac{1}{r}\frac{\partial}{\partial r}\left(r\frac{\partial c}{\partial r}\right) + \frac{1}{r^2}\frac{\partial^2 c}{\partial \theta^2} + \frac{\partial^2 c}{\partial z^2}\right) - B\frac{v_{\max} C}{K_M + C} \qquad (2)$$

In the above equation, the parameters are defined as: C the concentration of oxygen; B the viable cell density within the construct; D_{eff} the oxygen effective diffusivity; v_{max} the maximal rate of oxygen consumption by the βTC3 cells; and K_M the Michaelis kinetic constant for oxygen consumption by the cells. The boundary condition is that the concentration of oxygen at the surface, C_S, is the bulk oxygen concentration of the environment.

For βTC3 cells, published experiments found that oxygen tensions above 7 mmHg were required for the cells to retain their secretory capacity [6, 24, 25], therefore this tension was targeted as the minimum threshold ($C_{ox} = 9.5 \times 10^{-3}$ mM). Parameters v_{max} and K_M, which are inherent to the cell type, were estimated at 2 μmol/(10^9cells•min) and 0.01 mM, respectively, based on earlier published work [25-27]. The oxygen diffusivity in the agarose constructs was not experimentally measured in this study; however, the ratio of effective diffusivity in a matrix to the molecular diffusivity in water is reportedly not dependent on molecular size for low molecular weight molecules [28]. Therefore, the oxygen effective diffusivity could be estimated based on the percentages calculated in the glucose and insulin diffusivity experiments, which is approximately 100% for pure agarose and 75% for cell-containing agarose. Using the Wilke-Chang equation to estimate the oxygen diffusivity in water at 37°C to be 1.8×10^{-3} cm^2/min [25]. The oxygen effective diffusivities for the cell-free agarose and cell-containing agarose regions were assumed to be 100% (1.8×10^{-3} cm^2/min) and 75% (1.35×10^{-3} cm^2/min) of the oxygen diffusivity in water at 37°C, respectively.

Three dimensional profiles of dissolved oxygen concentration for three bulk oxygen concentrations, at three different cell densities, are shown in Fig. 2A-C for single agarose constructs and Fig. 3A-C for buffered agarose constructs. Fig. 2D and 3D illustrate the axial profiles for the given bulk oxygen tensions at the three different cells densities. The bulk oxygen concentrations used in the calculations correspond to normoxia (C_{ox} = 0.2 mM), arterial oxygen (C_{ox}= 0.1 mM) determined the maximum cell loading in each construct at the three bulk oxygen tensions, where all the cells are at or above the 7 mmHg oxygen threshold. These results are summarized in Table 1

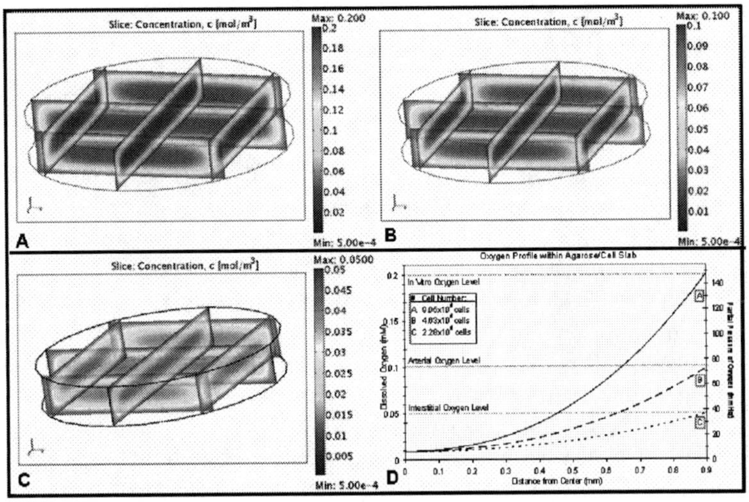

FIGURE 2. Model predicted oxygen profiles for single agarose disk constructs, where the oxygen level at center of construct is set to 7mmHg, is shown for normoxia (A), arterial (B) and approximated interstitial oxygen (C) conditions. (D) represents the oxygen axial profile through the construct.

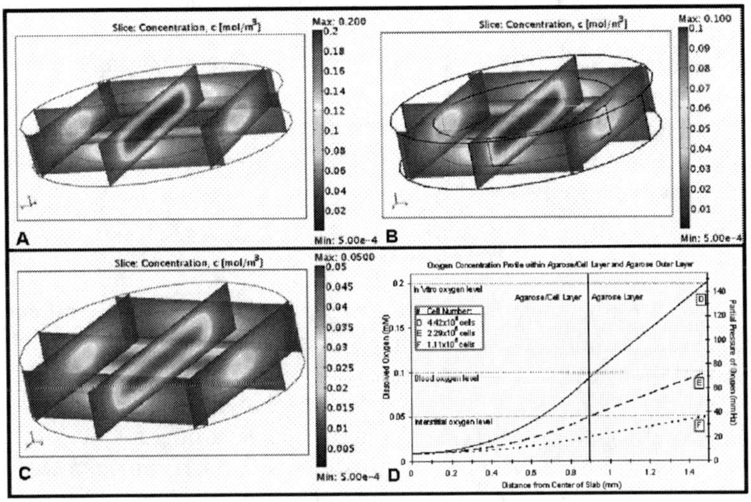

FIGURE 3. Model predicted oxygen profiles for buffered agarose disk constructs, where the oxygen level at center of construct is set to 7mmHg, is shown for normoxia (A), arterial (B) and approximated interstitial oxygen (C) conditions. (D) represents the oxygen axial profile through the construct.

TABLE 1. Predicted maximum cell densities (or total number) in constructs

Construct Type	Normoxia (0.2 mM)	Arterial oxygen (0.1 mM)	Interstitial oxygen (0.05 mM)
Single	$4.6x10^7$ cell/mL agar or $9.05x10^6$ cells	$2.35x10^7$ cell/mL agar or $4.63x10^6$ cells	$1.16x10^7$ cell/mL agar or $2.28x10^6$ cells
Buffered	$2.24x10^7$cell/mL agar or $4.42x10^6$ cells	$1.16x10^7$ cell/mL agar or $2.29x10^6$ cells	$5.65x10^6$ cell/mL agar or $1.11x10^5$ cells

3.3. *In vitro* construct characterization

The experimental *in vitro* temporal profiles of viable cell number, measured by MTT for single agarose constructs are shown in Fig. 4A. Three initial cell loadings were selected for study to correspond to approximately that supported by a normoxic environment ($10x10^6$ cells), followed by one lower ($7.0x10^6$ cells) and one higher ($14x10^6$ cells) than modeling predictions. Viable cell number results (Fig. 4A) show an increase in viable cell number for all of the constructs during the first 24 hrs, with a decline in viable cell number over the subsequent days towards an equilibrium value of approximately $7.5x10^6$ cells by Day 5. Comparable experiments using buffered agarose constructs involved two different initial cell loadings, one approximately equal to that calculated by the model to be supported by normoxia ($3x10^6$cells) and a higher one ($7x10^6$ cells), Fig. 4B. For the latter, the model predicts that only 58% of cells are exposed to oxygen levels at or above 7 mmHg. Constructs initially loaded with $7x10^6$ cells exhibited a decline in viable cell number, which stabilized to $\sim1.6x10^6$ cells at the end of the experiment. In constructs loaded with $3x10^6$ cells, the viable cell number slowly declined over the first 7 days before plateauing at $\sim1.6x10^6$ cells.

Histological sectioning of both construct types found changes in the viable cell regions from a homogeneous distribution on Day 0 to a preference for viable cells on the outer regions of the construct by Day 3. This remodeling was more pronounced in the buffered agarose constructs, where the band of viable cells on the periphery was reduced (images not shown).

3.4. *In vivo* construct characterization

All agarose implants were easily explanted from the peritoneal cavity and the host response was moderate and similar to that seen in the cell-free implants for the implantation times examined in this study (14 days). In all cases, the host cells were relegated to the periphery of the agarose material and did not penetrate into the gel. Furthermore, the entrapped βTC3 cells were unable to migrate out of the outer gel layer. Measurements of viable cell number within the buffered agarose disk implants were collected over the course of two weeks using the MTT assay. Three cell loadings within the buffered agarose disk constructs were selected for study ($4x10^6$ cells, $8x10^6$ cells, and $14x10^6$cells), all of which were higher than the cell loading predicted by the mathematical model. Temporal assessment of viable cell number for all three cell loadings tested, collected after explantation of the construct from normal C57BL/6J mice, is shown in Fig. 4C. The cell loadings plateaued to approximately the same viable cell number for all implants, equal to 1.3-2.0x10^6 cells. Histological cross-sections of explanted constructs showed that by Day 3 viable cells were primarily relegated to a small band at the periphery of the cell-containing region (image not shown). In vivo cell viability measurements were not performed on the single agarose disks implants, because host cells affixed to the outer interface could not be specifically detached without also removing a portion of the embedded βTC3 cells.

4. DISCUSSION

This study focused on the design, development and characterization of disk-shaped agarose constructs containing βTC3 cells as a pancreatic substitute and the use of modeling to predict viable cell number under varying experimental conditions. Initial studies examined the diffusion of critical solutes, in particular glucose and insulin, and determined effective diffusivities in pure 2% and deactivated cell agarose within the range of previously published values for pure and cell-containing hydrogels of similar concentrations [12, 29-33]. The presence of the agarose interface had an undetectable effect on the diffusion of glucose or insulin.

The oxygen profiles developed via mathematical modeling provided a useful estimation of the cellular environment within the agarose constructs. In vitro and in vivo monitoring of the insulin secretion and total viable cell number profiles tested the applicability of the modeling results. After an initial transient period, the ISR and MTT values for all constructs plateaued to statistically similar levels, regardless of the initial cell loading. These results indicate that, regardless of the initial cell seeding, the overall viable cell number with these agarose constructs will plateau at a level that is supportable by the surrounding environment. Therefore, it can be assumed that the agarose biomaterial is a permissive extracellular support that allows the cells to divide, when the environmental conditions are conducive to growth. Furthermore, given the strong agreement between the mathematical model predictions, which were based solely on oxygen diffusion and consumption, and the experimental results, it can be concluded that oxygen is the dominating factor in establishing this viable cell number plateau for these agarose constructs. However, when cellular remodeling is extensive, as seen for the buffered agarose disk constructs after Day 3, the mathematical modeling becomes less predictive.

Future work is focused on refining the construct architecture through the use of three dimensional modeling to increase *in vivo* efficacy. In addition, we will investigate the use of cells death models, as described in the literature [34-35], to further refine our models. In this case, we will further evaluate design parameters that can maximize function in diabetic rodent models, while maintaining the ease in implantation and retrievability afforded by the construct design. The data collected in this study will provide guidelines for future designs, such as appropriate cell loadings, construct architecture and expected secretion levels. Also, the models help refine and validate the modeling tools used for construct design.

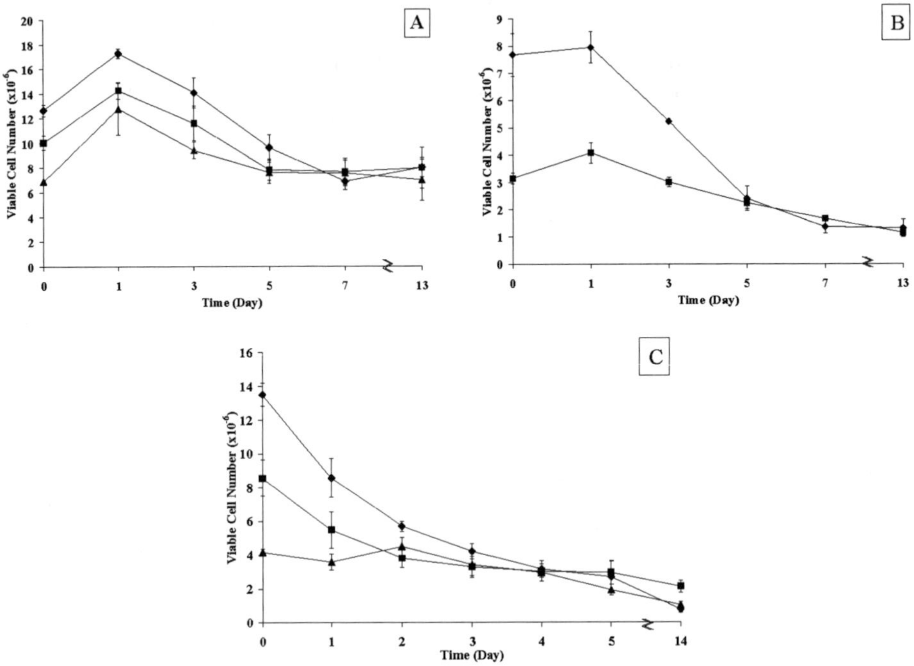

FIGURE 4. A. Temporal changes in MTT-assessed viable cell number for single agarose disk constructs with initial cell loadings of $14x10^6$ cells (diamonds), $10x10^6$ cells (squares), and $7x10^6$ cells (triangles). **B.** Temporal changes in MTT-assessed viable cell number for buffered agarose disk constructs with initial cell loadings of $7x10^6$ cells (diamonds) and $3x10^6$ cells (squares) (B). **C.** Temporal changes in MTT-assessed viable cell number for *implanted* buffered agarose disk implants with initial cell loadings of 4x106 cells (diamonds), 8x106 cells (squares) and 14x106 cells (triangles). Initial cell loading values were based on Trypan Blue measurements made during construct fabrication.

ACKNOWLEDGMENTS

This work was primarily supported by funding from the Georgia Tech/Emory Center (GTEC) for the Engineering of Living Tissues, an ERC Program of the National Science Foundation under Award Number EEC-9731643. Funding for the mathematical software was provided by the Diabetes Research Institute Foundation (DRIF). This financial support is greatly appreciated.

REFERENCES

1. (1993) The effect of intensive treatment of diabetes on the development and progression of long-term complications in insulin-dependent diabetes mellitus. The Diabetes Control and Complications Trial Research Group. *N Engl J Med* 329, 977-986.
2. Ryan, E. A., Lakey, J. R., Paty, B. W., Imes, S., Korbutt, G. S., Kneteman, N. M., Bigam, D., Rajotte, R. V., and Shapiro, A. M. (2002) Successful islet transplantation: continued insulin reserve provides long-term glycemic control. *Diabetes* 51, 2148-2157.

3. Ryan, E. A., Lakey, J. R., Rajotte, R. V., Korbutt, G. S., Kin, T., Imes, S., Rabinovitch, A., Elliott, J. F., Bigam, D., Kneteman, N. M., Warnock, G. L., Larsen, I., and Shapiro, A. M. (2001) Clinical outcomes and insulin secretion after islet transplantation with the Edmonton protocol. *Diabetes* 50, 710-719.
4. Efrat, S. (2001) Prospects for treatment of type 2 diabetes by expansion of the beta-cell mass. *Diabetes* 50, S189-190.
5. Efrat, S., Linde, S., Kofod, H., Spector, D., Delannoy, M., Grant, S., Hanahan, D., and Baekkeskov, S. (1988) Beta-cell lines derived from transgenic mice expressing a hybrid insulin gene-oncogene. *Proc Natl Acad Sci U S A* 85, 9037-9041
6. Papas, K. K., Long, R. C., Constantinidis, I., and Sambanis, A. (1996) Effects of oxygen on metabolic and secretory activities of beta TC3 cells. *Biochim Biophys Acta* 1291, 163-166.
7. De Vos, P., Hamel, A. F., and Tatarkiewicz, K. (2002) Considerations for successful transplantation of encapsulated pancreatic islets. *Diabetologia* 45, 159-173.
8. Iwata, H., Takagi, T., Amemiya, H., Shimizu, H., Yamashita, K., Kobayashi, K., and Akutsu, T. (1992) Agarose for a bioartificial pancreas. *J Biomed Mater Res* 26, 967-977.
9. Jain, K., Asina, S., Yang, H., Blount, E. D., Smith, B. H., Diehl, C. H., and Rubin, A. L. (1999) Glucose control and long-term survival in biobreeding/Worcester rats after intraperitoneal implantation of hydrophilic macrobeads containing porcine islets without immunosuppression. *Transplantation* 68, 1693-1700.
10. Lahooti, S., and Sefton, M. V. (2000) Agarose enhances the viability of intraperitoneally implanted microencapsulated L929 fibroblasts. *Cell Transplant* 9, 785-796.
11. van Schilfgaarde, R., and de Vos, P. (1999) Factors influencing the properties and performance of microcapsules for immunoprotection of pancreatic islets. *J Mol Med* 77, 199-205.
12. Yang, H., Iwata, J., Shimizu, H., Takagi, T., Tsuji, T., and Ito, F. (1994) Comparative studies of in vitro and in vivo function of three different shaped bioartificial pancreases made of agarose hydrogel. *Biomaterials* 15, 113-120
13. Stabler, C. L., Long, R. C., Jr., Constantinidis, I., and Sambanis, A. (2005) In vivo noninvasive monitoring of a tissue engineered construct using 1H NMR spectroscopy. *Cell Transplant* 14, 139-149
14. Stabler, C.L., Development of noninvasive methods for monitoring tissue engineered constructs using nuclear magnetic resonance, Ph.D. Thesis, Georgia Institute of Technology; 2004.
15. Stabler, C. L., Long, R. C., Sambanis, A., and Constantinidis, I. (2005) Noninvasive measurement of viable cell number in tissue-engineered constructs in vitro, using 1H nuclear magnetic resonance spectroscopy. *Tissue Eng* 11, 404-414
16. Carslaw, H., and Jaeger, J. (1959) *Conduction of Heat in Solids*, Clarendon
17. Crank, J. (1975) *The mathematics of diffusion*, JW Arrowsmith Ltd
18. Longsworth, L. (1953) Diffusion Measurements at 25C of Aqueous Solutions of Amino Acids, Peptides and Sugars. *J Am Chem Soc* 75, 5705
19. Ashcroft, F., and Ashcroft, S. (1992) *Insulin : molecular biology to pathology*, IRL Press, Oxford
20. Ochoa, E., and Vacanti, J. P. (2002) An overview of the pathology and approaches to tissue engineering. *Ann N Y Acad Sci* 979, 10-26
21. Kaihara, S., and Vacanti, J. P. (1999) Tissue engineering: toward new solutions for transplantation and reconstructive surgery. *Arch Surg* 134, 1184-1188.
22. Wu, H., Avgoustiniatos, E. S., Swette, L., Bonner-Weir, S., Weir, G. C., and Colton, C. K. (1999) In situ electrochemical oxygen generation with an immunoisolation device. *Ann N Y Acad Sci* 875, 105-125.
23. Dulong, J. L., Legallais, C., Darquy, S., and Reach, G. (2002) A novel model of solute transport in a hollow-fiber bioartificial pancreas based on a finite element method. *Biotechnol Bioeng* 78, 576-582
24. Papas, K. K., Long, R. C., Constantinidis, I., and Sambanis, A. (2000) Effects of short-term hypoxia on a transformed cell-based bioartificial pancreatic construct. *Cell Transplant* 9, 415-422.
25. Tziampazis, E., and Sambanis, A. (1995) Tissue engineering of a bioartificial pancreas: modeling the cell environment and device function. *Biotechnol Prog* 11, 115-126.
26. Tziampazis, E. (1993) Engineering functional, insulin-secreting cell systems: effect of entrapment on cellular environment and secretory response. In *Chemical Engineering* p. 115, Georgia Institute of Technology, Atlanta
27. Wohlpart, D., Kirwan, D., and Gainer, J. (1990) Effects of cell density and glucose and glutamine levels on the respiration rates of hybridoma cells. *Biotechnol Bioeng* 36, 630-635
28. Westrin, B., and Alxelsson, A. (1991) Diffusion in gels containing immobilized cells: a critical review. *Biotechnol Bioeng* 38, 439-445
29. Chresand, T., Dale, B., and Hanson, S. (1988) A Stirred Bath Technique for Diffusivity Measurements in Cell Matrices. *Biotechnol Bioeng* 32, 1029-1036
30. Lundberg, P., and Kuchel, P. W. (1997) Diffusion of solutes in agarose and alginate gels: 1H and 23Na PFGSE and 23Na TQF NMR studies. *Magn Reson Med* 37, 44-52.
31. Schantz, E., and Lauffer, M. (1962) Diffusion Measurements in Agar Gel. *Biochemistry* 1, 658-663
32. Hannoun, B., and Stephanopoulos, G. (1986) Diffusion coefficients of glucose and ethanol in cell-free and cell-occupired calcium alginate membranes. *Biotechnol Bioeng* 28, 829-835

33. Pu, H., and Yang, R. (1988) Diffusion of sucrose and yonimbine in calcium alginate ge beads with or without entrapped plant cells. *Biotechnol Bioeng* 32, 891-896
34. Gross JD, Long, RC Jr., Constantinidis, I and Sambanis, A, in press. Monitoring of dissolved oxygen and cellular bioenergetics within a pancreatic substitute Biotechnol. Bioeng.,
35. Gross, J. D., Constantinidis, I., and Sambanis, A. (2007) Modeling of encapsulated cell systems. *J Theor Biol* 244, 500-510

Presence of nonlinearity in intracranial EEG recordings: detected by Lyapunov exponents [1]

Chang-Chia Liu*,†, Deng-Shan Shiau**, W. Art Chaovalitwongse‡, Panos M. Pardalos†,* and J.C Sackellares**

*Department of Biomedical Engineering, University of Florida, Gainesville, FL 32611
†Department of Industrial and Systems Engineering, University of Florida, Gainesville, FL 32611
**Optima Neuroscience, Inc. 101 SE 2nd Place, Suite 201-A, Gainesville, FL 32601
‡Department of Industrial and Systems Engineering, Rutgers University, Piscataway, NJ 08854

Abstract. In this communication, we performed nonlinearity analysis in the EEG signals recorded from patients with temporal lobe epilepsy (TLE). The largest Lyapunov exponent (Lmax) and phase randomization surrogate data technique were employed to form the statistical test. EEG recordings were acquired invasively from three patients in six brain regions (left and right temporal depth, sub-temporal and orbitofrontal) with 28–32 depth electrodes placed in depth and subdural of the brain. All three patients in this study have unilateral epileptic focus region on the right hippocampus(RH). Nonlinearity was detected by comparing the Lmax profiles of the EEG recordings to its surrogates. The nonlinearity was seen in all different states of the patient with the highest found in post-ictal state. Further our results for all patients exhibited higher degree of differences, quantified by paired t-test, in Lmax values between original and its surrogate from EEG signals recorded from epileptic focus regions. The results of this study demonstrated the Lmax is capable to capture spatio-temporal dynamics that may not be able to detect by linear measurements in the intracranial EEG recordings.

Keywords: Lyapunov Exponents, Surrogate Data, Epilepsy
PACS: 89.20.Kk, 87.85.-d

1. INTRODUCTION

In general, EEG captures the spatio-temporal information of the underlying neurons activities nearby the recording electrodes. It is widely accepted and reasonable to hypothesize that the EEG recordings involved non-linear mechanisms at microscopic level. Many studies have shown the presences and the uses of of nonlinearity and nonlinear measurements. Casdagli et al. (1996) reported nonlinearity found using correlation intergal and surrogate data technique in intracranial EEG recordings from two patients with temporal lobe epilepsy(TLE)[1]. Andrzejak et al. (2006) showed by focusing on nonlinearity and a combination of nonlinear measures with surrogates appears as the key to a successful characterization of spatial distribution of epileptic process[2]. B. Weber et al. (1998) also shown evidences for the usefulness of nonlinear time series analysis for the characterization of the spatio-temporal dynamics of the primary epileptogenic area in patients with TLE. Le Van Quyen et al. (1998) showed that use of nonlinear dynamical interdependence for localization of the epileptogenic regions[3]. With correlation

[1] This work is partially supported by Optima Neuroscience, Inc. 101 SE 2 Place, Suite 201-A, Gainesville, FL 32601

CP953, *Data Mining, Systems Analysis, and Optimization in Biomedicine*
edited by O. Seref, O. E. Kundakcioglu, P. M. Pardalos

dimension K. Lehnertz et al. (1995) reported by extraction of a single value quantifying the dimension variance of interictal activity allowed the primary epileptogenic area to be laterized in exact agreement with the results of the presurgical work-up[4]. Despite of the above positive outcomes of using nonlinear dynamical measurements, one of the major drawbacks for applying nonlinear analysis is the computation complexity. Before launching to nonlinear analysis domain, it is important to address the fundamental motivations. The Lyapunov exponents has been introduced by Iasemides et al. (1990) to the field of epilepsy research, with proper implementation it has been shown to have ability of predicating incoming epileptic seizures[5, 6, 7, 8, 9, 10, 11, 12, 13]. To exam the applicability and sensitivity when using Lmax for complex signals(e.g. EEG signals). In this presented communication, phase randomization surrogate data technique was used for generating surrogate EEG recordings. By randomizing the phase of the Fourier amplitudes, all information which is not contained in the power spectrum is lost. The surrogate EEG will have the same linear properties thus the same power spectrum and equal coefficients of a linear autoregressive (AR) model. If Lmax of the EEG are significantly different from its surrogates, it is the evidence for presences of nonlinearity which is not captured by power spectrum. Moreover, presences of nonlinearity is not only meeting the needs of using nonlinear analysis tools but also demonstrate applicability of Lmax on the EEG analysis. The aim of this study was therefore to test existence of the differences between Lmax values from original EEG and its surrogates.

This chapter is organized as follows. In Section 2 the data information is given. In Section 3 we review the surrogate data technique and brief descriptions of Lyapunov exponents. In Section 4 statistical test on our numerical results are presented followed by conclusions in Section 5.

2. DATA INFORMATION

In this section we will present the background of EEG we used in this study. The EEG was acquired bilaterally placed depth and subdural electrodes. Patients with medically refractory partial seizures of mesial temporal origin were analyzed to investigate the existence of nonlinearity. EEG signals were recorded using amplifiers with input range of 0.6 mV, and a frequency range of 0.5–70Hz. The recordings were stored digitally on high fidelity videotape with a sample rate 200 Hz using an analog to digital (A/D) converter with 10 bit quantization. This multi-electrode (28–32 common reference channels) EEG recording setting was intend to obtain long-term 5 to 6 hours continuous recordings. Three patients were included in this study (see Table 1). Each of the EEG recordings contained 1 or 2 well-defined complex partial seizures with about 30s seizure duration. The EEG recordings were interpreted and reviewed by board–certified electroencephalograph technologists. A Video/EEG monitoring report file was made by electroencephalograph technologists for scientific study purposes. In report files, the state of the patient and EEG activities during inter ictal, ictal and post ictal periods were interpreted in detail (i.e. epileptic focus regions, type of seizure etc.) For analysis and demonstration purposes, two electrodes from each region (ROF, LOF, RST, LST, RTD, LTD) were included in the analysis.

TABLE 1. EEG Data description

	Gender	Age	Epileptic Focus Region RH/LH	Length of EEG
P1	M	19	RH	6.1
P2	M	60	RH	5.4
P3	F	41	RH	5.8

3. METHODS

We will first start from nonstationarity and follow up on two main tools of this study in next two subsections. Nonstationarity is an fundamental difficulty in field of time series analysis. As result of existing nonstationarity in a measured time series, there exist no invariant measures. Nonstationarity will cause problems for many algorithms to have difficulties when trying interpret the results of an invariant measure. Most of the time, one can try to remove the nonstationarity by using variety of filters or divided the time series into a number of shorter length and assume the underlying dynamics to be approximately stationary within each divided segment. In this study, prior to calculate Lmax the EEG recordings were first divided into non–overlapping window of 10.24 second duration to avoid the nonstationarity. The same segmentation method has been used by Iasemidis et al. such segmentation technique is often applied especially for medical time series[5]. In the following method section, we will first review the surrogate data technique and then brief Lmax calculate will be given in the following subsection.

3.1. Surrogate data technique

In this subsection, we reviewed the background of surrogate data techniques. The surrogate data techniques were first suggested by Theiler et al. (1992). Surrogate data will preserve major features of the original series but destroying their nonlinear components. Using surrogate data methods, one can test a time series for relationship of a specific class of dynamical systems. There are three different procedures for surrogate data generation[14]. They are described as follows.

1) Surrogates are realizations of independent identically distributed (*iid*) random variables with the same mean, variance, and probability density function as the original data. The *iid* surrogates were generated by randomly permuting in temporal order the samples of the original series. This shuffling process will destroy the temporal information and thus generated surrogates are mainly random observation drawn (without replacement) from the same probability distribution as original data. 2) Fourier transform (FT) surrogates are constrained realizations of linear stochastic processes with the same power spectra as the original data. FT surrogate series were constructed by computing the FT of the original series, by substituting the phase of the Fourier coefficients with random numbers in the range while keeping unchanged their modulus, and by applying the inverse FT to return to the time domain. To render completely uncoupled the surrogate pairs, two independent white noises where used to randomize the Fourier phases. 3)

Auto regressive (AR) surrogates are typical realizations of linear stochastic processes with the same power spectra as the original series. By generating a Gaussian time series with the same length as the data, and reordered it to have the same rank distribution. Take the Fourier transform of this and randomize the phases (FT). Finally, the surrogate is obtained by reordering the original data to have the same rank distribution as the inverse Fourier transform. These surrogates were obtained by fitting an AR model to each of the two original series, using pairs of independent white noises as model inputs to produce completely uncorrelated surrogate series.

In this presented study, we employed the second algorithm to generate the surrogate data. By shuffling the phases but keeping the amplitude of the complex conjugate pairs at the same time the surrogates will have the same power spectrum (autocorrelation) as the data, but will have no nonlinear determinism. Ten surrogates will be produced to insured Fourier phases are completely randomized. This surrogate algorithm has been applied and combining with correlation integral,correlation integral, a measure sensitive to a wide variety of non-linearities was used for detection for nonlinearity by Casdigali et al. (1996)[1].

3.2. Lyapunov exponents

The brief formulations and the parameters for obtaining the Lyapunov exponents (Lmax) in this study is given in this subsection. The Lmax calculation is describe elsewhere in Iasemides (1990)[5]. The first step in obtaining Lmax is phase space reconstruction using method of delay[15]. By applying the method of delay we obtain the time delay vector. The time-delay vector is as follows:

$$X_i = (x(t_i), x(t_i+\tau), \ldots x(t_i+(p-1)*\tau)). \tag{1}$$

Fig. 1 shows the embedded EEG time series to a higher dimensional space of dimension p, (p=7). $X(t)$, whose components consist of values of the recorded EEG signal $x(t)$ for p points in time separated by a time delay τ, (τ=5). The method for estimation of the largest Lyapunov Exponent for nonstationary data (e.g., EEG time series) is previously explained in [5, 16, 17]. In this section, we will only give a short description and basic notation of our mathematical models used to estimate STL_{max}. First, let us define the following notation.

- N_a is the number of local L_{max}'s that will be estimated within a duration T data segment. Therefore, if D_t is the sampling period of the time domain data, $T = (N-1)D_t = N_a\Delta t + (p-1)\tau$.
- $X(t_i)$ is the point of the fiducial trajectory $\phi_t(X(t_0))$ with $t = t_i$, $X(t_0) = (x(t_0), \ldots, x(t_0 + (p-1)*\tau))$, and $X(t_j)$ is a properly chosen vector adjacent to $X(t_i)$ in the phase space.
- $\delta X_{i,j}(0) = X(t_i) - X(t_j)$ is the displacement vector at t_i, that is, a perturbation of the fiducial orbit at t_i, and $\delta X_{i,j}(\Delta t) = X(t_i+\Delta t) - X(t_j+\Delta t)$ is the evolution of this perturbation after time Δt.

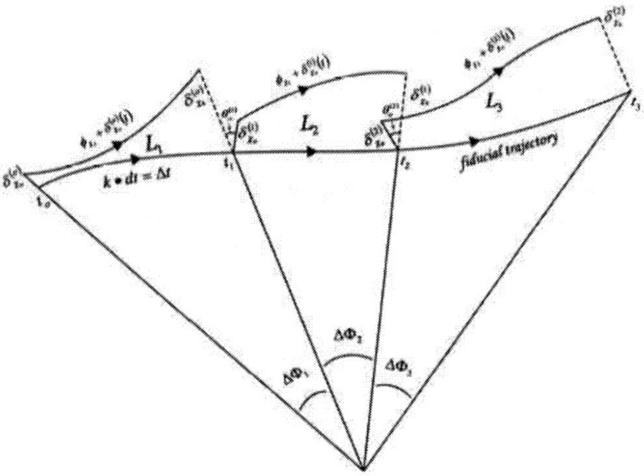

FIGURE 1. Diagram illustrates the embedded EEG time series in phase space used for the estimation of Lmax.

- $t_i = t_0 + (i-1) * \Delta t$ and $t_j = t_0 + (j-1) * \Delta t$, where $i \in [1, N_a]$ and $j \in [1, N]$ with $j \neq i$.
- Δt is the evolution time for $\delta X_{i,j}$, that is, the time one allows $\delta X_{i,j}$ to evolve in the phase space. If the evolution time Δt is given in second, then L is in bits per second.
- t_0 is the initial time point of the fiducial trajectory and coincides with the time point of the first data in the data segment of analysis. In the estimation of L, for a complete scan of the attractor, t_0 should move within $[0, \Delta t]$.

Let L be an estimate of the short term maximum Lyapunov exponent, defined as the average of local Lyapunov exponents in the state space. L can be calculated by the following equation:

$$L = \frac{1}{N_a \Delta t} \sum_{i=1}^{N_a} \log_2 \frac{|\delta X_{i,j}(\Delta t)|}{|\delta X_{i,j}(0)|}. \qquad (2)$$

4. STATISTICAL TEST AND RESULTS

The comparison between the STLmax value estimated from original EEG recordings and the mean of its 10 surrogates in three different time periods: 1. interictal (normal) 2. per-ictal (per-seizure) 3. postictal (post-seizure). The mean of Lmax value from 10 surrogates are taken to construct 95% confident interval. This hypothesis testing will exam whether there exits any significant difference in Lmax values between the original and surrogated EEG recordings. This 95% confident interval follows the t-distribution with degree of freedom equal to 9; the critical value of t–distribution is 2.26. The null hypothesis will be rejected if the Lmax values estimated from the original EEG recordings falls outside the 95% confident interval. Further if rejected, the original EEG recordings is considered lost information after performing the surrogate technique. The mean of T-index value for each 10.24 s epochs is served as an indicator of how far away (dissimilar) the original EEG recordings are from the surrogated EEG (null hypothesis). This statistical testing is derived from the well-known paired-t test for comparisons for each paired-dependent observation, it was employed here as a measure of statistical distance between pairs of Lmax profiles over a time window. [11] The T_{ij} index at time t between the Lmax profiles of i and j is defined as:

$$T_{ij}(t) = |\overline{D_{ij}^{t}}| \times \frac{\hat{\sigma_{ij}} t^{t}}{\sqrt{N}}. \tag{3}$$

where $||\overline{D_{ij}}^{t}||$ denotes the absolute value of the average of all paired differences

$$|\overline{D_{ij}}^{t}| = (L_i^t - L_j^t) \mid t \in w(t)) \tag{4}$$

over a moving window $w(t)$ defined as

$$w(t) = [\frac{t}{T} - N + 1, \frac{t}{T}] \tag{5}$$

where N denotes number of Lmax in the moving window and $\hat{\sigma_{ij}}^{t}$ denotes the standard deviation of the sample D_{ij} within $w(t)$. Asymptotically, $T_{ij}(t)$ follows the t-distribution with $N-1$ degree of freedom. We used $N = 30$ (i.e. averages of 30 paired differences of Lmax values per moving window). Since each Lmax value was derived from a 10.24 second of EEG epoch, the length of our moving window is approximately about 5 minutes. A critical value $T_{\frac{\alpha}{2}}$ from t-distribution with N-1 (=29) degrees of freedom at significance level $\alpha = 0.05$ was used to test the null hypothesis H_0. For the T-index to reject H_0 ,$T_{ij}(t)$ should be greater than 2.045 ($T_{(0.005,29)}$).

Our study results were plotted in three different figures for three different state. Fig. 2, 3, 4 show the Lmax profile from both original and its surrogates respectively, for analysis purposes, here we define interictal state is the time duration at least 60 mins prior to seizure onset, perictal is the time duration 10 mins prior to seizure onset, finally postictal is 5 mins after seizure onset. The results clear showed the nonlinearity was detected in three different states, this further clarified Lmax was capable to capture information which is not contained in linear AR model. The defferences between original Lmax and it surrogates was found largest in the EEG recorded from epileptic focus regions.

5. CONCLUSIONS

We compared Lmax values between original EEG and its surrogates for every 10.24 (2048 EEG points) second shorter segment of EEG recordings. The results of our approach successfully detected the presence of nonlinearity in our intracranial EEG recordings. Further more, EEG signals recorded from epileptic focus regions exhibited promising statistical differences from its surrogates. Detecting the nonlinearity in the scalp or intracranial EEG recordings is not unfamiliar in the field of epilepsy research, only

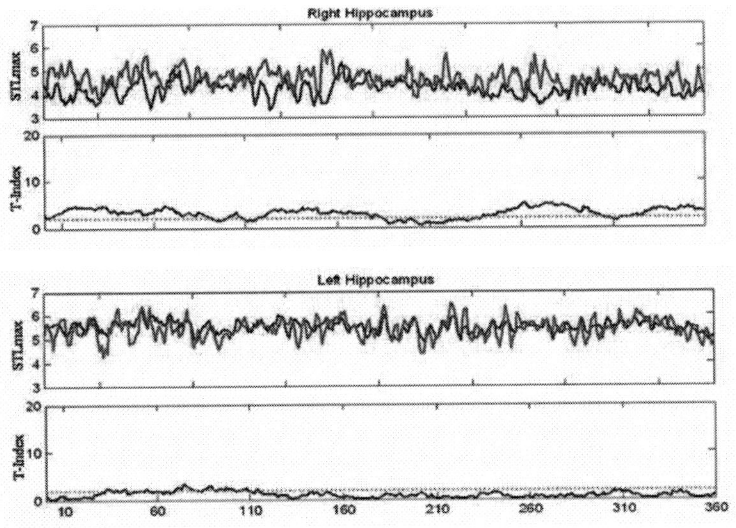

FIGURE 2. Lmax and T-index profiles during inter-ictal (360 points= 60minutes). EEG was recorded from right hippocampus (top) and left hippocampus (bottom). Red lines denotes the Lmax obtained from the original EEG, blue line denotes the mean of Lmax obtained from 10 surrogates. T-index curves demonstrate the statistical difference between two Lmax profiles over time.

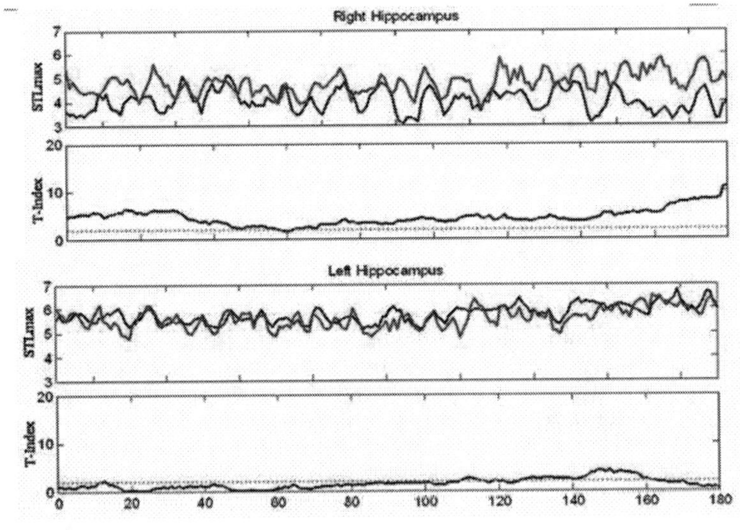

FIGURE 3. Lmax and T-index profiles during per-ictal (180 points= 30minutes). EEG was recorded from right hippocampus (top) and left hippocampus (bottom). Red lines denotes the Lmax obtained from the original EEG, blue line denotes the mean of Lmax obtained from 10 surrogates. T-index curves demonstrate the statistical difference between two Lmax profiles over time.

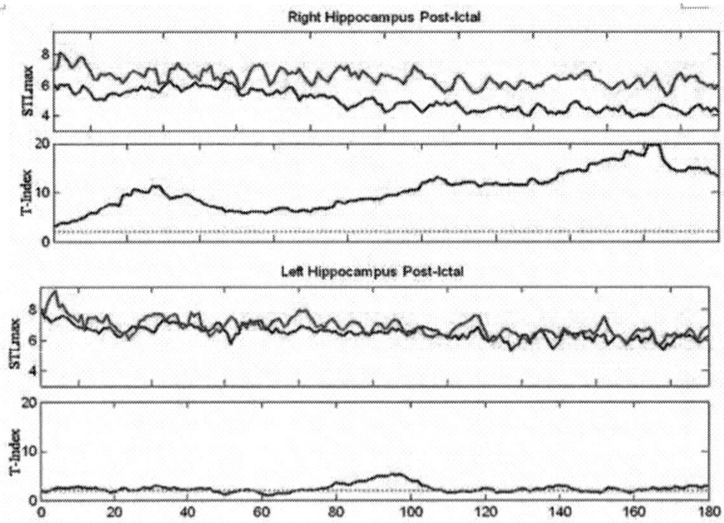

FIGURE 4. Lmax and T-index profiles during post-ictal (180 points= 30minutes). EEG was recorded from right hippocampus (top) and left hippocampus (bottom). Red lines denotes the Lmax obtained from the original EEG, blue line denotes the mean of Lmax obtained from 10 surrogates. T-index curves demonstrate the statistical difference between two Lmax profiles over time.

combining Lmax and surrogate data techniques for detecting nonlinearity in intracranial EEG recording is the first time. Our study results consistent with many pervious documented studies using varieties nonlinear or linear analysis tools [18, 19, 20, 21, 22, 23]. In conclusion of this study, we want to state that the nonlinear measurement Lmax can be significantly applicable with much favorable outcomes than linear measurements for the use of EEG analysis. Finally, Once the suitability is satisfied, the findings of this study could be use to serve for many therapeutic purposes (e.g. per-surgical evaluation, locate positions for deep brain stimulation). In order to achieve these objectives, further examination of our findings in a larger sample size is needed and samples should included not only confined with individuals have only unilateral seizure focus region but also bilateral subjects. More importantly, testing our proposed approach in the EEG recordings recorded from scalp is also an very important step, from practical point of view it makes our approach much valuable and useful.

REFERENCES

1. M. Casdagli, L. Iasemidis, J. Sackellares, S. Roper, R. Gilmore, and R. Savit, *Physica D* **99**, 381–399 (1996).
2. R. Andrzejak, F. Mormann, G. Widman, T. Kreuz, C. Elger, and K. Lehnertz, *Epilepsy Research* **69**, 30–44 (2006).
3. B. Weber, K. Lehnertz, C. E. Elger, and H. G. Wieser, *Epilepsia* **39(9)**, 922–927 (1998).
4. K. Lehnertz, and C. Elger, *Electroencephalography and Clinical Neurophysiology* **95(2)**, 108–117 (1995).

5. L. Iasemidis, *On the dynamics of the human brain in temporal lobe epilepsy*, Ph.D. thesis, University of Michigan, Ann Arbor (1991).
6. L. Iasemidis, H. Zaveri, J. Sackellares, and W. Williams, *Brain Topography* **2**, 187–201 (1990).
7. L. Iasemidis, A. Barreto, R. Gilmore, B. Uthman, S. Roper, and J. Sackellares, *Epilepsia* **35S**, 133 (1993).
8. L. Iasemidis, K. Pappas, R. Gilmore, S. Roper, and J. Sackellares, *Epilepsia* **37S**, 90 (1996).
9. L. Iasemidis, J. Principe, J. Czaplewski, R. Gilmore, S. Roper, and J. Sackellares, "Spatiotemporal transition to epileptic seizures: a nonlinear dynamical analysis of scalp and intracranial EEG recordings," in *Spatiotemporal Models in Biological and Artificial Systems*, edited by F. Silva, J. Principe, and L. Almeida, IOS Press, 1997, pp. 81–88.
10. L. Iasemidis, J. Sackellares, R. Gilmore, and S. Roper, *Epilepsia* **39S**, 207 (1998).
11. J. Sackellares, D.-S. Shiau, J. Principe, M. Yang, L. Dance, W. Suharitdamrong, W. Chaovalitwongse, P. Pardalos, and L. Iasemidis, *Journal of Clinical Neurophysiology* **23**, 509–520 (2006).
12. W. Chaovalitwongse, R. Sachdeo, P. Pardalos, L. Iasemidis, and J.C.Sackellares, *Epilepsia* (To appear).
13. J. Sackellares, L. Iasemidis, D.-S. Shiau, L. Dance, P. Pardalos, and W. Chaovalitwongse, *International Patent Application filed August 2003, Attorney Docket No. 028724–142* (2003).
14. J. Theiler, S. Eubank, A. Longtin, B. Galdrikian, and J. Farmer., *Physica D* **58**, 77–94 (1992).
15. F. Takens, "Detecting strange attractors in turbulence," in *Dynamical Systems and Turbulence, Lecture Notes in Mathematics*, edited by D. Rand, and L. Young, Springer-Verlag, 1981, vol. 898, pp. 366–381.
16. S. Nair, P. Pardalos, and V. Yatsenko, *Journal of Optimization Theory and Application* **134 (3)**, 533–547 (2007).
17. P. Pardalos, and V. Yatsenko, *Journal of Optimization Theory and Application* **128(1)**, 29–48 (2006).
18. A. Tuunainen, U. Nousiainen, A. Pilke, E. Mervaala, J. Partanen, and P. Riekkinen, *Epilepsia* **36**, 817–823 (1995).
19. A. Gambardella, J. Gotman, F. Cendes, and F. Andermann, *Epilepsia* **36**, 122–129 (1995).
20. J. Pijn, J. van Neerven, A. Noest, and F. L. da Silva, *Clin. Neurophysiol.* **79**, 371–381 (1991).
21. G. Widman, K. Lehnertz, H. Urbach, and C. Elger, *Epilepsia* **41**, 811–817 (2000).
22. F. Mormann, K. Lehnertz, P. David, and C. Elger, *Physica D* **144**, 358–369 (2000).
23. D. Panet-Raymond, and J. Gotman, *Electroencephalogr. Clin. Neurophysiol.* **75**, 474–481 (1990).

Quantification of the impact of vagus nerve stimulation parameters on electroencephalographic measures

Michael Bewernitz*, Georges Ghacibeh†, Onur Seref*,
Panos M. Pardalos*,**, Chang-Chia Liu** and Basim Uthman‡,§

*Department of Industrial and Systems Engineering, University of Florida, Gainesville, FL, 32611, USA.
†Northeast Regional Epilepsy Group 20 Prospect Ave, Suite 800 Hackensack, NJ, 07601, USA.
**Department of Biomedical Engineering, University of Florida, Gainesville, FL, 32611, USA.
‡Neurology Services, North Florida/South Georgia Veterans Health System, Gainesville, FL, 32608, USA.
§Departments of Neurology and Neuroscience, the University of Florida College of Medicine, Gainesville, FL, 32611, USA.

Abstract. This study presents an application of support vector machines (SVMs) to the analysis of electroencephalograms (EEG) obtained from the scalp of patients with epilepsy implanted with the vagus nerve stimulator (VNS) used in VNS Therapy®. The purpose of this study is to devise a physiologic marker using scalp EEG for determining optimal VNS parameters. Scalp EEG recordings were obtained from six patients with history of intractable partial onset epilepsy treated with VNS as adjunctive therapy to medicines. Averaged scalp EEG samples were used as features for separation. SVM classification accuracy was used as a measure of EEG similarity to separate a time segment during the beginning of stimulation from all the successive non-overlapping time segments within a full VNS on/off cycle. This analysis was performed for all the automated VNS cycles occurring during approximately twenty-four hours of 25 channels of scalp EEG. The patient that resulted in the lowest degree of EEG pattern similarity had the highest VNS stimulation frequency and experienced a monthly seizure rate among the lowest of all six patients included in this study. The patient with the greatest degree of pattern similarity had the lowest VNS stimulation frequency, shortest VNS pulse width, and experienced the greatest monthly seizure rate of all six patients included in this study. It is possible that VNS exerts its therapeutic effect by mimicking a theorized seizure effect for which a seizure has been observed to "reset" the brain from an unfavorable preictal state to a more favorable interictal state. These encouraging results suggest that data mining tools may be able to extract EEG patterns which could be used as an electrographic marker of optimal VNS stimulation parameters.

Keywords: Support Vector Machine, Vagus Nerve Stimulation, Data Mining, Quantitative EEG Analysis, Epilepsy
PACS: 89.20.Kk, 87.80.-y

1. INTRODUCTION

1.1. Epilepsy

Epilepsy is a disorder of the central nervous system that affects about 1% of the world's population with the highest incidence among infants and the elderly [1]. This disorder is characterized by recurrent seizures induced by the rapid development of syn-

CP953, *Data Mining, Systems Analysis, and Optimization in Biomedicine*
edited by O. Seref, O. E. Kundakcioglu, P. M. Pardalos

chronous neuronal firing. Seizure symptoms can include uncontrollable shaking, loss of awareness, visual and aural hallucinations, phantom odors, and other sensory disturbances depending on the location and spread of electrical activity in the brain [2]. Seizures can have a major effect on the patient's quality of life by imposing restricted driving privileges, adverse effects on social and career opportunities, self-esteem, education, and psychiatric issues [3]. A 1995 study estimated that epilepsy imposed an economic burden of $12.5 billion in associated health care costs and losses in employment, wages, and productivity in the U.S. [4].

Epilepsy treatment today is mostly treated with anti-epileptic drugs (AEDs), however approximately 30% of patients with epilepsy have seizures that are resistant to AED therapy [5]. Surgical treatment may be considered in extreme cases where the patient does not achieve adequate benefit from AED therapy [6]. The mean duration of inpatient hospital stay for pre-surgical EEG monitoring ranges from 4.7 to 5.8 days and costs over $200 million each year [4]. In order to be a candidate for surgery, the first criterion is that the patient must have seizures arising from a single seizure focus in the brain. In addition, the excision procedure cannot be performed on brain regions that are essential for normal functioning of the patient [7]. Even if as many as 50% of these patients were to benefit from surgical resection (an optimistic estimate), many patients still require new therapeutic approaches [5].

1.2. Vagus nerve stimulation

For many years there have been attempts to control medically intractable epileptic seizures by electrically stimulating the brain. Examples of electrical stimulation strategies include deep brain stimulation, transcranial magnetic stimulation, and chronic vagus nerve stimulation (VNS) [5]. VNS has been subject of much study since its approval for treatment of intractable seizures in 1997 [8, 9, 10]. It is suggested that VNS treatment has a much lower incidence of cognitive, neurological, and systemic effects than that of AED treatments [11]. At the end of 2004, over 32,000 epilepsy patients had VNS therapy® implants [8]. The device consists of an electric stimulator implanted subcutaneously in the left side of the chest and connected, via subcutaneous electrical wires, to the left cervical vagus nerve. The VNS is programmed to deliver electrical stimulation at a set intensity, duration, pulse width, and frequency for the treatment of epilepsy. Though VNS therapy has demonstrated therapeutic efficacy for seizure treatment, the mechanism of therapeutic action is still uncertain [9, 12, 13]. However, a physiologic basis for VNS therapeutic mechanism has been described [14]. The left vagus nerve is the 10th cranial nerve and composed of approximately 90% afferent fibers, whose function is to carry sensory information to the brain. The vagus nerve is involved in the control of numerous autonomic systems such as the intestines, the heart, and plays a role in respiration. Though the vagus nerve has widespread connections throughout the body, a study discovered no changes in physiologic markers for gastric, cardiac, or pulmonary function were found as a result of VNS [12].

Studies have suggested that metabolic activation of thalamic, brainstem, and limbic regions may be important in mediating VNS effects [12]. There is some evidence that

VNS exerts its effect via the locus coeruleus [13]. Also, a recent study discovered that depletion of norepinephrine in the locus coeruleus attenuates the AED-like effect of VNS [12].

Experimental stimulation of the vagus afferent neurons has caused cortical synchronization or desynchronization in cats, depending on the conduction velocity of the fibers [14]. Recent studies have determined that the activation of the vagus A and B fibers leads to synchronization of the EEG whereas vagus C fiber activation causes EEG desynchronization [13]. In short, these and other VNS effects have been reported but there is no consensus of the extent to which these empirical effects influence epilepsy. Though studies have explored how the stimulation of vagus A, B, and C fibers affect EEG synchronization, the overall VNS-induced effects of these fibers and other brain structures on EEG are not grossly visible in the time domain [12] or in frequency domain [15]. In general, there is no agreement on how the VNS affects EEG [15, 16].

With little knowledge of the mechanism of action or a rapid measure of efficacy, the current process of tuning VNS stimulation parameters in newly implanted patients is essentially based on the physician's experience and trial-and-error. This process consists of setting initial stimulation parameters then iteratively adjusting stimulation parameters based on patient reports of clinical efficacy (seizure frequency) and tolerability since the previous visit to the doctor. It is known that "high" settings for stimulation parameters (such as output current, frequency, and pulse width) are shown to result in greater seizure reduction than "low" settings [9]. However, there is currently no way to avoid the "guess-and-check" process for fine tuning stimulation parameters. Locating the optimal VNS parameters more rapidly would save money due to fewer doctor's office visits and would reduce the risk of seizures. Thus, it is desirable to devise a method to rapidly and objectively predict the efficacy of the numerous combinations of VNS stimulation parameters. A first step towards this goal is to identify a measurable quantity that is sensitive to the patient's therapeutic response to VNS therapy. There is reason to believe that such an effect could be present in the EEG signal.

1.3. EEG markers for treatment of neurological diseases and disorders

Recent research has demonstrated the utility of electrographic markers for the diagnosis and treatment of neurological disorders including the diagnosis, treatment, and evaluation treatment for epilepsy [17, 18, 19, 20]. Though there are no consistent electroencephalographic VNS effects reported in the literature [12, 16], [15], the non-invasive nature of scalp-EEG, state of the art of knowledge discovery in databases [21], and the potential for real-time application due to modern computer technology make EEG analysis a desirable approach to studying VNS effect. For this reason the current research aims to identify an electrographic marker that is sensitive to the configuration of stimulation parameters. If such an electrographic marker could be identified, and if the electrographic markers could also be correlated to clinical efficacy then such a marker could be used to determine optimal VNS stimulation parameters on a case-by-case basis.

1.4. Studying brain behavior in multidimensional feature space

One approach for studying aspects of brain behavior is to create a mathematical model of some brain function using measurable quantities, such as electrical signals. These models can allow predictions to be made about the behavior of sections of the brain. Due to the massive complexity of the interacting neurons in the brain, direct models have only been achieved for small groups of neurons and small brain structures [22, 23]. The enormous complexity of electrical brain signals often means exact modeling approaches are insufficient. The challenge of direct modeling is compounded by the fact that, as was stated earlier, no consistent VNS-induced effect in the raw EEG or time-frequency profile has been reported [15, 12].

Recent studies have suggested that therapeutic epilepsy treatments (e.g. anticonvulsants and electrical stimulation therapy) can mimic certain aspects of seizures in animal models of epilepsy. In particular, the brain resetting phenomenon accompanying epileptic seizures has been demonstrated after both anti-convulsant and electrical stimulation therapy [24, 25]. The term "resetting" is used to describe the theorized effect of a seizure to reset the brain from an unstable preictal state to a more stable interictal state [26, 27]. The electroencephalographic correlate of the theorized resetting effect manifests as a disentrainment of critical electrode sites (where disentrainment refers to two critical electrodes reaching a threshold "statistical distance" quantified by the t-index).

Another method to quantify the similarity of patterns in the EEG data is the similarity index. Similarity index has demonstrated success for neural state classification using EEG signals inputs [28, 29]. While the similarity index and STLmax t-index measures have identified preictal transitions by measuring the level of difference between two EEG time series vectors, recent studies have extended these ideas to include multidimensional (multichannel) pattern similarity measures.

A reasonable approach to extracting complex patterns from multidimensional datasets (such as EEG) is to utilize data mining techniques. Data mining refers to the application of algorithms to extracting patterns from large databases [21] and provides reasonable tools for extracting any patterns which may be sensitive to electric stimulation parameters for the VNS implant. Data mining techniques have demonstrated successful detection of scalp EEG patterns which may be difficult for the human eye to visualize [30, 18, 19], [26, 31, 20], [32, 33]. Often such algorithms can be utilized in real time, providing a basis for real-time EEG analysis tool [17, 31], [27, 32].

Given the degree of success in determining similarity of different EEG segments in real time [18, 32] it is possible that support vector machines may be able to provide a computationally efficient measure of similarity which can quantify any stimulation-induced effect on EEG signals. As was stated earlier, evidence exists indicating that electrical stimulation mimics the electroencephalographic resetting effect of a seizure. Thus, a reasonable approach for examining VNS therapy is from the framework that it mimics the effect of a seizure. For these reasons, the use of support vector machines to calculate a measure of similarity of an EEG segment during stimulation to subsequent EEG segments may provide a rapid means to investigate and quantify electroencephalographic effects of VNS.

The goal of this study is to extract EEG patterns which could be used as an electroencephalographic marker of optimal VNS stimulation parameters. Our hypothesis is "electroencephalographic effects accompanying vagus nerve stimulation are correlated with stimulation parameters".

2. METHODS

2.1. EEG acquisition

The scalp EEG analyzed in preliminary studies was acquired at 512 Hz analog-to-digital conversion rate with a 0.16 Hz high pass filter and 105 Hz low pass filter hardwired into the amplifier. The 25 EEG channels are fp1, fp2, f3, f4, c3, c4, p3, p4, o1, o2, f7, f8, t3, t4, t5, t6, a1, a2, fz, cz, pz, leye, reye, lmn, and rmn ("mn" refers to an electrode placed in the mandibular notch located between the skull and jaw bone, which provides information about the nearby temporal lobe) as shown in Fig. 1.

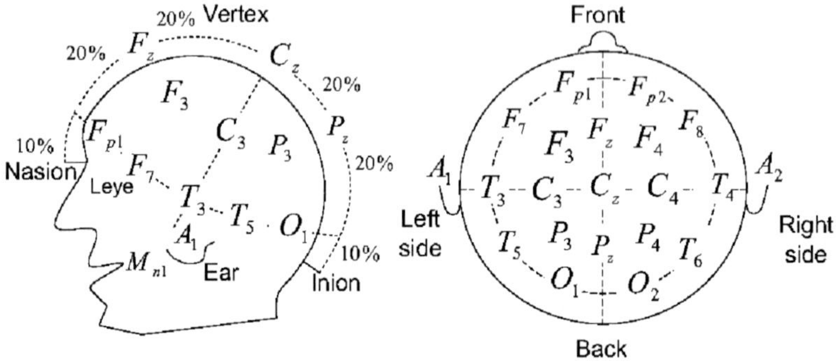

FIGURE 1. The international 10-20 system for electrode placement. Electrodes are spaced proportionally (e.g. 10%-20%) to the size of the patient's head.

The PG electrode, located between Cz and Fz, is the reference for the EKG and 25 EEG channels. The study utilized EEG from six epileptic patients with VNS implants at Shands Hospital General Clinical Research Center (GCRC protocol # 614, "Neurophysiologic Measures of Vagus Nerve Stimulation") at the University of Florida, Gainesville.

2.2. Support vector machines

Support Vector Machines (SVMs) are a class of data classification algorithms used to model and classify large volumes of multivariate data [34]. The SVM algorithm determines the optimal separating hyper surface between two multidimensional datasets. SVMs have been demonstrated success in numerous applications where neural states were classified from EEG data [30, 18, 32].

The SVM formulation for linearly separable data quantifies the distance from the hyper surface as expression (1) for each data point in the positive class and expression (2) for each data point in the negative class.

$$\langle \mathbf{w}, \mathbf{x}^+ \rangle + b \tag{1}$$

$$-(\langle \mathbf{w}, \mathbf{x}^- \rangle + b) \tag{2}$$

In this context, the term "margin" will refer to the distance between the separating hyper surface and the nearest point for either class. The margin of the function output is referred to as the functional margin. The geometric margin is the functional margin of a normalized weight vector. Hence the geometric margin can be optimized by fixing the functional margin to be equal to one and then minimizing the norm of the weight vector [35]. If w is the weight vector for a functional margin of distance equal to one from a positive (reference) class point, x^+ and a distance equal to one from a negative (comparison) class point, x^-, then the geometric margin can be determined as follows. For a functional margin equal to one:

$$\langle \mathbf{w}, \mathbf{x}^+ \rangle + b = +1 \tag{3}$$

$$\langle \mathbf{w}, \mathbf{x}^- \rangle + b = -1 \tag{4}$$

To calculate the geometric margin, w must be normalized. The geometric margin, γ, is then the functional margin of the resulting classifier

$$\gamma = \frac{1}{2}(\langle \frac{\mathbf{w}}{||\mathbf{w}||_2}, \mathbf{x}^+ \rangle - \langle \frac{\mathbf{w}}{||\mathbf{w}||_2}, \mathbf{x}^- \rangle) \tag{5a}$$

$$= \frac{1}{2||\mathbf{w}||_2}(\langle \mathbf{w}, \mathbf{x}^+ \rangle - \langle \mathbf{w}, \mathbf{x}^- \rangle) \tag{5b}$$

$$= \frac{1}{||\mathbf{w}||_2} \tag{5c}$$

Thus, the interclass margin is maximized by

$$\min \quad ||\mathbf{w}||_2 \tag{6a}$$

$$\text{subject to} \quad y_i(\langle \mathbf{w}, \mathbf{x}^+ \rangle + b) \geq 1 \qquad \forall i = 1, \ldots, n \tag{6b}$$

where the constraints are that all points are properly classified for positive and negative classes, as stated in (6b). For non-separable data, the objective function (7a) combines the term for maximizing interclass margin with a term for minimizing misclassification error [35].

$$\min \quad ||\mathbf{w}||^2 + C \sum_{i=1}^{n} \xi_{2,i} \qquad \forall i = 1, \ldots, n \tag{7a}$$

$$\text{subject to} \quad y_i(\langle \mathbf{w} \cdot \mathbf{x}_i \rangle + b) \geq 1 - \xi_i \qquad \forall i = 1, \ldots, n \tag{7b}$$

$$\xi_i \geq 0 \qquad \forall i = 1, \ldots, n \tag{7c}$$

The cost term, C, in (7a) is the weight assigned to the error. Expressions (7b) and (7c) represent the constraints that each misclassified point is assigned a linear penalty (slack variable) and that all slack variables are non-negative, respectively. The SVM formulation in (7a), (7b), and (7c) utilizes a linear hyper surface for discrimination. A common practice for improving separation accuracy is to perform a transformation to remap the data from input space into feature space using a kernel function. The inner product is then performed on the transformed feature space data (see equation (8)).

$$K(\mathbf{x}_i, \mathbf{x}_j) = \langle \phi(\mathbf{x}_i), \phi(\mathbf{x}_j) \rangle \quad (8)$$

The purpose of kernel mapping is to view the data in a transformed space to allow class distinction which may not have been possible in input space. The radial basis function (9) is a standard choice as a kernel function [36].

$$K(\mathbf{x}_i, \mathbf{x}_j) = \exp(-\frac{||\mathbf{x}_i - \mathbf{x}_j||^2}{\sigma}) \quad (9)$$

The SVM is trained using a data set with a known classification. The SVM's performance is measured in terms of the accuracy at which it is able to classify a test data set. In other words, SVMs utilize optimization characteristics to generate a sophisticated and robust measure the amount of separation between two clusters of points in a multidimensional feature space representation. In contrast to the traditional SVM application where the goal is to obtain maximal separation on a case-by-base basis by adjusting any and all parameters, this study fixes the cost(C) and "bandwidth" of the radial basis function (σ) in an effort to provide objectivity to this measure which can be used to study the evolution of EEG feature space similarity over time. SVM distinguishing accuracy of 50% is interpreted as maximum similarity between reference class and comparison class. SVM distinguishing accuracy of 100% is interpreted as the minimum similarity between the reference class and comparison class.

2.3. Experimental design

EEG pattern similarity was quantified as the amount of separation achieved with support vector machines between two EEG segments both mapped into feature space using the radial basis function kernel. This exploratory study examines how the evolution of EEG feature space pattern similarity throughout VNS stimulation is related to stimulation parameters. This study was performed on scalp EEG recordings from six epileptic patients. Patient information can be found in Tables 1 and 2.

Approximately 24 hours of 512 Hz scalp EEG was recorded and analyzed for six epileptic patients treated with AEDs and VNS therapy. The VNS stimulation times were obtained by examining the EKG channel. The EKG electrode was positioned on the skin over the VNS pulse generator so that VNS stimulations appeared as a high frequency waveform superimposed on the EKG rhythm. The EKG channel was excluded from SVM analysis. The following procedure was applied to each stimulation epoch (an epoch defined as a VNS "on" cycle and one "off" cycle) for each patient. A segment of EEG occurring from 10 seconds to 20 seconds after the start of the VNS stimulation was

TABLE 1. VNS Stimulation Parameters at Time of Study

Patient	A	B	C	D	E	F
Output Current (mA)	1.75	1.5	2.5	2	1.25	0.75
Stimulation Frequency (Hz)	30	20	25	30	20	30
Pulse Width (μs)	500	250	500	500	500	750
On time (seconds)	30	30	30	30	30	30
Off time (minutes)	5	5	5	3	5	3
Magnet Output Current (mA)	2	1.75	2.75	2	1.5	0.75
Magnet On Time (seconds)	60	60	60	30	60	60
Magnet Pulse width (μs)	500	500	500	500	500	750

selected as the positive class (from here on this class will be referred to as the "reference class"). The duration of each class was set at 10 seconds because this time frame is considered long enough to be able to quantify brain dynamics yet short enough to be approximately stationary [37, 38]. The reference segment was selected to start at one full window length after the start of stimulation because that allows the stimulator to have completed the ramp-up phase. The EEG segment ranging from 20 seconds to 30 seconds was selected as the first negative class (from here on this class will be referred to as the "comparison class"). The first preprocessing step was to down sample both classes to decrease SVM classification time. This was achieved by calculating the mean value of a moving-average non-overlapping window [32]. The smoothing window was set at 10 points in duration based on prior experience. Next, each channel was normalized to zero mean and unit standard deviation to prevent channels with greater mean amplitudes from dominating the classification decision [30]. LIBSVM software package for Matlab [39] was used to train and test the SVM classifier. An RBF kernel with σ=39 was utilized for pre-classification feature space transformation and a cost parameter of C=1000 were applied based on studies utilizing RBF SVM to distinguish neural states from scalp EEG data [30, 40, 41]. The SVM was trained and tested using v-fold cross validation, as described in [42]. Briefly, the first step is to divide the training set into v subsets of equal size. The SVM is trained using v-1 segments and is tested using the remaining subset v. This process is repeated until all v subsets have been tested. The accuracy is the percentage of properly classified points from each of the v trials. Based on prior experience, v=2 folds was used for SVM training and testing. The training accuracy was then recorded for the particular reference class and comparison class combination. The comparison class was then advanced one full window length and the SVM was trained to classify the original reference segment from the new classification segment. This process of computing SVM accuracy then advancing the comparison class was repeated until the time point at the end of the comparison segment is $\geq$ the time of next stimulation (into the next epoch). This process is completed for each VNS epoch in the 24 hour recording period for each patient.

3. RESULTS

The results for patients A, B, C, D, E, and F are shown in Fig. 2,3,4,5, and Table 3.

TABLE 2. Seizure frequency and types, concurrent medications, EEG discharge localization, and duration of VNS therapy in participating patients (CPS=Complex Partial Seizure; 2° GTCS=secondarily generalized tonic-clonic seizure)

Patient	A	B	C	D	E	F
Age	38	53	54	29	54	54
Region of Focal EEG Discharge	Right Frontal Left Frontal	Right Frontal	Right Temporal	Left Frontal	Right Fronto-Parito-Temporal	Right Temporal Left Temporal
Mean seizures per month	3	30	2	3	0	0
# of seizures during recording	0	14	0	0	0	0
Seizure Type	CPS with occasional 2° GTCS	CPS with occasional 2° GTCS	CPS with occasional 2° GTCS	CPS with occasional 2° GTCS	CPS with occasional 2° GTCS	CPS with occasional 2° GTCS
Duration of VNS therapy (years)	> 1 year	> 1 year	> 1 year	> 1 year	> 1 year	> 1 year
Medications	gabapentin lamotrigine levetiracetam pregabalin	levetiracetam phenobarbital topiramate	gabapentin topiramate	carbamazepine lamotrigine levetiracetam zonisamide	clorazepate gabapentin	pregabalin levetiracetam

4. DISCUSSION

These results suggest a correlation between pulse width and SVM separation accuracy (Fig. 2) as well as between stimulation frequency and SVM separation accuracy (Fig. 3). Furthermore, there is less similarity (greater SVM separation accuracy) between the reference class (10 second EEG segment during stimulation) and each subsequent comparison class (all subsequent non-overlapping 10 second EEG segments prior to the next stimulation) for higher values of pulse width and stimulation frequency than occurred with the lower values. As Table 3 shows, these patients with the lowest similarity following stimulation also have fewer seizures per month. The results of this study did not demonstrate any correlation with output current (Fig. 4), a rather important stimulation parameter which along with pulse width is has been reported to be correlated with level

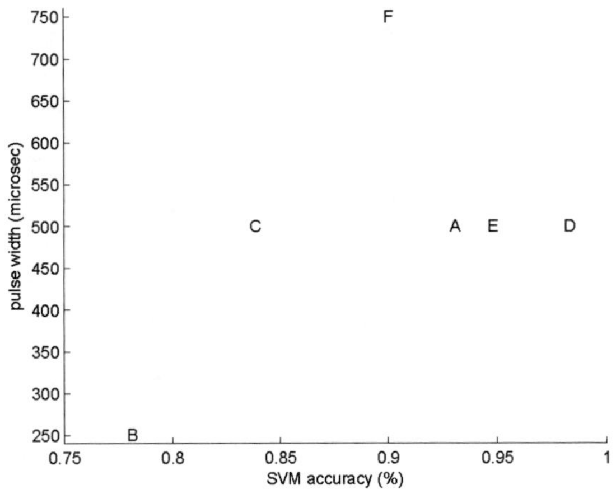

FIGURE 2. VNS pulse width (microseconds) plotted against mean SVM accuracy over 24 hours.

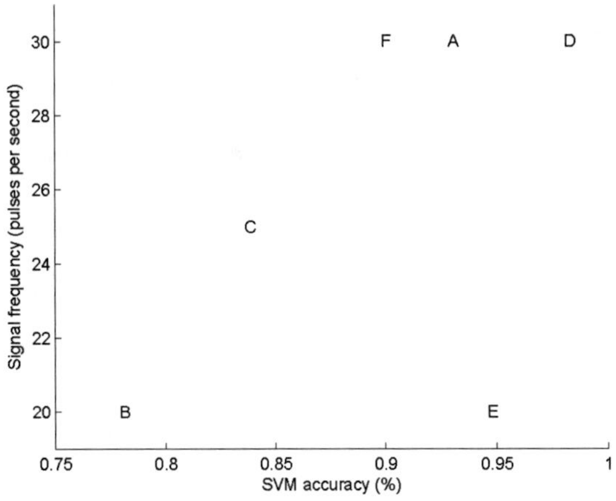

FIGURE 3. VNS stimulation frequency (pulses per second) plotted against mean SVM accuracy over 24 hours.

of seizure protection [9].

From the perspective that the VNS mimics the seizure effect of "resetting" the brain from an unfavorable preictal state to a more favorable interictal state [26, 27], the decrease in EEG similarity (signified by higher SVM separation accuracy) between the reference class and all subsequent comparison classes in an epoch could be analogous to disentrainment of STLmax t-indicies after epileptic seizures. Thus, an explanation for the observed feature space similarity phenomena could be that these measures are

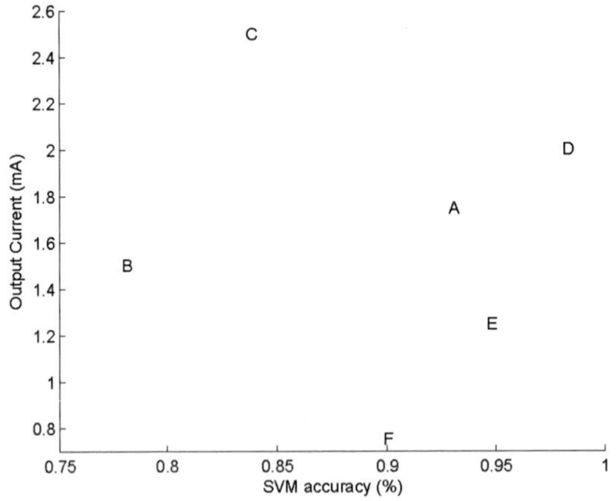

FIGURE 4. VNS output current (milliamps) plotted against mean SVM accuracy over 24 hours.

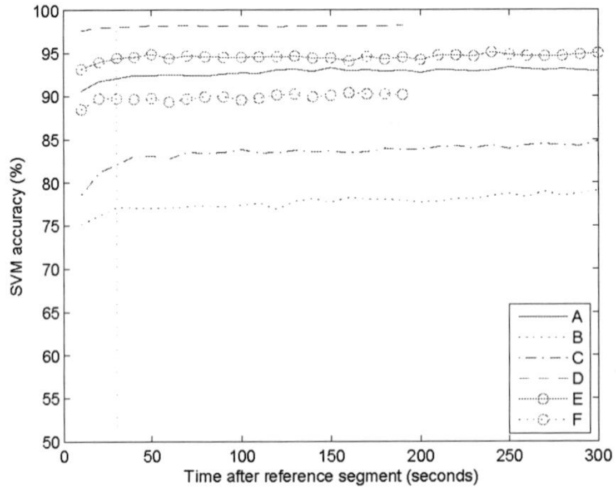

FIGURE 5. SVM accuracy averaged across the corresponding comparison segment delay for all stimulations over the duration of the recording (24 hours for each patient). The vertical line at 30 seconds represents the time when the VNS stimulator deactivates. Patients D and F have fewer comparison segments as their off time is 3 minutes (thus the time between stimuli is shorter than the rest of the patients which have an off time of 5 minutes). Patients E and F are seizure free.

electroencephalographic evidence that VNS is resetting the brain more effectively (as denoted by the lower similarity measures after stimulation) in the patients with fewer seizures per month.

TABLE 3. Mean SVM separation accuracy and patient seizure information

Patient	Mean SVM separation	Seizures per month	Seizures during recording
A	0.9284	3	0
B	0.7788	30	14
C	0.8359	2	0
D	0.9807	3	0
E	0.9454	0	0
F	0.8982	0	0

It should be noted the time each patient's last seizure prior to the EEG recording session for this study is unknown. The time of last seizure could prove to be important for the non-seizure free patients (A,C and D), as there is a period of reduced seizure susceptibility immediately after experiencing a generalized tonic-clonic seizure. If patients A, C, and D had experienced a generalized tonic clonic seizure shortly before the recording session for this study, it is possible that they may still have been under the influence of the reduced seizure susceptibility phenomenon which may explain why they did not experience any seizures during this study. Furthermore, if patient B had not experienced a generalized tonic clonic seizure recently then that could be a factor contributing to the large number of seizures reported this patient underwent during the study. It is also worth noting that all six VNS patients are considered responders to the therapy defined as having at least 50% seizure reduction after one year of VNS therapy [43].

Differences in drug medications could also have an effect on the results, as patient D (whom demonstrated the lowest average similarity) is the only patient using either carbamazepine or zonisamide and patient B (whom demonstrated the highest average similarity) is the only patient taking phenobarbital. Thus, it is possible that these differences in antiepileptic medications could alter the observed electroencephalographic effects.

According to Fig. 5, there was no significant EEG effect of acute stimulation detected with this measure of EEG similarity. This could be due to the fact that the SVM computational parameters used to generate this EEG similarity measure are not optimized for this application. It should also be noted that all patients have have undergone VNS therapy for >1 year thus their vagus nerve/brain may have adapted such that they do not demonstrate a significant acute EEG response to individual stimulations.

5. FUTURE DIRECTION

Additional EEG features should be utilized for possible improvements in EEG similarity measurements. For example, complexity measures such as the short-term maximum Lyapunov exponent and entropy measures have been shown to be useful features when using data mining tools to extract information about brain dynamics [18].

A necessary step for furthering the understanding of EEG effects accompanying VNS is to track these measures in patients before, immediately after, and at least six months after implantation. This would provide insight into the evolution of EEG similarity patterns before and after numerous stimulation parameter adjustments. Such an analysis would help determine if there were any electroencephalographic characteristics present prior to VNS implantation which may influence feature space similarity measures.

Also, studying the effect of epileptic seizures on this measure of EEG feature space similarity would help determine how robust this method is. In addition, a study of the contribution of the various EEG channels to the outcome, e.g. a focal electrode group versus a non-focal electrode group, could provide insight into the spatial effect of VNS. Patients whom are receiving VNS for depression treatment could serve as useful control subjects to determine the effect of VNS stimulation on a non-epileptic brain. The potential knowledge gained from these suggested studies could lead us one step closer to the creation of an EEG marker for optimal VNS parameters.

ACKNOWLEDGMENTS

General Clinical Research Center, Grant # MO1-RR00082
The Evelyn F. & William L. McKnight Brain Institute of the University of Florida
Neurology Services, North Florida / South Georgia Veterans Health System, Gainesville, FL, 32608, USA.

REFERENCES

1. J. Engel Jr., and T.A. Pedley, *Epilepsy: a Comprehensive Textbook*, Philadelphia, PA: Lippincott-Raven, 1997.
2. D. Weaver, *Epilepsy and Seizures: Everything You Need to Know*, Buffalo, NY: Firefly Books Inc., 2001.
3. A. N. Wilner, *Epilepsy: 199 Answers: A Doctor Responds to His Patients' Questions*, New York: Demos Vermande, 1996.
4. C. E. Begley, M. Famulari, J. F. Annegers, D. R. Lairson, T. F. Reynolds, S. Coan, S. Dubinsky, M. E. Newmark, C. Leibson, E. L. So, and W. A. Rocca, *Epilepsia* **41**, 342–351 (2000).
5. W. H. Theodore, and R. S. Fisher, *Lancet Neurol* **3**, 111–118 (2004).
6. J. Engel Jr., P. C. Van Ness, T. B. Rasmussen, and L. M. Ojemann, "Outcome with Respect to Epileptic Seizures" in *Surgical Treatment of the Epilepsies*, edited by J. Engel Jr., 1993, pp. 609–622.
7. S. N. Roper, and A. L. Rhoton, *Neurosurgery clinics of North America* **4**, 223–231 (1993).
8. Cyberonics®, Inc., Houston, Texas, *Epilpesy Patient's Manual for Vagus Nerve Stimulation with the VNS Therapy™ System*, Cyberonics®, 2006.
9. E. Ben-Menachem, R. Manon-Espaillat, R. Ristanovic, B. J. Wilder, H. Stefan, W. Mirza, W. B. Tarver, and J. F. Wernicke, *Epilepsia* **35**, 616–626 (1994).
10. B. M. Uthman, B. J. Wilder, J. K. Penry, C. Dean, R. E. Ramsay, S. A. Reid, E. J. Hammond, W. B. Tarver, and J. F. Wernicke, *Neurology* **43**, 1338–1345 (1993).
11. M. S. George, *Epilepsy and Behavior* **2**, 95–100 (2001)
12. R. S. Fisher, and A. Handforth, *Neurology* **53**, 666–669 (1999).
13. D. A. Groves, and V. J. Brown, *Neuroscience and Biobehavioral Reviews* **29**, 493–500 (2005).
14. A. R. M. Upton, and A. M. White, *Pacing and Clinical Electrophysiology* **14**, 50 (1991).
15. M.C. Salinsky, and K. J. Burchiel, *Epilepsia* **34**, 299–304 (1993).
16. E. J. Hammond, B. M. Uthman, S. A. Reid, and B. J. Wilder, *Epilepsia* **33**, 1013–1020 (1992).
17. W. A. Chaovalitwongse, L. D. Iasemidis, P. M. Pardalos, P. R. Carney, D. S. Shiau, and J. C. Sackellares, *Epilepsy Res.* **64**, 93–113 (2005).

18. W. A. Chaovalitwongse, O. A. Prokopyev, and P. M. Pardalos, *Ann Oper Res* **148**, 227–250 (2006).
19. L. D. Iasemidis, and J. C. Sackellares, *Neuroscientist* **2**, 118–126 (1996).
20. P. M. Pardalos, V. Yatsenko, J. C. Sackellares, D. S. Shiau, W. A. Chaovalitwongse, and L. D. Iasemidis, *Computational Statistics and Data Analysis* **44**, 391–408 (2003).
21. A. Flexer, Statistical *Methods in Medical Research* **9**, 395–413 (2000).
22. G. A. Chauvet, and T. W. Berger, *Hippocampus* **12**, 698–712 (2002).
23. A. L. Hodgkin, and A. F. Huxley, *J. Physiol.* **7**, 500–544 (1952).
24. L. B. Good, S. Sabesan, L. D. Iasemidis, K. J. Garvey, and D. M. Treiman, *Epilepsia* **45**, 42–43 (2004).
25. L. B. Good, S. Sabesan, S. T. Marsh, L. D. Iasemidis, and D. M. Treiman, *Epilepsia* **46**, 330–330 (2005).
26. L. D. Iasemidis, D. S. Shiau, J. C. Sackellares, P. M. Pardalos, and A. Prasad, *IEEE transactions on biomedical engineering* **51**, 2004.
27. J. C. Sackellares, and L. D. Iasemidis, *Epilepsia* **38**, 1–284 (1997).
28. M. Le Van Quyen, J. Martinerie, M. Baulac, and F. Varela, *Neuroreport* **10**, 2149–2155 (1999).
29. V. Navarro, J. Martinerie, M. Le Van Quyen, M. Baulac, F. Dubeau, and J. Gotman, *Epilepsia* **46**, 385–396 (2005).
30. N. Acir, and C. Guzelis, *Neural computing and applications* **14**, 56–65 (2005).
31. L. D. Iasemidis, D. S. Shiau, P. M. Pardalos, W. A. Chaovalitwongse, K. Narayanan, A. Prasad, K. Tsakalis, P. R. Carney, and J. C. Sackellares, *Clin Neurophysiol* **116**, 532–544 (2005).
32. M. Thulasidas,C. Guan, and J. Wu, *IEEE Transactions On Neural Systems And Rehabilitation Engineering* **14**, (2006).
33. J. C. Sackellares, D. S. Shiau, J. C. Principe, M. C. K. Yang, L. K. Dance, W. Suharitdamrong, W. A. Chaovalitwongse, P. M. Pardalos, and L. D. Iasemidis, *Journal of clinical neurophysiology* **23**, 509–520 (2006).
34. C. J. C. Burges, *Data Mining and Knowledge Discovery* **2**, 121–167 (1998).
35. N. Cristianini, and J. Shawe-Taylor, *An introduction to Support Vector Machines and other kernel-based learning methods*, Cambridge University Press, 2000.
36. O. Seref, O. E. Kundakcioglu, and P. M. Pardalos, “Selective linear and nonlinear classification”, in *CRM Proceedings and Lecture Notes*, (forthcoming).
37. M. C. Casdagli, L. D. Iasemidis, R. S. Savit, R. L. Gilmore, S. N. Roper, and J. C. Sackellares, *Epilepsia* **36**, 142 (1995).
38. M. C. Casdagli, L. D. Iasemidis, J. C. Sackellares, S. N. Roper, R. L. Gilmore, and R. C. Savit, *Physica D* **99**, 381–399 (1996).
39. C. C. Chang, and C. J. Lin, LIBSVM : a library for support vector machines, `http://www.csie.ntu.edu.tw/~cjlin/libsvm`, (2001).
40. F. Lotte, M. Congedo, A. L'ecuyer, F. Lamarche, and B. Arnaldi, *J. Neural Eng.* **4**, 1–13 (2007).
41. M. Kaper, P. Meinicke, U. Grossekathoefer, T. Lingner, and H. Ritter, *IEEE transactions on biomedical engineering* **51**, (2004).
42. C. W. Hsu, C. C. Chang, and C. J. Lin, A practical guide to support vector classification, *Technical report*, `http://www.csie.ntu.edu.tw/~cjlin/libsvm/`, Department of Computer Science and Information Engineering, National Taiwan University, Taipei, (2003).
43. M. Morell, *Current Opinion in Neurology* **19**, 164–168 (2006).

Biclustering EEG data from epileptic patients treated with vagus nerve stimulation

Stanislav Busygin*, Nikita Boyko*, Panos M. Pardalos*, Michael Bewernitz† and Georges Ghacibeh**

*Department of Industrial and Systems Engineering, University of Florida, Gainesville, FL, 32611, USA.
†Department of Biomedical Engineering, University of Florida, Gainesville, FL, 32611, USA.
**Norteast Regional Epilepsy Group 20 Prospect Ave, Suite 800 Hackensack, NJ, 07601, USA.

Abstract. We present a pilot study of an application of consistent biclustering to analyze scalp EEG data obtained from epileptic patients undergoing treatment with a vagus nerve stimulator (VNS). The ultimate goal of this study is to develop a physiologic marker for optimal VNS parameters (e.g. output current, signal frequency, etc.) using measures of scalp EEG signals. A time series of $STL_{\max}$ values was computed for each scalp EEG channel recorded from two epileptic patients and used as a feature of the two datasets. The averaged samples from stimulation periods were then separated from averaged samples from non-stimulation periods by feature selection performed within the consistent biclustering routine.

The obtained biclustering results allow us to assume that signals from certain parts of the brain consistently change their characteristics when VNS is switched on and could provide a basis for desirable VNS stimulation parameters. A physiologic marker of optimal VNS effect could greatly reduce the cost, time, and risk of calibrating VNS stimulation parameters in newly implanted patients compared to the current method of clinical response.

Keywords: Consistent biclustering, vagus nerve stimulation, data mining, biomedicine, epilepsy research
PACS: 87.85.D-, 87.85.-d, 87.55.de, 87.55.kd, 87.55.kh

1. INTRODUCTION

Epilepsy is one of the most common nervous system disorders. It affects about 1% of the world's population with the highest incidence among infants and the elderly [6, 8]. The hallmark of epilepsy is recurrent seizures caused by the sudden development of synchronous neuronal firing in the cerebral cortex. Electroencephalograph (EEG) recordings show that these discharges begin either locally in portions of the cerebral hemisphere or simultaneously in both cerebral hemispheres [13, 20, 1]. These seemingly unpredictable seizures can cause a variety of motor symptoms and have major effects on the patient's quality of life if they are not controlled [12].

Epilepsy treatment today is for the most part pharmacological, however surgical treatment is used in medically refractory cases [7]. For many years there have been attempts to control epileptic seizures by electrically stimulating the brain [10]. This alternate method of treatment is the subject of much study since the approval of the chronic Vagus Nerve Stimulation (VNS) implant for treatment of intractable seizures [23, 9, 21]. The device consists of an electric stimulator implanted subcutaneously in the chest and connected, via subcutaneous electrical wires, to the left cervical vagus nerve.

CP953, *Data Mining, Systems Analysis, and Optimization in Biomedicine*
edited by O. Seref, O. E. Kundakcioglu, P. M. Pardalos

The VNS is programmed to deliver electrical stimulation at a set intensity, duration, pulse width, and frequency. Optimal parameters are determined on a case by case basis, depending on clinical efficacy (seizure frequency) and tolerability.

The anticonvulsant mechanism of VNS is unknown. The many studies performed to determine the mechanism have produced inconsistent results. Studies by Lesser and co-workers have, however, suggested that certain types of pulse stimulation reduce the excitability of the cerebral cortex of epileptic patients [18, 19]. It is also suggested that VNS treatment has a much lower incidence of cognitive, neurological, and systemic effects than that of anticonvulsant drug treatments [11].

Though various VNS-induced EEG effects have been reported, there is no general agreement on how the VNS affects EEG [15]. One explanation for the wide range of reported EEG effects is that activation of the vagus *A* and *B* fibers leads to synchronization of the EEG whereas vagus C fiber activation causes EEG desynchronization [14]. These opposing effects on the EEG by the various vagus nerve fibers may effectively cancel each other out, which could explain the lack of a consistent empirical EEG effect. Nevertheless, expression of the VNS effect in terms of scalp-EEG recordings is desirable due to widespread availability at medical facilities and usefulness in diagnosing epilepsy.

The Lyapunov exponent is a measure of signal complexity which when applied to EEG recordings from epileptic patients has demonstrated effectiveness in predicting epileptic seizures [4, 16, 26]. Thus, the successful application of the Lyapunov exponent to EEG data for seizure prediction purposes makes it a great candidate for detecting the EEG effects of VNS stimulation.

Biclustering has been successfully applied to determine the existence of supervised binary classifications in multivariate data. A major application of this method is to establish connections between disease symptoms (e.g., certain types of cancer) and accompanying gene expression patterns [5, 17, 24]. This ability of biclustering to distinguish between two predefined states makes it a potentially useful tool to distinguish between VNS stimulation and VNS deactivation states using EEG data (or measures of EEG data, such as the Lyapunov exponent) as the input.

Though biomedical research is now one of the major applications of data mining, previous biomedical applications of biclustering were limited mostly to DNA microarray analysis [3, 29]. However, it should be noted that deliverables of biclustering are especially valuable for feature extraction purposes which is the major concern of many biomedical problems including the present study. Furthermore, we would like to emphasize the novelty of our studies representing the first application of biclustering to analyzing EEG data.

The current method of adjusting VNS parameters based on the clinical response is time consuming, costly, and delays reaching the optimal parameters, putting patients at increased risk of recurrent seizures. Therefore, there is a need to develop a reliable, objective and rapid method of determining the optimal stimulation parameters for each individual patient, without having to wait for seizures to recur or for side effects to develop. The ultimate goal of this pilot study is to develop a physiologic marker for optimal VNS parameters using measures of scalp EEG signals. The combination of Lyapunov exponents with biclustering classification to model the VNS-induced effects on scalp-EEG recordings could be a useful biomarker of the VNS's therapeutic effect. Clinically, such marker can be used to create an online EEG analysis system to help

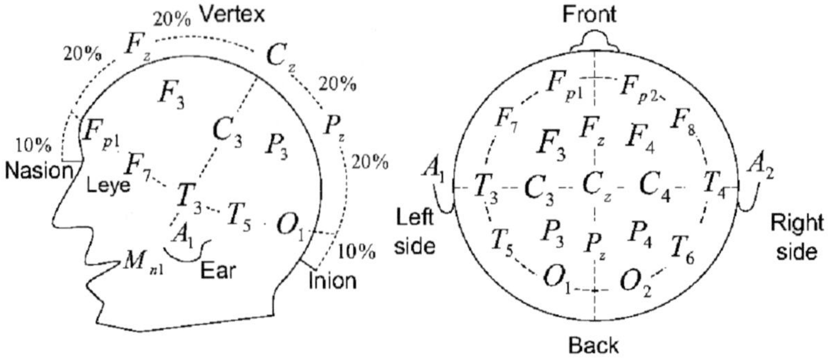

FIGURE 1. Montage for scalp electrode placement

neurologists make more direct, targeted VNS parameter adjustments in newly-implanted patients.

2. ELECTROENCEPHALOGRAM ACQUISITION

This study was conducted on scalp-EEG recordings obtained from two epileptic patients implanted with a functioning VNS system in the General Clinical Research Center (GCRC) in Shands Hospital at The University of Florida. EEG was acquired under GCRC protocol # 614, "Neurophysiologic Measures of Vagus Nerve Stimulation". EEG data was obtained at 512 Hz sampling rate from 26 scalp-EEG channels arranged in the standard international 10-20 system (see Fig. 1). EEG channels included in the study were: fp1, fp2, f3, f4, c3, c4, p3, p4, o1, o2, f7, f8, t3, t4, t5, t6, a1, a2, fz, cz, pz, leye, reye, lmn, and rmn ("mn" refers to an electrode placed in the mandibular notch located between the skull and jaw bone, which provides information about the nearby temporal lobe). ECG (electrocardiogram) channel was excluded from the analysis since it has no connection to the brain activity.

3. VAGUS NERVE STIMULATION SETTINGS

The VNS paradigm is set to 30 seconds stimulation, and a 5 minute non-stimulation epoch, which repeated continuously throughout the entire recording for both patient A and patient B. Patient A's VNS stimulation parameters were 1.75 mA output current, 30 Hz signal frequency, 500 μsec pulse width, 30 second signal duration, and a 5 minute rest duration. Patient B's VNS stimulation parameters were 1.5 mA output current, 20 Hz signal frequency, 250 μsec pulse width, 30 second signal duration, and a 5 minute rest duration. VNS implants allow manual activation of the device in the event a patient requires immediate stimulation (e.g. the patient can sense an imminent seizure). Patient A's manual activation stimulation parameters were 2 mA output current, 30 Hz signal frequency, 500 μsec pulse width, and a 60 second signal duration. Patient A did not undergo any seizures during the recording session, and did not manually activate the stimulator. Patient B's manual activation stimulation parameters were 1.75 mA output current, 20 Hz signal frequency, 500 μsec pulse width, and a 60 second signal duration. Patient B experienced 14 seizures during the recording and manually activated the stimulator approximately 19 to 37 seconds after onset of every seizure. Patient B manually activated the stimulator a total of 14 times during the EEG recording period (once manual activation per seizure for every seizure). The ECG electrode was positioned over the VNS pulse generator which introduced a high-frequency high amplitude artifact to the ECG channel. This explicit artifact introduced into the ECG channel by the VNS pulse generator provides a simple mechanism to visually identify epochs during which the VNS device is active. Each patient was recorded continuously for approximately 24 hours. Patient A had a total of 255 VNS cycles used in the analysis. Patient B had a total of 237 VNS cycles used in the analysis. Some of the VNS cycles were excluded from the study due to occasional severe EEG artifact. Also, the final VNS cycle was excluded from the study in both patients because the cycle began closer than a full cycle length to the end of the recording.

4. LYAPUNOV EXPONENT CALCULATION

Lyapunov Exponents are considered to be one of the most useful dynamical descriptors of chaotic systems. A famous practical application is the Short Term Largest Lyapunov Exponent ($STL_{\max}$) of EEG signal time series. The STLmax measure of EEG has demonstrated effectiveness in predicting epileptic seizures, see [25, 26, 27].

Let the following set of nonlinear of equations describe the behavior of the chaotic system,

$$\dot{X}(t) = F(X), \tag{1}$$

where $X : R \mapsto R^n$, $F : R^n \mapsto R^n$. Lyapunov Exponents are the average exponential rates divergence or convergence of nearby orbits of the system in the phase space. More formally

$$\lambda = \lim_{t \to \infty} \frac{1}{t} \log_2 \frac{\delta X(\Delta t)}{\delta X(0)} \tag{2}$$

In practice, however, explicit equations modeling the system's behavior are not usually available. In these situations it is necessary to compute Lyapunov Exponents from a time series of discrete measurements of one or more of the system's characteristics.

Wolf *et al* [28] proposed an algorithm for estimation largest Lyapunov Exponent from a time series. Iasemidis *et al* [25] developed a modification of Wolf's algorithm for estimating short term largest Lyapunov exponent ($STL_{\max}$) that can cope with nonstationary and nonuniform EEG data. A sketch of the algorithm is provided below. A more detailed explanation and justification can be found in [28, 25].

Input: Time series data $x(t_i)$. Time window size N. Dimension of the reconstructed phase space p. Discretization time Δt. Selected time lag between the component of each vector in the phase space τ. Propagation time ΔT.

Output: Array L of $STL_{\max}$ for each time window.

Define $X_i = (x(t_i), x(t_i+\tau), \ldots, x(t_i+(p-1)\tau))$;

$k = 1$;

for $wnd = 0;\ wnd < size(X) - N;\ wnd += N$ **do**

Set $r = wnd$;

Select s, $\delta_0 = X_r - X_s$ such that

$$|\delta_0| = \min_{\substack{s\in[wnd,wnd+N],\\ |r-s|>(p-1)*\tau}} |X_r - X_s|$$

Set $L_k = 0$;

repeat

Set $\delta_1 = X_{r+\Delta T} - X_{s+\Delta T}$

Set $L_k = L_k + \log_2(|\delta_1|/|\delta_0|)$;

Find $\Delta_{j,\max}$ for $j = 1,\ldots,p$ such that

$$\Delta_{j,\max} = \max_{\substack{i\in[wnd,wnd+N],\\ \tau<|r-i|<(p-1)*\tau}} |X_{r,j} - X_{i,j}|$$

Find a δ_0^{new} vector with relatively small magnitude which is also close to parallel to the previous vector δ_0That is, we need to find s and $\delta_0^{new} = X_r - X_s$ such that $(\delta_0^{new}, \delta_0) < a \cdot V_{\max}$ and $l \cdot \Delta_{j,\max} < |\delta_{0,j}^{new}| < u \cdot \Delta_{j,\max}$ If δ_0^{new} is not found then loosen u and a and search again

Set $\delta_0 = \delta_0^{new}$, $r = r + \Delta T$

until $r > wnd + N - \Delta T$;

Set $L_k = L_k/(N\Delta t)$, $k = k+1$;

end

The following values of problem parameters have been used: reconstructed dimension $p = 7$, lag step $\tau = 7$ (14 msec), evolution time $\Delta T = 21$ (41 msec), window size $N = 2048$ (4 sec). $STL_{\max}$ time series were computed for all 26 channels for the entire recording duration (approximately 24 hours) for both Patient A and B.

5. BICLUSTERING THE EEG DATA

5.1. Biomedical research and biclustering

Biclustering is a data mining methodology allowing for clustering/classification of not only data samples, but also data attributes (features) of which the samples are composed. It is performed in such a way that each class of data features created within the biclustering is related to a class of data samples by a certain property distinguishing it from other classes and believed to be "responsible" for its creation.

As we mentioned above, though biomedical research is now one of the major applications of data mining, previous biomedical applications of biclustering were limited to DNA microarray analysis, drug design, and few others [3, 29, 30, 31]. However, deliverables of biclustering are especially valuable for feature extraction purposes which is the major concern of many biomedical studies including the present study of EEG channels most indicative in VNS treatment. Indeed, similarly to the interest in upregulated and downregulated genes in each sampled condition in a microarray dataset, we were interested in revealing subsets of channels whose Lyapunov exponents consistently change with switching the VNS stimulation on and off. Therefore, representing each EEG channel as a data feature and analyzing data samples taken within the stimulation periods versus others taken outside of these periods with the biclustering setup, we were able to achieve our goals.

Let a dataset of n samples and m features be given as a rectangular matrix $A = (a_{ij})_{m \times n}$, where the value a_{ij} is the expression of i-th feature in j-th sample. We consider classification of the samples into classes

$$\mathcal{S}_1, \mathcal{S}_2, \ldots, \mathcal{S}_r, \; \mathcal{S}_k \subseteq \{1 \ldots n\}, \; k = 1 \ldots r,$$

$$\mathcal{S}_1 \cup \mathcal{S}_2 \cup \ldots \cup \mathcal{S}_r = \{1 \ldots n\},$$

$$\mathcal{S}_k \cap \mathcal{S}_\ell = \emptyset, \; k, \ell = 1 \ldots r, \; k \neq \ell.$$

This classification should be done so that samples from the same class share certain common properties. Correspondingly, a feature i may be assigned to one of the feature classes

$$\mathcal{F}_1, \mathcal{F}_2, \ldots, \mathcal{F}_r, \; \mathcal{F}_k \subseteq \{1 \ldots m\}, \; k = 1 \ldots r,$$

$$\mathcal{F}_1 \cup \mathcal{F}_2 \cup \ldots \cup \mathcal{F}_r = \{1 \ldots m\},$$

$$\mathcal{F}_k \cap \mathcal{F}_\ell = \emptyset, \; k, \ell = 1 \ldots r, \; k \neq \ell,$$

in such a way that features of the class $\mathcal{F}_k$ are "responsible" for creating the class of samples $\mathcal{S}_k$. Such a simultaneous classification of samples and features is called *biclustering* (or *co-clustering*).

Definition 1 *A* biclustering *of a dataset is a collection of pairs of sample and feature subsets* $\mathcal{B} = ((\mathcal{S}_1, \mathcal{F}_1), (\mathcal{S}_2, \mathcal{F}_2), \ldots, (\mathcal{S}_r, \mathcal{F}_r))$ *such that the collection* $(\mathcal{S}_1, \mathcal{S}_2, \ldots, \mathcal{S}_r)$ *forms a partition of the set of samples, and the collection* $(\mathcal{F}_1, \mathcal{F}_2, \ldots, \mathcal{F}_r)$ *forms a partition of the set of features. A pair* $(\mathcal{S}_k, \mathcal{F}_k)$ *will be called a* bicluster.

The criteria used to relate clusters of samples and clusters of features may have different nature. Most commonly, it is required that the submatrix corresponding to a bicluster either is overexpressed (i.e., mostly includes values above average), or has a lower variance than the whole dataset, but in general, biclustering may rely on any kind of common patterns among elements of a bicluster. We employed the consistent biclustering framework with feature selection based on 0–1 fractional programming [2].

Let each sample be already assigned somehow to one of the classes $\mathcal{S}_1, \mathcal{S}_2, \ldots, \mathcal{S}_r$. Introduce a 0–1 matrix $S = (s_{jk})_{n \times r}$ such that $s_{jk} = 1$ if $j \in \mathcal{S}_k$, and $s_{jk} = 0$ otherwise. The sample class *centroids* can be computed as the matrix $C = (c_{ik})_{m \times r}$:

$$C = AS(S^T S)^{-1}, \tag{3}$$

whose k-th column represents the centroid of the class $\mathcal{S}_k$.

Consider a row i of the matrix C. Each value in it gives us the average expression of the i-th feature in one of the sample classes. As we want to identify the checkerboard pattern in the data, we have to assign the feature to the class where it is most expressed. So, let us classify the i-th feature to the class $\hat{k}$ with the maximal value $c_{i\hat{k}}$:

$$i \in \mathcal{F}_{\hat{k}} \Rightarrow \forall k = 1 \ldots r, \, k \neq \hat{k} : \; c_{i\hat{k}} > c_{ik} \tag{4}$$

Now, provided the classification of all features into classes $\mathcal{F}_1$, $\mathcal{F}_2$, ..., $\mathcal{F}_r$, let us construct a classification of samples using the same principle of maximal average expression and see whether we will arrive at the same classification as the initially given one. To do this, construct a 0–1 matrix $F = (f_{ik})_{m \times r}$ such that $f_{ik} = 1$ if $i \in \mathcal{F}_k$ and $f_{ik} = 0$ otherwise. Then, the feature class centroids can be computed in form of matrix $D = (d_{jk})_{n \times r}$:

$$D = A^T F (F^T F)^{-1}, \tag{5}$$

whose k-th column represents the centroid of the class $\mathcal{F}_k$. The condition on sample classification we need to verify is

$$j \in \mathcal{S}_{\hat{k}} \Rightarrow \forall k = 1 \ldots r, \, k \neq \hat{k} : \; d_{j\hat{k}} > d_{jk} \tag{6}$$

The framework for feature selection and supervised biclustering proposed in [2] is based on the following definition:

Definition 2 *A biclustering $\mathcal{B}$ will be called* consistent *if both relations (4) and (6) hold, where the matrices C and D are defined as in (3) and (5).*

In contrast to other biclustering schemes, this definition of consistent biclustering is justified by the fact that *consistent biclustering implies separability of the classes by convex cones* [2]:

Theorem 1 *Let $\mathcal{B}$ be a consistent biclustering. Then there exist convex cones $\mathcal{P}_1, \mathcal{P}_2, \ldots, \mathcal{P}_r \subseteq \mathbb{R}^m$ such that all samples from $\mathcal{S}_k$ belong to the cone $\mathcal{P}_k$ and no other sample belongs to it, $k = 1 \ldots r$.*

Similarly, there exist convex cones $\mathcal{Q}_1, \mathcal{Q}_2, \ldots, \mathcal{Q}_r \subseteq \mathbb{R}^n$ such that all features from $\mathcal{F}_k$ belong to the cone $\mathcal{Q}_k$ and no other feature belongs to it, $k = 1 \ldots r$.

It also follows from the conic separability that convex hulls of classes are separated, i.e, they do not intersect.

We also say that a dataset is *biclustering-admitting* if some consistent biclustering for it exists. Furthermore, the data set will be called *conditionally biclustering-admitting* with respect to a given (partial) classification of some samples and/or features if there exists a consistent biclustering preserving the given (partial) classification.

Assuming that we are given the training set $A = (a_{ij})_{m \times n}$ with the classification of samples into classes $\mathcal{S}_1, \mathcal{S}_2, \ldots, \mathcal{S}_r$, we are able to construct the corresponding classification of features according to (4). Next, if the obtained biclustering is not consistent, our goal is to exclude some features from the dataset so that the biclustering with respect to the residual feature set is consistent.

Formally, let us introduce a vector of 0–1 variables $x = (x_i)_{i=1\ldots m}$ and consider the i-th feature selected if $x_i = 1$. The condition of biclustering consistency (6), when only the selected features are used, becomes

$$\frac{\sum_{i=1}^{m} a_{ij} f_{i\hat{k}} x_i}{\sum_{i=1}^{m} f_{i\hat{k}} x_i} > \frac{\sum_{i=1}^{m} a_{ij} f_{ik} x_i}{\sum_{i=1}^{m} f_{ik} x_i}, \; \forall j \in \mathcal{S}_{\hat{k}}, \; \hat{k}, k = 1 \ldots r, \; \hat{k} \neq k. \tag{7}$$

We will use the fractional relations (7) as constraints of an optimization problem selecting the feature set. It may incorporate various objective functions over x, depending on the desirable properties of the selected features, but one general choice is to select the maximal possible number of features in order to lose minimal amount of information provided by the training set. In this case, the objective function is

$$\max \sum_{i=1}^{m} x_i \tag{8}$$

The optimization problem (8), (7) is a specific type of *fractional 0–1 programming problem*, which can be solved using the approach described in [2]. When it has a feasible solution, the corresponding feature selection makes the dataset conditionally biclustering-admitting with respect to the given classification of samples.

5.2. Biclustering experiment

After the Lyapunov exponent's time series of two patients' EEG have been computed, we employed consistent biclustering framework to analyze whether we can separate EEG fragments corresponding to times of stimulation versus regular EEG. As each stimulation lasted 30 sec and a 4 sec time window was used to compute one element of the Lyapunov exponent time series, each stimulation provided 7 data points. Since the EEG patterns of a patient may have been changing throughout the observed period due to changes in his/her conditions not relevant to the investigated phenomenon, we averaged each of the 7 samples across all stimulation cycles. Thus, 7 Lyapunov exponent samples have been created to represent the positive class.

To create the negative class, we considered 10 Lyapunov exponent data points 250 sec after each stimulation. In the similar way, we averaged these 10 samples across all stimulation cycles. So, the created negative class contains 10 averaged Lyapunov exponent data samples from non-stimulation time intervals.

There were 25 features of the data each of which represents an EEG channel. So, with the data dimensionality of 25×17 or even 25×17, we were able to solve the problem without any relaxation using CPLEX [32]. We note that the previously published computational experiments with consistent biclustering were performed with help of an

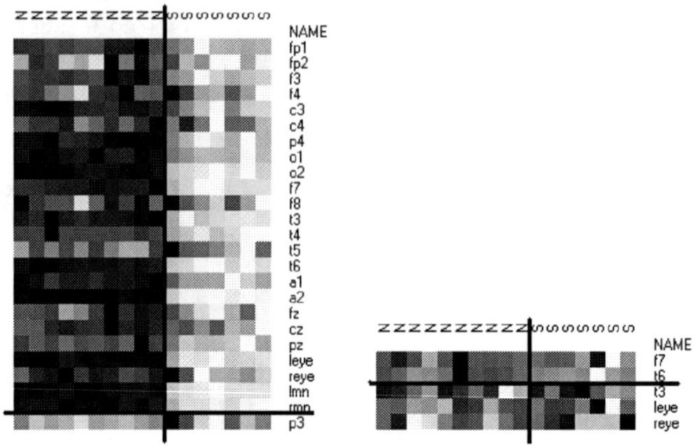

FIGURE 2. Heatmaps for patients A and B

iterative algorithm relaxing the mixed 0-1 programming formulation to make the formulation tractable by existing integer programming packages [2].

6. RESULTS

The patient A data were conditionally biclustering-admitting with respect to given stimulation and non-stimulation classes without excluding any features. All but one feature (channel p3) were classified into the non-stimulation class, which indicates that for almost all EEG channels the Lyapunov exponent was consistently decreasing during the stimulation with p3 being the only exception. We performed cross-validation of the obtained biclustering by leave-one-out method examining for each sample whether it would be classified in the appropriate class if the feature selection was performed without it. It turned out that all classes of all 17 samples are confirmed by this method.

To make the patient B dataset conditionally biclustering-admitting with respect to given stimulation and non-stimulation classes only five features were selected. Channels f7 and t6 were classified into the non-stimulation class, while channels t3, leye and reye were classified into the stimulation class. The one-leave-out experiment classified correctly all but 4 samples. Two of them were from the non-stimulation class (samples 1 and 9) and the other two were from the stimulation class (samples 15 and 16).

The biclustering heatmaps are presented in Fig. 2.

7. DISCUSSION

The successful application of biclustering with Lyapunov exponents demonstrates the potential to distinguish VNS stimulation from VNS deactivation epochs using scalp-EEG recordings.

Scalp EEG recordings, though relatively easy to obtain are challenging to work with. Scalp EEG recordings are inherently limited as each channel detects a massive summation of excitatory and inhibitory post-synaptic potentials [22] and the signal is highly filtered by the skull, muscle, and skin between the electrode and neurons. In addition, scalp EEG is highly susceptible to muscle artifacts (e.g. a1, a2, leye, and reye) and signal filtering effects of skull and scalp. However, one justification for the mathematical analysis of scalp EEG is that the signals originate from inside the brain and contain redundant physiological or pathological information about the brain [22]. Also, scalp-EEG recording is a widely available non-invasive measurement tool available in almost every hospital.

It should be noted that Patient *A* had no seizures during the recording, whereas Patient *B* did have seizures, which were accompanied by a manually-induced extra VNS stimulation, which could alter results. Perhaps the seizures and/or the manual stimulations were the reason why so few features were selected in Patient *B*. Despite the seizures and manual stimulations, it appears that signals arising from the frontal lobe (f7, leye, and reye) and temporal lobe (t3 and t6) of the brain were sufficient to cluster the stimulation class (from t3, leye and reye) and non-stimulation class (from f7 and t6) for Patient *B*. This could mean that the electrographic effect of the VNS is the most pronounced in the frontal and temporal lobes.

Future studies may run a separate experiment where manual stimulations are included along with the standard periodic VNS stimulations, or where stimulations occurring close to seizures are excluded. Perhaps additional studies will demonstrate a connection between the ability of the biclustering algorithm to separate stimulation and non-stimulation and the amount of seizures a patient experiences. At this time, we cannot determine if the successful biclustering separation is the cause or effect of a seizure-free recording session.

The ability to distinguish between stimulation and non-stimulation states by processing scalp-EEG is an important first step in this line of research. The next goal of this research is to be able to distinguish between various VNS stimulation parameter settings. Ultimately, classification method would provide a physiologic marker of optimal VNS therapeutic action, and lead to the creation of a bedside device which allow the physician a rapid measure of VNS effectiveness by applying these classification analyses in real-time. A physiologic marker of this type could greatly expedite the VNS calibration process, and minimize the amount of time and doctor office visits required for calibration in newly implanted VNS patients.

ACKNOWLEDGMENTS

General Clinical Research Center, Grant # MO1-RR00082
The Evelyn F. & William L. McKnight Brain Institute of the University of Florida

REFERENCES

1. Binnie CD, Mizrahi EM. The epilepsy monitoring unit. in Epilepsy: A comprehensive textbook, Eds. J. Engel Jr. and T. A. Pedley, Philadelphia: Lippincott-Raven, 1997.

2. Busygin S, Prokopyev OA, Pardalos PM, Feature Selection for Consistent Biclustering via Fractional 0–1 Programming, *Journal of Combinatorial Optimization*, Vol. 10/1 (2005), pp. 7–21.
3. Busygin S, Prokopyev OA, Pardalos PM, Biclustering in Data Mining, to appear in *Computer and Operations Research*, 2006.
4. Chaovalitwongse W, Iasemidis LD, Pardalos PM, Carney PR, Shiau DS, Sackellares JC. Performance of a seizure warning algorithm based on the dynamics of intracranial EEG. Epilepsy Research. 2005 May;64(3):93-113.
5. Cheng Y, Church GM. Biclustering of expression data. Proceedings from the International Conference on Intelligent Systems for Molecular Biology. 2000;8:93-103.
6. Engel J. Jr., Seizures and Epilepsy. Philadelphia, PA: F. A. Davis Co., 1989
7. Engel J. Jr., P. C. Van Ness, T. B. Rasmussen and L. M. Ojemann. Outcome with Respect to Epileptic Seizures. In Surgical Treatment of the Epilepsies, J. Engel Jr., Ed., New York: Raven Press, 1993, pp. 609-622.
8. Engel J. Jr., T.A. Pedley, Epilepsy: a Comprehensive Textbook. Philadelphia, PA: Lippincott-Raven, 1997.
9. Fisher RS, Krauss GL, Ramsay E, Laxer K, Gates J. Assessment of Vagus Nerve Stimulation for Epilepsy: Report of the Therapeutics and Technology Assessment Subcommittee of the American Academy of Neurology. Neurology, vol. 49, pp. 293-297, 1997.
10. Fisher RS, Theodore WH. Brain stimulation for epilepsy. Lancet Neurol. 2004 Feb;3(2):111-8. Review. Erratum in: Lancet Neurol. 2004 Jun;3(6):332.
11. George MS. Summary and Future Directions of Therapeutic Brain Stimulation: Neurostimulation and Neuropsychiatric Disorders. Epilepsy & Behavior. 2 (2001), pp. S95–S100
12. Goldstein MA, Harden CL. Continuing Exploration of the Neuropsychiatry of Seizures: a Review of Anxiety and Epilepsy. Epilepsy & Behavior., vol. 2 pp. 311-317, 2001.
13. Gotman J, Ives J, Gloor P. Long-Term Monitoring in Epilepsy. EEG suppl. Vol. 37, Amsterdam: Elsevier, 1985.
14. Groves DA, Brown VJ. Vagal Nerve Stimulation: a Review of its Applications and Potential Mechanisms that Mediate its Clinical Effects. Review. Neuroscience & Biobehavioral Reviews. Volume 29, Issue 3 , May 2005, Pages 493-500.
15. Hammond EJ, Uthman BM, Reid SA, Wilder BJ. Epilepsia. Electrophysiological Studies of Cervical Vagus Nerve Stimulation in Humans: I. EEG effects. 1992 Nov-Dec;33(6):1013-20.
16. Iasemidis LD, Shiau DS, Pardalos PM, Chaovalitwongse W, Narayanan K, Prasad A, Tsakalis K, Carney PR, Sackellares JC. Long-term prospective on-line real-time seizure prediction. Clinical Neurophysiology. 2005 Mar;116(3):532-44. Epub 2005 Jan 6.
17. Kluger Y, Basri R, Chang JT, Gerstein M. Spectral biclustering of microarray data: coclustering genes and conditions. Genome Res. 2003 Apr;13(4):703-16.
18. Lesser RP, Kim SH, Beyderman L, et al. Brief Bursts of Pulse Stimulation Terminate After Discharges Caused by Cortical Stimulation. Neurology, vol. 53, pp. 2073-2081, 1999.
19. Motamedi GK, Lesser RP, Miglioretti DL, et al. Optimizing Parameters for Terminating Cortical Afterdischarges with Pulse Stimulation. Epilepsia, vol. 43, pp. 836-846, 2002.
20. Niedermeyer E, Lopes da Silva FH. Electroencephalography. Baltimore: Williams & Wilkins, 1993.
21. Schachter SC, Wheless JW. Eds. Vagus Nerve Stimulation Therapy 5 Years After Approval: A Comprehensive Update. Neurology, S4, vol. 59, 2002.
22. Thakor NV, Tong SB. Advances in quantitative electroencephalogram analysis methods. Annual review of biomedical engineering. 6: 453-495 2004.
23. Uthman BM, Wilder BJ, Penry JK, Dean C, Ramsay RE, Reid SA, Hammond EJ, Tarver WB, Wernicke JF. Treatment of Epilepsy by Stimulation of the Vagus Nerve. Neurology, vol. 43, pp. 1338-1345, 1993.
24. Yoon S, Nardini C, Benini L, De Micheli G. Discovering coherent biclusters from gene expression data using zero-suppressed binary decision diagrams. IEEE/ACM Transactions on Computational Biology and Bioinformatics. 2005 Oct-Dec;2(4):339-54.
25. Iasemidis LD, Principe JC, Sackellares JC, Measurement and Quantification of Spatiotemporal Dynamics of Human Epilepic Seizures. Nonlinear Signal Processing in Medicine, Ed. M. Akay, IEEE Press, 1999
26. Pardalos PM, Chaovalitwongse WA, Iasemidis LD, Sackellares JC, Shiau DS, Carney PR, Prokopyev OA, Yatsenko VA, Seizure Warning Algorithm Based on Optimization and Nonlinear Dynamics,

Mathematical Programming, Vol. 101/2 (2004), pp. 365–385.
27. Pardalos PM, Chaovalitwongse WA, Prokopyev OA, Electroencephalogram (EEG) Time Series Classification: Application in Epilepsy, Annals of Operations Research, 2006.
28. Wolf A, Swift JB, Swinney HL, Vastano JA, Determining Lyapunov Expononets from a Time Series, Phisica 16D (1985), pp. 285 – 317.
29. Madeira SC, Oliveira AL, Biclustering Algorithms for Biological Data Analysis: A Survey, *IEEE Transactions on Computational Biology and Bioinformatics*, 1 (2004), pp. 24–45.
30. Tanay A, Sharan R, Shamir R, Biclustering Algorithms: A Survey, to appear in the Handbook of Bioinformatics, 2004. Available at `http://www.cs.tau.ac.il/~rshamir/papers/bicrev_bioinfo.ps`, last accessed November 2006.
31. Biclustering - Wikipedia, the Free Encyclopedia, `http://en.wikipedia.org/wiki/Biclustering`, last accessed August 2006.
32. ILOG Inc. CPLEX 9.0 User's Manual, 2004.

Numerical limitations in application of vector autoregressive modeling and Granger causality to analysis of EEG time series

Alla Kammerdiner, Petros Xanthopoulos and Panos M. Pardalos

Department of Industrial and Systems Engineering, University of Florida, Gainesville, FL, 32611, USA.

Abstract. In this chapter a potential problem with application of the Granger-causality based on the simple vector autoregressive (VAR) modeling to EEG data is investigated. Although some initial studies tested whether the data support the stationarity assumption of VAR, the stability of the estimated model is rarely (if ever) been verified. In fact, in cases when the stability condition is violated the process may exhibit a random walk like behavior or even be explosive. The problem is illustrated by an example.

Keywords: EEG, time series, vector autoregression
PACS: 01.10.Fv

1. INTRODUCTION

The problem of determining the connectivity among different areas in the brain has recently become one of the vital problems in neuroscience. Indeed, the ability to extract the information about the organization and interaction of the neurons in the different parts of the brain from neurophysiological data allows one to draw the conclusions about the functional relations of various areas in the brain.

Among existing brain imaging methods, the electroencephalography, EEG, is one of the most widely used techniques. Although EEG measurements, which go back as far as 1875 [1], are used to determine both the relative strength and the position of electrical activity at the different sites in the brain, EEG provides less spatial resolution than such brain imaging techniques as magnetic resonance imaging (MRI) and positron emission tomography (PET). However, the recent technological advances made it possible to record the EEG signals from numerous (up to 30) channels simultaneously and, therefore, produce more spatial information. As a result, the techniques that facilitate joint analysis of the EEG data in both the temporal and the spatial domains have received special attention in recent years.

Several methods for joint spatio-temporal analysis of multichannel EEG recordings based on the idea of Granger causality were presented in the last decade. The concept of Granger causality was first introduced by Clive Granger for measuring linear dependence and feedback in economic time series. Later, this idea was further extended by John Geweke who proposed an equivalent measure based on the spectral representation of time series. Both Granger's and Geweke's approaches employ the vector autoregressive modeling to derive estimates of underlying casual relations in the data. However,

CP953, *Data Mining, Systems Analysis, and Optimization in Biomedicine*
edited by O. Seref, O. E. Kundakcioglu, P. M. Pardalos

the latter approach is found particularly useful for analysis of EEG time series, since it investigates the casual relation in the frequency domain, instead of the time domain as in the former approach. In particular, the spectral measure of Granger causality proposed by John Geweke was employed on intracortical local field potentials recorded from 8 electrodes during go/no-go trials of cat's visual responses [2]. Another study [3] utilized a similar method of directed transfer function (which is equivalent to the spectral measure of Granger causality) to examine casual influences in the primate visual cortex during the task of visual patterns recognition. The direct transfer function approach to Granger causality was also applied to analyzing brain connectivity patterns on human EEG data recorded during stage 2 sleep [4].

Michael Eichler proposed a graphical approach for modeling Granger-causal relationships in multivariate time series [5] and later applied this method to studying connectivity in neural systems [6, 7]. Luiz Baccalá and Koichi Sameshima introduced a concept of the partial directed coherence for inference of Granger causality in the frequency domain based on the linear vector autoregressive modeling, and applied it to investigating the functional interactions among different brain structures [8, 9].

Vector autoregressive modeling is a common basis in various approaches to estimating Granger causality. In this chapter, we review the details of the multivariate autoregressive modeling and investigate some limitations in its application to the EEG data.

The remaining of this chapter is organized as follows. In Section 2 we define the concept of Granger causality in the time domain and its extension to the frequency domain. We also present some simple tests of Granger causality relations in the data. Section 3 deals with various aspects of vector autoregressive modeling such as estimation of model parameters, criteria for selecting a suitable order for the model and imposed assumptions. In Section 4, we present the numerical example illustrating the violation of the assumptions of the vector autoregression (VAR). The chapter ends with conclusions and possible directions of further research in Section 5.

2. GRANGER CAUSALITY

Mathematically, the data obtained from EEG measurements can be viewed as the time series. Specifically, the EEG data recorded from multiple channels are represented by the multidimensional time series, and so, various multivariate statistical techniques can be applied to such data.

Frequently, when analyzing multivariate time series, one wants to know whether or not the information about the past values of some variables helps forecast another variable. One way to determine this was proposed by Clive Granger [10, 11] based on the Wiener's original idea [12] that since a cause precedes its consequence in a causal relationship then random processes are causally related if one process is better predicted by incorporating knowledge of previous realizations of another process. Granger was able to formalize the Wiener's idea in the case of linear systems by applying a linear autoregressive modeling. More precisely, the first process (Granger) causes the second process if inclusion of the available information about the past realizations of the first process improves the prediction given by the linear regressive model by reducing the residual error. Granger's concept of causality was soon adopted and popularized by

Christopher Sims in the study of the causal ordering between money and income [13]. Due to simplicity of its application and interpretation, the notion of Granger causality became widely used in economic time series analysis.

In 1982 John Geweke extended the notion of Granger causality into the frequency domain by introducing a measure of linear dependance and feedback between two time series based on the spectral density representation [14]. Geweke noted that the general measure of linear dependence between two series is a sum of three components, the measure of linear feedback from the first series to the second, the linear feedback from the second to the first series, and the instantaneous linear feedback. To guarantee the existence of the spectral density matrix for the derivation of the measures, a moving average representation of the multiple time series is adopted. In addition, the consideration is limited to the cases that admit representation of the inverted series in the form of linear regression. The Fourier transform and the inverse Fourier transform are applied to alternate between the time domain and the frequency domain. The significance of Geweke's measures is that the causal relations are expressed as functions of the frequency, which allows for additional interpretation of the results.

Many different tests of Granger Causality are developed. Some versions of the test are based on a vector autoregressive model, others are based on a multivariate moving average representation. For simplicity, we present the alternative based on the bivariate regression. For a detailed review of Granger causality tests, see the book on time series by Hamilton [15].

Let $X(t)$ and $Y(t)$, $t \in \mathbb{Z}$ denote two time series (or stochastic processes) with the corresponding realizations x_t and y_t, $t \in \mathbb{Z}$. Suppose that $\Omega_{X,t}$ and $\Omega_{Y,t}$ denote all the information about the realizations of processes X and Y, respectively, up to time t. Then, the relationship of Granger causality between such series can be formally defined as follows:

The time series $X(t)$ is said to *Granger-cause* $Y(t)$ if there exists $p = 1,2,\ldots$ such that the mean squared error (MSE) of the p-step forecast of $Y(t)$ based on the information $\Omega_{X,t}$ and $\Omega_{Y,t}$ is smaller than the MSE of the p-step forecast of $Y(t)$ based on $\Omega_{Y,t}$ alone, i.e.

$$\exists p = 1,2,\ldots \quad : \quad \Sigma_Y(p|\Omega_{X,t},\Omega_{Y,t}) < \Sigma_Y(p|\Omega_{Y,t}), \tag{1}$$

where $\Sigma_Y(p|\Omega)$ is the MSE of the p-step forecast of $Y(t)$ based on information Ω.

Using the above definition 1, we now present the test for Granger causality based on the bivariate autoregressive model. Suppose that for some integer lag parameter $p > 0$, the realizations of time series Y(t) are given by the model

$$y_t = \nu + \sum \alpha_i y_{t-i} + \sum \beta_i x_{t-i} + \varepsilon_t, \tag{2}$$

where ε_t is a *standard white noise* (or *innovation process*, i.e. has zero mean and zero autocorrelation).

We test the hypothesis

$$H_0: \quad \beta_i = 0 \quad \forall i = 1,2,\ldots,p \tag{3}$$

against the alternative

$$H_1: \quad \exists j, 1 \leq j \leq p \quad \text{s.t.} \quad \beta_j \neq 0 \tag{4}$$

Note that if the null hypothesis is accepted, then the time series $Y(t)$ believed to be NOT Granger-caused by $X(t)$. Meanwhile, rejecting the null hypothesis (i.e. accepting the alternative) means that $X(t)$ is believed to cause $Y(t)$ in Granger's sense.

Let T be the sample size parameter. The model parameters for the null hypothesis H_0 and the parameters for the alternative H_1 are estimated from the sample data using the ordinary least squares method (or other methods) to obtain the estimates of the forecast errors $\hat{\varepsilon}_{0t}$ and $\hat{\varepsilon}_{1t}$, respectively, $t = 1,2,\ldots,T$. Then the sum of squared residuals RSS_0 under the assumption of null hypothesis H_0 is

$$RSS_0 = \sum_{t=1}^{T} \hat{\varepsilon}_{0t}^2, \tag{5}$$

and the sum of squared residuals RSS_1 under the alternative H_1 is

$$RSS_1 = \sum_{t=1}^{T} \hat{\varepsilon}_{1t}^2. \tag{6}$$

By conducting the F-test of the null hypothesis, one can find the test statistic

$$S_1 = \frac{(RSS_0 - RSS_1)/p}{RSS_1/(T-2p-1)} \sim F_{p,T-2p-1}. \tag{7}$$

If the test statistic S_1 exceeds the specified critical value, then the null hypothesis is rejected that $X(t)$ does not Granger-causes $Y(t)$. Otherwise H_0 is accepted.

An asymptotically equivalent test of Granger causality is given by the following statistic

$$S_2 = \frac{T(RSS_0 - RSS_1)}{RSS_1} \sim \chi^2(p). \tag{8}$$

It is noteworthy to point out that the tests of Granger causality are very sensitive to the choice of the lag length parameter p, and to the methods utilized for handling any non-stationarity in the time series.

The bivariate approach to testing Granger causality can be naturally extended to the multivariate case by partitioning the vector autoregressive process $Z(t)$ into two components $X(t)$ and $Y(t)$, so that $Z(t) = (X(t),Y(t))$, and then testing the suitable zero constraints on the coefficients of vector autoregression. For the derivation of the Wald statistic and the F-statistic for testing Granger causality in the multivariate case, see the book on multiple time series by Lütkepohl [16].

3. VECTOR AUTOREGRESSIVE LINEAR MODELS

This section introduces the vector autoregressive (VAR) model of finite order that serves as a foundation for establishing Granger causal relationships in multidimensional time series.

Let p denote a positive integer, and let y_t denote the K-variate time series (i.e. realizations of K-dimensional process $Y(t)$). Formally, the vector autoregressive model of order p, denoted VAR(p), is defined as follows:

$$y_t = \nu + A_1 y_{t-1} + \ldots + A_p y_{t-p} + \varepsilon_t, \quad t = 0, \pm 1, \pm 2, \ldots, \tag{9}$$

where $y_t = (y_{1t}, \ldots, y_{Kt})'$ is a $(K \times 1)$ random vector, $\nu = (\nu_1, \ldots, \nu_K)'$ is a fixed $(K \times 1)$ vector representing a non-zero mean $EY(t)$, the A_i, $i = 1, \ldots, p$ are fixed $(K \times K)$-dimensional coefficient matrices, and $\varepsilon_t = (\varepsilon_{1t}, \ldots, \varepsilon_{Kt})'$ is a K-dimensional *white noise* process (i.e. $E\ [\varepsilon_t] = 0$, $E\ [\varepsilon_s \varepsilon_t'] = 0$, for $s \neq t$, and $E\ [\varepsilon_s \varepsilon_t'] = \Sigma_\varepsilon$). It is assumed that the covariance matrix Σ_ε is nonsingular. In addition, three important conditions are usually imposed on the time series in the VAR model. The first condition is stability of the process $Y(t)$, the second is stationarity of $Y(t)$, while the third one supposes that the underlying white noise process ε_t is *Gaussian*.

3.1. Methods for parameter estimation

Next, we review some common approaches to estimating the model parameters.

Suppose that the the lag length parameter p is specified. Although in the above definition of the VAR(p) model the process mean ν, the coefficient matrices A_i, and the covariance matrix Σ_ε were assumed to be known, in practice, these parameters must be derived from the sample data. There are three main approaches to estimating the parameters of a VAR(p) time series, namely, the multivariate least squares method, the Yule-Walker estimator, and the maximum likelihood estimation[16]. Under the assumptions of stability and Gaussian distribution, these approaches lead to estimators with the same asymptotic properties. However, the asymptotic results should be used cautiously in inference from small samples. As a result, different approaches may sometimes lead to different results when estimating the model parameters using small samples.

Let us now briefly present the multivariate least squares estimation, which is a higher dimensional extension of the well-known method of ordinary least squares. For more detailed discussion, refer to [16].

Suppose that the available data include $(T + p)$ successive realizations of estimated multiple time series represented by K-dimensional vectors

$$y_{-p+1}, \ldots, y_0, y_1, \ldots, y_T$$

where p is the fixed lag length, and T is the sample size parameter. For convenience, we partition the data into the pre-sample $y_{-p+1}, \ldots, y_0$ and the sample values $y_1, \ldots, y_T$. In addition, the following notation is introduced:

$$
\begin{aligned}
Y &:= (y_1,\ldots,y_T) && (K \times T), \\
B &:= (\nu, A_1,\ldots,A_p) && (K \times (Kp+1)), \\
Z_t &:= \begin{bmatrix} 1 \\ y_t \\ \vdots \\ y_{t-p+1} \end{bmatrix} && ((Kp+1) \times 1), \\
Z &:= (Z_0,\ldots,Z_{T-1}) && ((Kp+1) \times T), \\
\varepsilon &:= (\varepsilon_1,\ldots,\varepsilon_T) && (K \times T),
\end{aligned}
\tag{10}
$$

where $t = 1,\ldots,T$. Then using this notation in (9), the vector autoregressive model of order p can be represented in the compact form:

$$
Y = BZ + \varepsilon, \tag{11}
$$

and the coefficients B of the model are given by the least squares estimator:

$$
\hat{B} = YZ'(ZZ')^{-1}. \tag{12}
$$

The covariance matrix can be estimated in various ways. Since $\Sigma_\varepsilon = E\,[\varepsilon_t \varepsilon_t']$, the estimator

$$
\tilde{\Sigma}_\varepsilon = \frac{1}{T}\sum_{t=1}^{T} \hat{u}_t \hat{u}_t' = \frac{1}{T}(Y - \hat{B}Z)(Y - \hat{B}Z)' \tag{13}
$$

is consistent. However, this estimator of the covariance matrix Σ_ε is not unbiased. Therefore, it is often replaced by the following unbiased estimator

$$
\hat{\Sigma}_\varepsilon = \frac{1}{T - Kp - 1}\sum_{t=1}^{T} \hat{u}_t \hat{u}_t' = \frac{1}{T - Kp - 1}(Y - \hat{B}Z)(Y - \hat{B}Z)' \tag{14}
$$

Obviously, both estimators are consistent estimators of the covariance matrix, and they are asymptotically equivalent.

3.2. Order selection criteria

When estimating the coefficients of the vector autoregressive model from data, we assumed the order p of the VAR(p) to be known. In practice, however, it is unknown, and therefore, needs to be derived from the data. Since zero coefficient matrices are allowed, one could simply set p to some upper bound on the VAR order. On the other hand, selecting an unnecessary large p will affect the forecast precision of the estimated model. Therefore, it is advantageous to apply some suitable criteria for optimal selection of the lag length parameter p.

Various criteria for choosing the optimal model order are developed. Some of the most commonly used are the final prediction error (FPE) criterion, Akaike's information criterion (AIC), Hannan-Quinn criterion (HQ), and Schwarz or Bayesian information criterion (SC) [16].

Let $\tilde{\Sigma}_u(m)$ denote the maximum likelihood estimator of Σ_u computed by fitting the VAR model of order m. The FPE criterion proposed by Akaike in 1969 is based on the idea that minimizing the mean square error improves the forecast of the model. For a VAR(p) time series, the FPE criterion is defined as

$$FPE(m) = \left(\frac{T+Km+1}{T-Km-1}\right)^K \det\left(\tilde{\Sigma}_u(m)\right) \tag{15}$$

Using the FPE criterion, the estimate $\hat{p}_{FPE}$ of the model order p is selected so that

$$FPE\left(\hat{p}_{FPE}\right) = \min_{m=1,\ldots,M} FPE(m), \tag{16}$$

where M denote some upper boundary on the model order. In other words, first, for each $m = 1,\ldots,M$, the vector autoregressive model of order m is estimated from the data, and the respective values of the $FPE(m)$ are calculated using (15); then the order producing the smallest value of $FPE(m)$ is chosen among the possible orders $m = 1,\ldots,M$.

AIC is another popular order selection criteria that was also introduced by Akaike. Given a VAR(m) model, the Akaike information criteria is defined as follows:

$$AIC(m) = \frac{2mK^2}{T} + \ln\left(\tilde{\Sigma}_u(m)\right) \tag{17}$$

Similarly to the FPE criterion, the VAR(m) models are estimated for different $m = 1,\ldots,M$ to obtain the corresponding $AIC(m)$ values for each order. Then the estimate $\hat{p}_{AIC}$ of the model order p with the smallest $AIC(m)$ is selected.

The Hannan-Quinn order selection criterion, HQ, is given by:

$$HQ(m) = \frac{2mK^2 \ln\ln T}{T} + \ln\left(\tilde{\Sigma}_u(m)\right) \tag{18}$$

As before, among the model parameters $m = 1,\ldots,M$, the one p having the smallest value of $HQ(m)$ is chosen as the estimator $\hat{p}_{HQ}$ of the true model order.

Last, but not least, we present Schwarz criterion, which was derived using Bayesian arguments. The SC is formulated as:

$$SC(m) = \frac{mK^2 \ln T}{T} + \ln\left(\tilde{\Sigma}_u(m)\right), \tag{19}$$

and the order minimizing $SC(m)$ is chosen among $m = 1,\ldots,M$ as the estimator $\hat{p}_{SC}$ of the model order p.

Some interesting statistical properties of the above criteria are proved in [16]. In particular, it is shown that AIC and FPE criteria for VAR order selection are asymptotically equivalent, although these estimators of the model order are not consistent. On the other hand, the other two criteria provide consistent estimators of the order parameter p. More precisely, in the univariate case ($K = 1$), the Hannan-Quinn criterion is consistent (i.e.

$\lim_{T\to+\infty}\Pr\{\hat{p}=p\}=1$). In addition, the HQ criterion is strongly consistent for $K\geq 2$ (i.e. $\Pr\{\lim_{T\to+\infty}\hat{p}=p\}=1$). The SC is shown to be strongly consistent for any dimension K.

It is important to keep in mind that even though FPE and AIC do not provide consistent estimators, they are not necessarily inferior to HQ and SC. Actually, in small samples, and even in larger samples, FPE and AIC may produce better forecast, although they may not estimate the model order correctly.

3.3. Stability condition and other assumptions

As mentioned above, the conditions of stability, stationarity and Gaussian distribution are usually imposed on time series when dealing with the VAR models. Below we define these conditions, and discuss their role. A K-dimensional VAR(p) time series (9) are called *stable*, if

$$\det(I_K - A_1 z - \ldots A_p z^p) \neq 0 \qquad \text{for} \quad \text{complex} z: \quad |z|\leq 1. \tag{20}$$

In other words, the VAR(p) process (9) satisfies the *stability condition* when its *reverse characteristic polynomial* (given by $\det(I_K - A_1 z - \ldots A_p z^p)$) has no roots on and inside the complex unit circle.

The stability condition guarantees that there exists a moving average (MA) representation for the VAR(p) process. Also stability ensures that the process is a well-defined stochastic process with the distributions and joint distribution of the process y_t uniquely determined by the innovation process ε_t. For a stable VAR(p)process, both the process mean and the autocovariance are time-invariant (which, according to the definition below, implies stationarity).

When the stability condition is violated, the process variance is increasing with time and unbounded. Specifically, if the inverse characteristic polynomial of the time series has a single unit root and all the other roots are outside the complex unit circle, then the time series behavior is similar to a *random walk*. In this special case, the variance increases linearly with time, the correlation between y_t and y_{t+h} approaches 1, and the process mean $E\left[Y(t)\right]$ exhibits linear trend for $\nu\neq 0$. In addition, if one of the roots of the inverse characteristic polynomial lies strictly inside the complex unit circle, then such process is explosive, i.e. the process variance grows exponentially. Various approaches are developed in the time series literature to address the time series with the unit roots. For example, the unit roots can be removed by taking differences. However, the explosive time series are not as well-studied, because it is believed that an exponential increase in the variance of the economic time series is not well founded. As one can see the stability assumption plays important role in VAR(p).

A wide-sense stationarity for stochastic processes is imposed on the VAR time series as follows. A stochastic process $Y(t)$ is considered *stationary* if

1. $E\left[Y(t)\right]=\nu$ for all t;
2. $E\left[(Y(t+h)-\nu)(Y(t)-\nu)'\right]=\Gamma_y(h)=\Gamma_y(-h)$ for all t and $h=0,1,\ldots$.

In other words, the stationarity condition supposes that the first and second moments are time invariant. Also note that the process mean ν and the autocovariance matrix $\Gamma_y(h)$ are finite. It is shown (see Proposition 2.1 in [16]) that

A stable VAR(p) time series y_t, $t = 0, \pm 1, \pm 2, \ldots$ are stationary.

Since stability of time series implies that the series are stationary, the stability condition (20) is sometimes cited in the literature as the stationarity condition. However, it is important to remember that these two conditions are not equivalent. In fact, although a stable vector autoregressive series are always stationary, the converse is not true, i.e. an unstable time series are not necessary non-stationary.

The Gaussian distribution assumption is introduced into the VAR(p) model through ε_t. Specifically, given representation (9) of the VAR (p), the innovation process ε_t is assumed to be *Gaussian white noise*, i.e.

1. $\varepsilon_t \sim N(0, \Sigma_u)$,
2. ε_s and ε_t are independent for $s \neq t$.

This condition implies that y_t is a Gaussian process, i.e. any subcollection $y_t, \ldots, y_{t+h}$ follows a multivariate normal distribution for all possible values of t and h.

4. NUMERICAL EXPERIMENTS

In the applications of Granger causality and related measures to the EEG, the directional dependencies in neural data are analyzed based on autoregressive modeling. Although, Bernasconi and König [2] applied statistical testing to verify the stationarity of the data, and established the duration of the stationary interval for EEG to be approximately 1 second. The statistical testing of underlying assumptions of the VAR (discussed in the previous section) are often omitted in the later studies. As shown above, the stability condition is a very important assumption of vector autoregression. In this study, we examine how the parameters of the model order and sample size influence the stability of the derived VAR model.

In order to estimate VAR(p) model from data and investigate the properties of the derived model, the rodent intracranial EEG data was used. The data set consisted of the electroencephalographic recordings from 6 electrodes (implanted in left frontal, right frontal, two left hippocampal and two right hippocampal parts of the rodent's brain) sampled at 200 Hz.

To examine the applicability of the vector autoregression to EEG data, we estimated the VAR(p) model parameters for different values of model order p and different sample sizes T. The sample sizes $T \in [1, 300]$ were used, and the lag length parameter p varied between 1 and 30. In addition, we filtered data using a Rectangular band pass Hamming window with 100 coefficients into the frequency bands of 0 – 30 Hz, 30 – 60 Hz, 60 – 90 Hz, 90 – 120 Hz, and 120 – 150 Hz. The raw data and the five differently filtered data represented separate datasets. For each dataset, we ran the model estimation procedure with $T = 1, 2, \ldots, 300$ and $p = 1, 2, \ldots, 30$. The procedure for estimating coefficients of the model was implemented in MATLAB environment based on the multivariate least squares method presented above. For every VAR model derived from the data, the

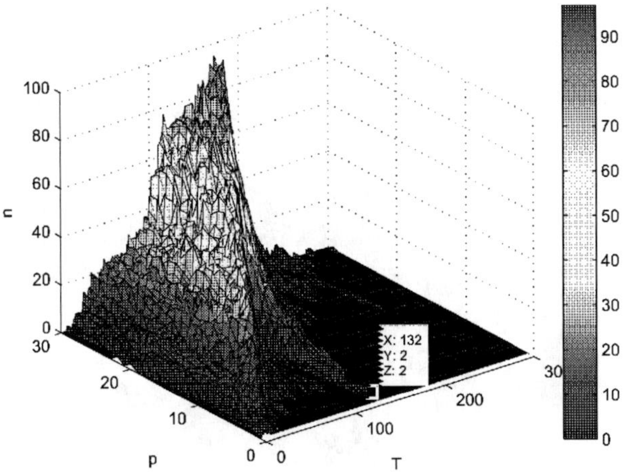

FIGURE 1. Numbers n of roots of the inverse characteristic polynomial (ICP) for VAR(p), which lie on and inside the complex unit circle, computed for different sample sizes T and for different model orders p using the raw data

stability condition was checked, and the number of the roots of the inverse characteristic polynomial (ICP) on and inside the complex unit circle was stored. For each dataset, the 3-dimensional surface plot was produced by graphing (T,p,n), where n is the numbers n of the ICP roots on and inside the unit circle, and T and p represent the sample size and the model order (or lag length), respectively. The surface plots are displayed in Fig. 1 – 6.

The surface plot in Fig. 1 supports the fact [16] that the sample size parameter should significantly exceed the lag length p, i.e. $T \gg p$. On the other hand, as can be seeing from Fig. 1, even for $T = 132$ considerably larger than $p = 2$, the stability assumptions may be violated (the numbers of the ICP roots on and inside the unit circle $n = 2$).

Fig. 2 – 6 shows that for the filtered data the (T,p) region where the stability condition of the estimated VAR(p) model is violated covers almost the whole domain, while the (T,p) region that corresponds to stable VAR(p) models is much smaller than the stable region in Fig. 1 and characterized by large T and very small p.

The experiment was repeated with consistent results on various samples from the six datasets. The results of our experiments clearly shows that the stability condition imposed on the simple VAR models is often violated even for the parameters $T \gg p$. Furthermore, filtering the data within some restricted frequency band often leads to reduction of the (T,p) domain where the estimated VAR(p) models remain stable. In practice, the EEG data are often filtered within a certain frequency band. In many studies that utilize the vector autoregression to study Granger causality among series, the verification of the conditions assumed by the VAR model is often omitted. From our point of view, the relevant statistical tests justifying the suitability of the model are always required when estimating the model from data.

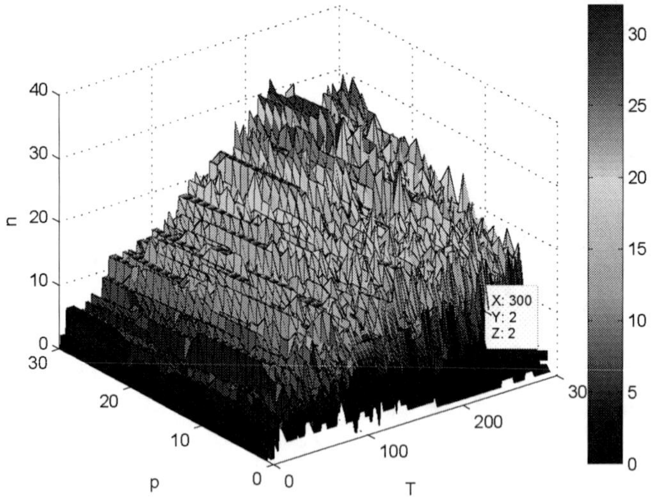

FIGURE 2. Numbers n of roots of the inverse characteristic polynomial (ICP) for VAR(p), which lie on and inside the complex unit circle, computed for different sample sizes T and for different model orders p using the 0 – 30 Hz band filtered data

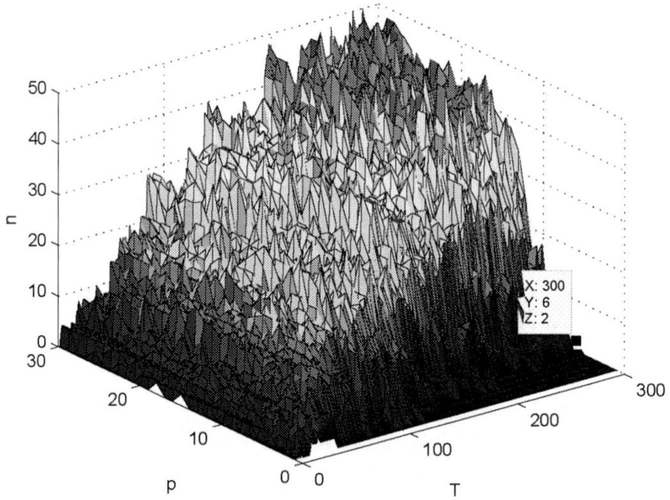

FIGURE 3. Numbers n of roots of the inverse characteristic polynomial (ICP) for VAR(p), which lie on and inside the complex unit circle, computed for different sample sizes T and for different model orders p using the 30 – 60 Hz band filtered data

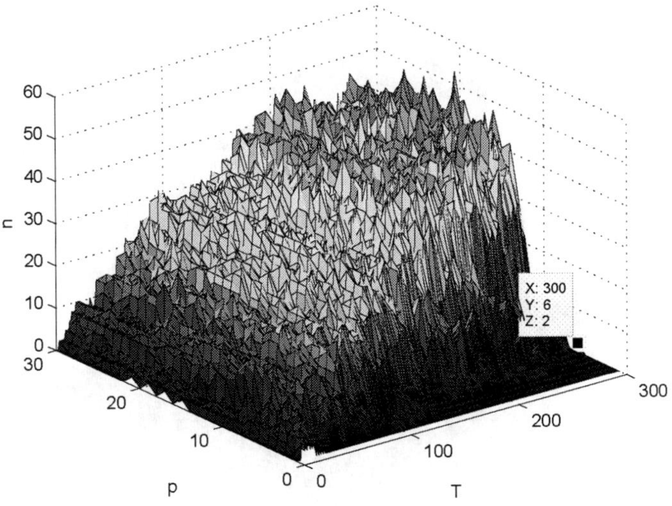

FIGURE 4. Numbers n of roots of the inverse characteristic polynomial (ICP) for VAR(p), which lie on and inside the complex unit circle, computed for different sample sizes T and for different model orders p using the 60 – 90 Hz band filtered data

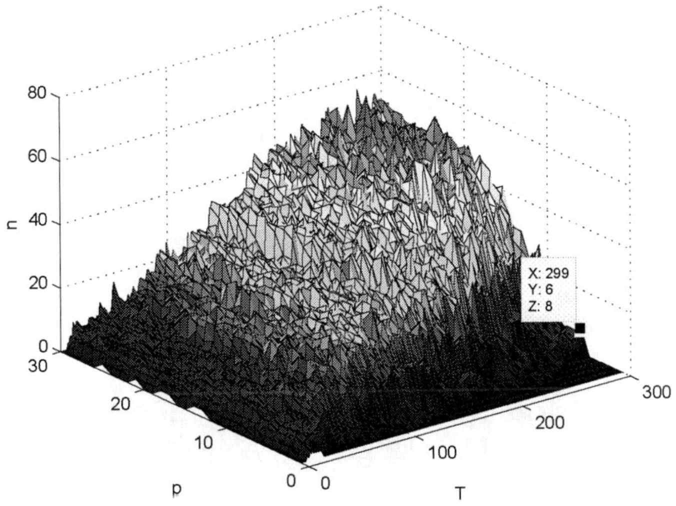

FIGURE 5. Numbers n of roots of the inverse characteristic polynomial (ICP) for VAR(p), which lie on and inside the complex unit circle, computed for different sample sizes T and for different model orders p using the 90 – 120 Hz band filtered data

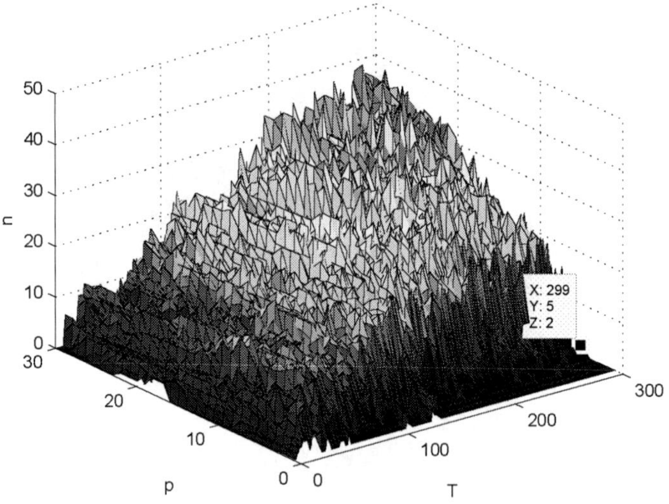

FIGURE 6. Numbers n of roots of the inverse characteristic polynomial (ICP) for VAR(p), which lie on and inside the complex unit circle, computed for different sample sizes T and for different model orders p using the 120 – 150 Hz band filtered data

5. CONCLUSIONS

Various modifications of vector autoregression (which relax the stability condition - no roots inside unit circle) are developed for analysis of economic time series. Although Bernasconi and König [2] tested the stationarity of different EEG data, and concluded that a time interval of 1 second is the interval on which the EEG time series can be considered stationary, the underlying assumptions of the VAR modeling such as stationarity and stability are rarely statistically tested in applications to EEG data. The results of our experiments shows that the stability assumptions of vector autoregressive model are often violated in application to the EEG data. This observation suggests that suitable extensions of multivariate autoregression to unstable processes may fit the data better, and so, be more appropriate for the EEG time series analysis than linear vector autoregressive modeling.

Additional statistical testing is required in order to make conclusions on what multivariate models are the most suitable for extracting the directional dependencies from multichannel EEG data.

ACKNOWLEDGMENTS

The experimental data were obtained in collaboration with P.R. Carney and S.P. Nair. We thank them very much for providing the EEG data for this study.

REFERENCES

1. M. Teplan, *Measurement science review* **2**, 112–23 (2002).
2. C. Bernasconi and P. Koënig, *Biol. Cybern.* **81**, 199–210 (1999).
3. H. Liang, M. Ding, R. Nakamura, and S. L. Bressler, *Neuroreport* **11**, 2875–2880 (2000).
4. M. Kaminski, M. Ding, W. A. Truccolo, and S. L. Bressler, *Biol. Cybern.* **85**, 145–157 (2001).
5. M. Eichler, *Graphical modelling of multivariate time series*, University of Heidelberg: Preprint, 2001.
6. M. Eichler, *Philos. Transact. R. Soc. B* **360**, 953–967 (2005).
7. M. Eichler, R. Dahlhaus, J. Sandkühler, *Biol. Cybern.* **89**, 289–302 (2003).
8. L. Baccala, K. Sameshima, *Biol. Cybern.* **84**, 463–474 (2001).
9. K. Sameshima, L. Baccala, *J. Neurosci. Methods* **94**, 93–103 (1999).
10. C. W. J. Granger, *Information and Control* **6**, 28–48 (1963)
11. C. W. J. Granger, *Econometrics* **37**, 424–438 (1969).
12. N. Wiener, "The Theory of Prediction," in *Modern Mathematics for the Engineer*, ed. E.F. Beckenbach, New York: McGraw-Hill, 1956.
13. C. A. Sims, *American Economic Review* **62**, 540–552 (1972).
14. J. Geweke, *J. American Statis. Association* **77**, 304–313 (1982).
15. J. D. Hamilton, *Time Series Analysis*, Princeton, 1994.
16. H. Lütkepohl, *Intoduction to multiple time series analysis*, Springer-Verlag Berlin, Heidelberg, 1991.

Quantitative analysis on electrooculography (EOG) for neurodegenerative disease [1]

Chang-Chia Liu[*,†], W. Art Chaovalitwongse[**], Panos M. Pardalos[†,*], Onur Seref[†], Petros Xanthopoulos[†], J.C. Sackellares[‡] and Frank M. Skidmore[§,¶]

[*]*Department of Biomedical Engineering, University of Florida, Gainesville, FL 32611*
[†]*Department of Industrial and Systems Engineering, University of Florida, Gainesville, FL 32611*
[**]*Department of Industrial and Systems Engineering, Rutgers University, Piscataway, NJ 08854*
[‡]*Optima Neuroscience, Inc. 101 SE 2nd Place, Suite 201-A, Gainesville, FL 32601*
[§]*Malcolm Randle VA Medical Center NF/VAMC, 1601 S.W Archer Road, Gainesville, FL 32608*
[¶]*Department of Neurology, University of Florida, Gainesville, FL 32611*

Abstract. Many studies have documented abnormal horizontal and vertical eye movements in human neurodegenerative disease as well as during altered states of consciousness (including drowsiness and intoxication) in healthy adults. Eye movement measurement may play an important role measuring the progress of neurodegenerative diseases and state of alertness in healthy individuals. There are several techniques for measuring eye movement, Infrared detection technique (IR), Video-oculography (VOG), Scleral eye coil and EOG. Among those available recording techniques, EOG is a major source for monitoring the abnormal eye movement. In this real-time quantitative analysis study, the methods which can capture the characteristic of the eye movement were proposed to accurately categorize the state of neurodegenerative subjects. The EOG recordings were taken while 5 tested subjects were watching a short (>120s) animation clip. In response to the animated clip the participants executed a number of eye movements, including vertical smooth pursued (SVP), horizontal smooth pursued (HVP) and random saccades (RS). Detection of abnormalities in ocular movement may improve our diagnosis and understanding a neurodegenerative disease and altered states of consciousness. A standard real-time quantitative analysis will improve detection and provide a better understanding of pathology in these disorders.

Keywords: Eye movement, Parkinson Disease, Motor Fluctuation, Electro-oculography
PACS: 89.20.Kk, 87.85.-d

1. INTRODUCTION

Continuous EOG recordings, similar to continuous EEG recordings, provide a continuous spatially and temporally precise stream of data related to eye movements. Due to the higher metabolic rate of the retina compared to the cornea, the eye maintains a voltage of 0.40 to 1.0 millivolts with respect to the retinal. This corneoretinal potential, which is roughly aligned with the optic axis and rotates with the direction of gaze, can be measured by surface electrodes on the skin around eyes. Recorded potentials, in the range of 15 to 200 microvolts, are amplified before further processing. With proper

[1] This work is partially supported by Optima Neuroscience, Inc. 101 SE 2 Place, Suite 201-A, Gainesville, FL 32601.

CP953, *Data Mining, Systems Analysis, and Optimization in Biomedicine*
edited by O. Seref, O. E. Kundakcioglu, P. M. Pardalos

calibration, the orientation of the electric dipole can be used to specify the angular position of the eyeball to within 2 degrees vertically and 1.5 degrees horizontally [1]. Even without careful calibration, pattern and amplitude of eye movements can also be determined over time using current clinical EOG recording devices. Bedside measurement of eye movements is an important part of the neurologic examination of patients. For example, abnormalities of eye movements are commonly noted in multiple neurodegenerative disease such as Parkinson's disease, Multiple Systems Atrophy, Progressive Supranuclear Palsy, and other disorders affecting the human basal ganglia [2, 3, 4] as well as in cerebellar ataxias. Differentiating between various types of Parkinsonism or ataxia syndromes can be difficult, however many disorders have different characteristics of eye movements that may allow distinction with the appropriate measurement techniques. Similarly, altered states of consciousness such as drowsiness[3] or intoxication [5, 6, 7] may also cause changes in eye movements. Measurement of changes in eye movements may further be useful in following neurodegenerative disease in which there is a fluctuation in status such as is found in Parkinson's disease (PD). Multiple authors have noted abnormalities in both saccadic and smooth pursuit eye movements in levodopa responsive PD [8, 9, 10, 11, 12, 13, 14]. Versino et al. have also demonstrated that binocular control of eyes is also impaired in PD[15]. In an experiment of nature paralleling animal models, Hotson and Langson (1986)[9] studied humans drug users who used an illicit drug formulation containing MPTP. In these patients, dopamine agonists and precursors improved both accuracy and reaction times of saccades. In an EOG study of patients with idiopathic Parkinson's disease, Rascol et al. noted alterations of saccade latency and smooth pursuit peak velocity in individuals with PD that were in part improved with levodopa[16]. More recently, Hood and Sereno 2006 have shown that levodopa increases response time for reflexive prosaccades (stimulus driven saccades) and reduces error rate in voluntary antisaccades [14]. It is therefore possible that measurement of eye movements in the future may be used to track disease progression or motor fluctuations. Therefore, detection of abnormalities in ocular movement may improve our diagnosis and understanding of multiple neurodegenerative disease and altered states of consciousness. A standard real-time quantitative analysis will improve detection and provide a better understanding of pathology in these disorders, and may be useful in screening, diagnosis, and management of these disorders in the future. In this study, we attempted to utilize pattern classification methods to investigate the relationship between eyemovements and neurodegenerative diseases.

The remainder of the chapter is organized as follows. In Sections 2, 3 and 4, we introduce different types of eye movements, the data acquisition process, and the screening paradigms, respectively. In Section 5 we present the details of the EOG analysis, and finally in Section 6 we give a discussion on the results and future directions.

2. EYE MOVEMENTS

In this section, different types of eye movement is presented. 1. Saccades are the main method for moving eyes to different areas of the visual scene, and are sudden, rapid movements of the eyes. It takes about 100-300ms to initiate a saccade and another 30-120ms to complete the saccade, depending on-among other things-the visual angle

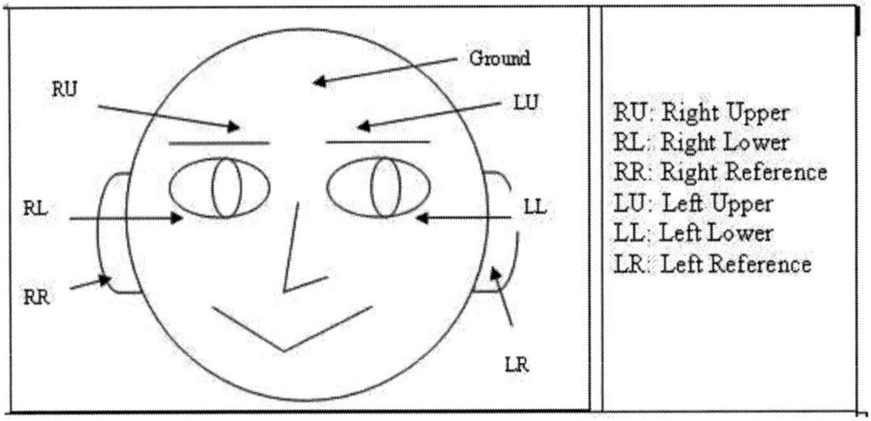

FIGURE 1. EOG electrode placement

traversed. Saccades can be initiated voluntarily, but are ballistic: that is, once they are initiated, their path of motion and destination cannot be replaced. During saccadic movement, it is actually from about 50ms before saccades are initiated–processing of the visual image is suppressed but not entirely inhibited. Thus, processing of the retinal image takes place mainly between the saccades, during the so-called fixations, that last for about 200-600ms. 2. Pursuit motion is a much smoother, slower movement than a saccade; it acts to keep a moving object focused. It cannot be induced voluntarily, but requires a moving object in the visual field. 3. Nystagmus is a pattern of eye movements that occur as a response to the turning of the head (acceleration detected by the inner ear) or the viewing of a moving, repetitive pattern (the train window phenomenon). It consists of smooth pursuit motion in one direction to follow a position in the scene, followed by a fast motion in the opposite direction to select a new position. 4. Drift and microsaccades occur during fixations and consist of slow drifts followed by very small saccades (microsaccades). These movements are involuntary. Steinman et al., 1973 suggested that microsaccades do not have any function. These movements are involuntary[17].

3. DATA ACQUISITION

There are several modalities for monitoring eye movements. The ElectroOculography (EOG) records the potential differences between the cornea and the fundus while eyes are moving. It is non-invasive, and is the least expensive and most available technology as it utilizes standard electroencephalographraphy (EEG) technoloques that are commonly available in most physician clinics. For this study, the EOG recordings were acquired using a standard clinical Nicolet BMSI 5000 EEG acquisition system at a sampling rate 250Hz using an analog to digital (A/D) converter with 16bits quantization, and amplifier with input range of -2.5 to +2.5mV and frequency range of 0.005Hz to 70Hz

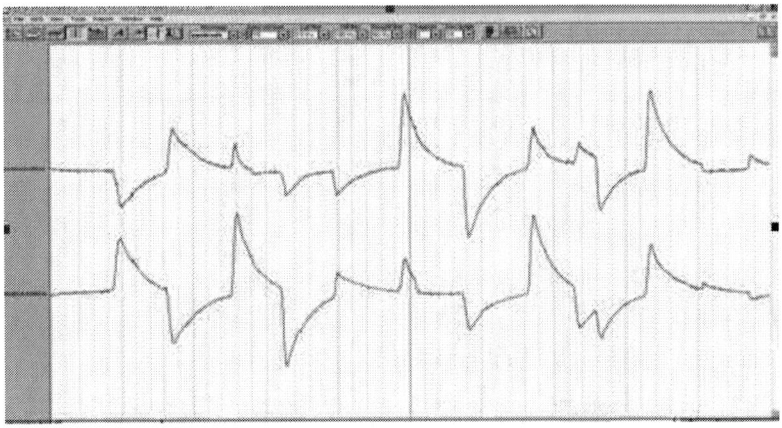

FIGURE 2. 10 seconds EOG from Left(upper) and Right(lower) eye.

4. SCREENING PARADIGMS

In order to capture the desired eye movements, an animation clip (>120s)Table 1 was designed for this research study. The animation contains different eye movement scenarios designed to stimulate participants to engage in vertical smooth pursuit movements (SVP), horizontal smooth pursuit movements (HVP) and random saccades (RS). Three separate experimental sessions are performed. In the first two sessions, (testing vertical smooth pursuit and horizontal smooth pursuit), participants are asked to focus and follow the moving object appearing on a computer screen. These sessions test the ability of the eyes to smoothly follow a moving object. Smooth pursuit can be divided into two separated sections - initiation and continuance. These two sections of smooth pursuit can be quantified using the following 3 parameters: 1. Acceleration: this parameter is the rate of change of eye velocity defines 25ms after onset of the movement. 2. Peak velocity: After onset of the movement, the velocity increases to a maximum point define as the peak velocity of the movement. 3. Latency: The time between the object appearances to the beginning of movement of the eye. In the third session, we evaluate random saccades. A saccade is an abrupt rapid small movement of both eyes, such as when the eyes scan a line of print. The underline dynamics of saccadic eye motion gives insights into the complexity of the mechanisms that control the motion of the eye. The parameters used for describe the saccade movement are the following 4 parameters: 1. The duration of a saccade is the time for completion a saccade movement captured by EOG. 2. The amplitude of a saccade is the angular distance that the eye needs to travel during the movement. It is also called the size of the saccade. 3. The peak velocity is the maximum velocity attainable by the eye. 4. The latency is the time taken from the appearance of an object to the beginning of a saccade movement. The saccade movements are highly predictable, and there exits a linear correlation between peak velocity and the amplitude (needs a reference). This linear correlation can use quantitative measurements of saccades to evaluate the function of the oculomotor system, examine the level of alertness

TABLE 1. Eye movement tasks

	Duration	Repetition
Vertical smooth pursued	50 sec	3
Horizontal smooth pursued	50 sec	3
Random saccades	20 sec	10

and diagnosis the neurological diseases.

5. QUANTITATIVE EOG ANALYSIS

In the quantitative analysis we utilize a pattern recognition algorithm namely the Artificial neural network (ANN)[18, 19]. ANN is being successfully designed and used for pattern classification on bio-signals. We employed ANN here to provide a metrics to differentiate normal subjects from individuals who suffer from neurological disorders or altered states of consciousness, and determine the level of dysfunction based on comparison with the normative data we are developing. The features that we selected and demonstrate here is the latency between eye movement and movement of the moving target object. To extract this feature, we preprocess the EOG recordings associated with saccadic eye movements and smooth pursuit. Structures of pattern recognition for these two types of eye movement are different in the trace/shape in the EOG recordings. The saccadic eye movements analysis algorithm is based on a self-organized processing scheme, a type of artificial neural network that is trained using unsupervised learning to produce low-dimensional representation of the training samples, while the smooth pursuit analysis algorithm reconstructs eye movement trace from EOG signal in a supervised learning framework. The smooth pursuit algorithm operates in two phases. In the first phase EOG data is from collected subjects for training purposes and the system will calibrate itself and neural network will be trained to project EOG recording to a visual cue. If the system maps a set of input variables X_n to output variable Y_k, then forward classification is obtained by a feedforward neural network whose inputs correspond to X and outputs correspond to Y, the network is trained to minize an error measure that captures the difference between the netwoork outputs and actual control system inputs. In the second phase, this neural network will be used to reconstruct the eye movement trajectory from EOG recordings in real time. The self calibration feature of this system makes it independent of the variations in electrode positions. In the saccadic eye movement the important feature is the latency between onset of eye movement and visual cue so we have used an edge detector to detect sharp changes in the EOG signal resulting from eye movement responses to visual cues. The edge detector is designed for image processing and computer vision, and here we also use it as a tool for feature extraction. The variability in different trials mean and standard deviation of latency between visual cue and saccadic eye movement can be used as a score in this task. This analysis has been tested on the recorded data from 5 normal subjects. Fig. A shows the symmetry of the Fig. B shows the output reflecting the difference between different saccadic eye movements. In other words, the difference between the time stamp of the detector and

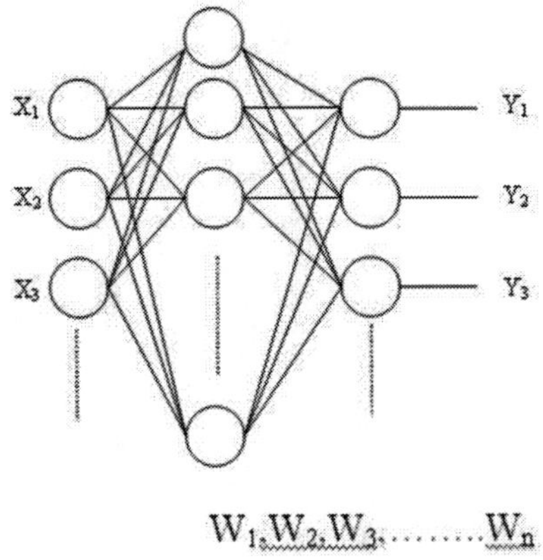

FIGURE 3. Structure of Artificial neural network (ANN).X_n is the input vector.Y_k is the output vector.W_n is the weight matrix.

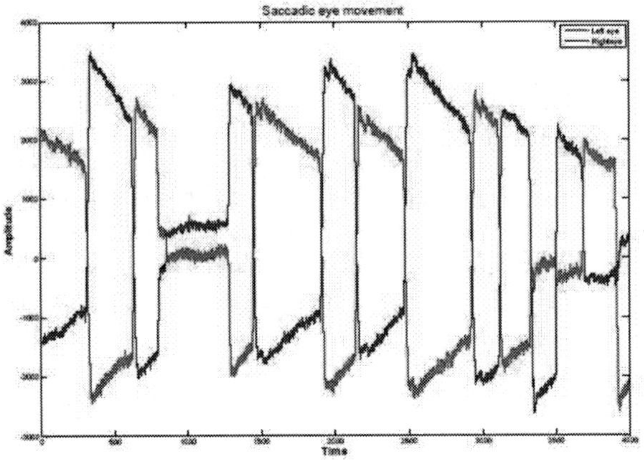

FIGURE 4. The saccadic EOG response of the left and right eye for the normal subject.

visual cue is a signature of the saccadic latency and the amplitude of the detector output is proportional to the speed of saccade. The results with normal subjects here we showed that this analysis is capable of providing a diagnostic measure that neurologists can use it in a clinical setting.

6. DISCUSSION AND FUTURE WORK

In this presented communication, we presented artificial neural network based pattern identification approach for distinguishing normal (control subjects) EOG signal from subjects with neurological disorders. We presented our results from our neural network

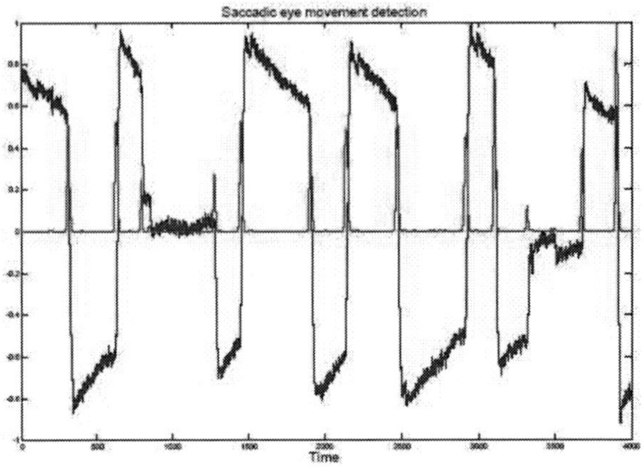

FIGURE 5. The output of the edge detector. The figure shows the output of the amplitude from edge detector

for normal subjects with selected features. Our neural network successful captured the desired patterns from our normal control data sets. A standard automated diagnostic tool that has the ability to quantify critical spacio-temporal characteristics of eye movement that directly relate to disease state, disease severity, level of arousal, and dopaminergic status would be immediately applicable in multiple areas, including diagnosis of neurologic disease (such as differential diagnosis in differentiation of PD from other conditions with similar neurologic findings),alertness monitoring, and intoxication evaluation. Similar ideas have been proposed previously with no clear means of automated real-time analysis. For example Neurokinetics uses a video-based system to determine saccades by identifying the center movement of the iris, and by Eye Dynamics, Inc uses an invasive scleral eye coil. The current methods of diagnoses in the clinic require direct involvement of physicians and neurologists and are subject to human error and judgment. A physician can either visually observe patients' eye movements or an experienced neurologist can directly interpret real-time or recorded EOG data. Both of these methods take extra time from an appointment slot that is usually already restricted to fifteen minutes or less. With the purposed classification algorithm for EOG, we propose to develop a quick screening procedure to classify normal and abnormal eye movement patterns. This studies will help us to select some important features that show prominent changes and give us the best classification results, which might lead us to the solution to further classifications. Development of a larger normative data pool is in process. Also, further studies on a larger sample of subjects to begin comparing individuals with varying disease states to our normative data pool are warranted. The success of this study will provide more much-needed information to guide patients and physician to improve the likelihood of accurate exanimation. We also anticipate we will learn about the interaction between disease state and alterations in consciousness and the oculomotor system. As oculomotor changes parallel dopaminergic state, and eye movement measurements compared to a normative database may be useful for defining optimal dopaminergic state, we anticipate that lessons learned from our study will not only be useful for screening and diagnosis, but also, potentially, for intervention in the future using a feedback-loop system. Portable devices, therefore, using automated eye movement tracking and feed back control loops may provide the basis for major advances in the treatment of neurodegenerative disease in the future.

ACKNOWLEDGMENTS

The authors want to thank Dr. Deng-Shan Shiau for his helpful discussion about data analysis.

REFERENCES

1. Young, and Sheena, *Behavior Research Methods and Instrumentation* **7 (5)** (1975).
2. K. Rottach, D. Riley, A. DiScenna, A. Zivotofsky, and R. Leigh, *Ann Neurol.* **39(3)**, 368–77 (1996).
3. L. Crevits, J. Versijpt, M. Hanse, and K. De Ridder, *Neuropsychobiology* **42(4)**, 202–6 (2000).
4. S. Rivaud-Pechoux, M. Vidailhet, J. Brandel, and B. Gaymard, *Brain* **130(Pt1)**, 256–9 (2007).
5. M. Nawrot, B. Nordenstrom, and A. Olson, *Psychol Sci.* **15(12)**, 858–65 (2004).
6. A. Degia, R. Meadows, S. Johnsen, I. Hindmarch, and J. Boyle, *Percept Mot Skills* **101(2)**, 383–92 (2005).
7. B. Abroms, L. Gottlob, and M. Fillmore, *Psychopharmacology (Berl)* **188(3)**, 324–34 (2006).
8. O. White, J. Saint-Cyr, R. Tomlinson, and J. Sharpe, *Brain* **106(Pt3)**, 571–587 (1983).
9. J. Hotson, E. Langston, and J. Langston, *Ann Neurol.* **20(4)**, 456–463 (1986).
10. F. Degl'Innocenti, A. Ginanneschi, and L. Cipparrone, *Riv Neurol.* **58(5)**, 210–5 (1988).
11. T. Nakamura, R. Kanayama, R. Sano, M. Ohki, Y. Kimura, M. Aoyagi, and Y. Koike, *Acta Otolaryngol Suppl* **481**, 559–62 (1991).
12. F. Chan, I. Armstrong, G. Pari, R. Riopelle, and D. Munoz, *Neuropsychologia* **43(5)**, 784–96 (2005).
13. S. Amador, A. Hood, M. Schiess, R. Izor, and A. Sereno, *Neuropsychologia* **44(8)**, 1475–82 (2006).
14. A. Hood, S. Amador, A. Cain, K. Briand, A. Al-Refai, M. Schiess, and A. Sereno, *J Neurol Neurosurg Psychiatry* **78(6)**, 565–70 (2006).
15. M. Versino, C. Zavanone, S. Colnaghi, G. Beltrami, C. Pacchetti, R. Zangaglia, and V. Cosi, *Ann N Y Acad Sci* **1039**, 588–92 (2005).
16. O. Rascol, M. Clanet, J. Montastruc, M. Simonetta, M. Soulier-Esteve, B. Doyon, and A. Rascol, *Brain* **112(pt5)**, 1193–214 (1989).
17. R. M. Steinman, G. M. Haddad, A. A. Skavenski, and D. Wyman, *Science* **181 no.4102**, 810–819 (1973).
18. G. Irwin, K. Warwick, and K. Hunt, *IEEE Control Engineering Series* **53**, Short Run Press Ltd. (1995).
19. N. Steele, E. Gaura, and J. Godjevac, *The 3rd European IEEE Workshop on Computer-Intensive Methods in Control and Data Processing* **Proceedings**, 187–199 (1998).

Inverse source localization for EEG using system identification approach

Petros Xanthopoulos*, Vitaliy Yatsenko†, Alla Kammerdiner* and Panos M. Pardalos*

*Department of Industrial and Systems Engineering, University of Florida, Gainesville, FL, 32611, USA.

†Institute of Space Research, 40 Glushkov Ave, Kiev 03022, Ukraine.

Abstract. The reconstruction of the brain current sources from scalp electric recordings (Electroencephalogram) also known as the inverse source localization problem is a highly underdetermined problem in the field of computational neuroscience, and this problem still remains open . In this chapter we propose an alternative formulation for the inverse electroencephalography (EEG) problem based on optimization theory. For simulation purposes, a three shell realistic head model based on an averaged magnetic resonance imaging (MRI) segmentation and Boundary Element method (BEM) is constructed. System identification methodology is employed in order to determine the parameters of the system. In the last stage the inverse problem is solved using the computed forward model.

Keywords: Source localization, systems identification, bilinear systems, optimization
PACS: 01.10.Fv

1. INTRODUCTION

The inverse source localization problem can be stated as a problem of finding the brain current given the electric scalp recordings (Electroencephalography - EEG). Let M denote the number of electrodes, N be the number of time points, and L denote the number of grid points of the head model. In terms of matrices the problem can be stated:

Determine appropriate matrix $J(t)$ such that

$$F(t) = K \cdot J(t) \tag{1}$$

where $F(t)$ is the matrix of scalp recordings ($M \times N$), K is the lead field matrix ($M \times 3L$) and $J(t)$ is the current distribution matrix ($3L \times N$ matrix). Since the number of the electrodes (N) is smaller (usually 32-128) than the number L (approximately some thousands) the source localization problem is a highly underdetermined problem.

Most solutions for the inverse problem consider a matrix W, which is a generalized pseudoinverse matrix of K and estimates $\hat{J}(t)$ in the following way:

$$F(t) = K \cdot J(t) \Rightarrow \hat{J}(t) = W \cdot F(t) \tag{2}$$

Such algorithms are minimum norm solution (MN) [2], weighted minimum norm (WMN), LORETA (weighted minimum norm with Laplacian operation) [1], Recursive minimum norm solution (FOCUSS) [3] and LAURA (pseudoinversing method that embodies biophysical constraints into the pseudoinversion) [5]. An alternative approach to

CP953, *Data Mining, Systems Analysis, and Optimization in Biomedicine*
edited by O. Seref, O. E. Kundakcioglu, P. M. Pardalos

the source localization problem is represented by the so-called MUSIC algorithms (Multiple signal classification) [4], in which the data covariance matrix is eigen-decomposed and the uncorrelated components are back-projected and localized separately. On the other hand, these algorithms assume that the brain current sources are uncorrelated, which may not always be the case.

2. FORWARD MODEL

Before introducing the methodology for the inverse problem, we should consider a forward model. The forward problem is to determine the scalp potentials that satisfy the equation:

$$F(t) = K \cdot J(t), \tag{3}$$

given the current distribution matrix $J(t)$. The difficulty of the forward problem lies in describing the human brain in terms of conductivity and producing the lead field matrix K.

There are two major approaches in the literature to estimation of the head conductivity matrix. The first one uses spherical models that are computed via analytical expressions for single or multiple concentric spheres . Although analytical expressions of spherical models are easily implementable they are considered very simplistic. The second approach to estimating the conductivity matrix uses more realistic models of the human head. Even though, realistic head models are closer to the reality, they are computationally hard to implement and sometimes involve estimation of parameters that are unknown (such as anisotropic conductivities of brain tissues).

In this study, we consider a realistic 3 isotropic shell BEM model based on MRI segmentation. We segment the MRI slice using image processing techniques in order to extract the brain,the scalp and the skin surface [7]-[11]. Each one of the three volumes is considered to be a homogenous conductor material.

2.1. BEM model

The electric potential created by some current configuration $j(\underline{r})$, (where $\underline{r}$ represents the Cartesian coordinates vector) in a infinite homogenous matter with conductivity σ_0 is given by:

$$V_0(\underline{r}) = \frac{1}{4\pi} \int_G j(\underline{u}) \frac{\underline{d}}{\|\underline{d}\|^3} d\underline{u}, \tag{4}$$

where $\underline{d}$ is the Euclidean distance vector between the current source and the corresponding $\underline{r}$ and G is the support of $\mathbf{j}(\underline{r})$. According to [6], the potential for a closed surface that consists of k isotropic compartments, each having fixed conductivity σ_i^{in}, $i = 1, \ldots, k$, is given by:

$$\left(\frac{\sigma_k^{in}+\sigma_k^{out}}{2}\right)V(r)=\sigma_0 V_0+\frac{1}{4\pi\sigma_0}\sum_{i=1}^{k}\left(\sigma_k^{in}-\sigma_k^{out}\right)\oiint_{S_i} V(u)n(u)\frac{u-r}{\|u-r\|^3}dS_i \tag{5}$$

In order to approximate the double integral, we segment the head volume in a triangular grid and therefore the double integral is replaced by a double sum. We use the sBEM methodology described in [6] and the spherical electrode coordinates have been transformed into the PAN (Pre Auricular points and nasion) system. The MRI dataset used is an averaged MRI dataset available at [7].

3. DISCRETE BILINEAR FORWARD MODEL AND INVERTABILITY

Consider the multi input-output bilinear forward model described by the equations

$$x(t+1)=Ax(t)+\sum_{i=1}^{m} B_i x(t)u_i(t), \tag{6}$$

$$y(t)=Cx(t), \tag{7}$$

where $x(t)$ is the n-dimensional state vector that corresponds to the matrix $F(t)$, $n=M$, $u_i(t)$, $i=1,\ldots,m$ are the currents, $m=3L$, and $y(t)$ is the m-dimensional measurement vector ($m\leq M$). The matrices A, B_i $(i=1,\ldots,m)$, C are real constant matrices of appropriate dimension and

$$\operatorname{rank} B_i=1, i=1,\ldots,m. \tag{8}$$

Consider another system of the form

$$\begin{aligned}\tilde{x}(t+1)&=\widetilde{A}\tilde{x}(t)+\widetilde{B}\tilde{u}(t),\\ \tilde{y}(t)&=H(\tilde{x}(t),\tilde{u}(t)),\end{aligned} \tag{9}$$

where $\tilde{x}(t)$ is the n-dimensional state vector, $\tilde{u}(t)$ is the m-dimensional control vector, $\tilde{y}(t)$ is the m-dimensional output vector, matrices $\widetilde{A}$ and $\widetilde{B}$ are constant matrices of appropriate dimension, and H is a nonlinear map.

Definition 1 *System (9) is called the inverse for the bilinear system (6)–(7) if, when driven by the output of the original system $y(t+1)$, it produces the input $u(t)$ of the original system as an output.*

The purpose of this chapter is to construct an inverse system for the bilinear system (6)–(7), i.e. to find the matrices $\widetilde{A}$, $\widetilde{B}$ and the form of the map H.

3.1. Construction of inverse system

We shall now derive the equations for the inverse system. When the unit shift operator is applied to equation (7), and then equation (6) is substituted into it, we obtain:

$$y(t+1) = CAx(t) + \sum_{i=1}^{m} CB_i x(t) u_i(t). \tag{10}$$

Equations (10) and (6) combined give a pair of algebraic equations relating $x(t+1)$, $x(t), y(t+1)$ and $[u_1(t) \ldots u_m(t)]^T$

$$\begin{bmatrix} I_n & -B_1x(t) \ldots & -B_mx(t) \\ 0 & CB_1x(t) \ldots & CB_mx(t) \end{bmatrix} \begin{bmatrix} x(t+1) \\ u(t) \end{bmatrix} = \begin{bmatrix} Ax(t) \\ y(t+1) - CAx(t) \end{bmatrix}. \tag{11}$$

The desired inverse system is a pair of equations for $x(t+1)$ and $u(t)$ in terms of $x(t)$ and $y(t+1)$. It is known that the equations (11) can be solved for $x(t+1)$, $u(t)$ uniquely if and only if the matrix

$$R = \begin{bmatrix} I_n & -B_1x(t) & \ldots & -B_mx(t) \\ 0 & CB_1x(t) & \ldots & CB_mx(t) \end{bmatrix} \tag{12}$$

has an inverse. Now consider the following fact from matrix theory.

Lemma 1 *Any n-dimensional square matrix of rank one can be uniquely (within a scalar factor) expressed as a product of a column and a row n-vector.*

Consequently, the matrices B_i can be expressed as

$$B_i = g_i h_i^T, \quad i = 1, \ldots, m.$$

Now let us introduce the following notation:

$$h_I^T x(t) = \gamma_i(x), \quad \Gamma(x) = \operatorname{diag}\{\gamma_1(x)\ldots\}, \quad G = [g_1 \ldots g_m].$$

With this notation matrix R can be written as:

$$R = \begin{bmatrix} I_n & -G\Gamma(x) \\ 0 & CG\Gamma(x) \end{bmatrix}.$$

Matrix R has an inverse if and only if matrix CG is of full rank and

$$\gamma_i(x) = h_i^T x(t) \neq 0, \quad i = 1, \ldots, m, \quad t = 0, 1, \ldots.$$

If the matrix R is invertible, then applying the well-known matrix inversion lemma, we get

$$R^{-1} = \begin{bmatrix} I_n & G(C\Gamma)^{-1} \\ 0 & \Gamma^{-1}(CG)^{-1} \end{bmatrix}$$

and the solution of equations (11) is

$$x(t+1) = [I - G(CG)^{-1}C]Ax(t) + G(CG)^{-1}y(t+1),$$
$$u(t) = -\Gamma^{-1}(x)(CG)^{-1}CAx(t) + \Gamma^{-1}(x)(CG)^{-1}y(t+1). \tag{13}$$

Now, denote $H_i = \{x \mid h_x^T = 0\}$, $X = R^n - \bigcup_{i=1}^{m} H_i$. For any $x(t_0) \in X$ introduce a set $\mathcal{U}_{x(t_0)}^{t_0,t_1}$, containing the sequences $\{u(t_0),\ u(t_0+1),\ldots,u(t_1)\}$ that yield $x(t) \in X$, $t = t_0,\ldots,t_1$.

We have proved the following theorem.

Theorem 1 *The bilinear system (6)–(8) has the inverse on the time interval* $[t_0,t_1]$ *at* $x(t_0) \in X$ *with respect to* $\{u(t_0),\ldots,u(t_1)\} \in \mathcal{U}_{x(t_0)}^{t_0,t_1}$ *if and only if the matrix CG has a full rank. If the bilinear system (6)–(8) has an inverse, then the inverse system is a linear system with nonlinear output, defined by*

$$x(t+1) = \widetilde{A}x(t) + \widetilde{B}y(t+1),$$
$$u(t) = \widetilde{C}x(t) + \widetilde{D}y(t+1), \tag{14}$$

where

$$\widetilde{A} = [I - G(CG)^{-1}C]A, \quad \widetilde{B} = G(CG)^{-1},$$
$$\widetilde{C} = -\Gamma^{-1}(x)(CG)^{-1}CA, \quad \widetilde{D} = \Gamma^{-1}(x)(CG)^{-1}. \tag{15}$$

4. IDENTIFICATION OF A BILINEAR FORWARD MODEL

Now we present the framework for systems identification of the forward model based just on the input-output map. In the continuous-time case these simple models are bilinear systems (BS), given by:

$$\dot{x}(t) = \left[A_0 + \sum_{i=1}^{p} u_i(t)A_i\right] x(t),$$
$$y(t) = \lambda x(t), \tag{16}$$

where the state is $x \in \mathbb{R}^n$ and the (scalar) output is $y(t)$; $u_1(t)$ is the measured signal; $u_2(t),\ldots,u_p(t)$ are the control inputs of the forward model. The approximation result is as follows:

Let $J \subset [0,\infty)$ and $C \subset [C^*(J)]^p$ be compact sets, where $C^*(J)$ denotes the space of continuous functions $J \to \mathbb{R}$, with the supreme norm. A functional $J \times C \to \mathbb{R}$ is said to be causal if and only if its value in $t \in J$, for $(u_1,\ldots,u_p) \in C$, does not depend on $u_1(r),\ldots,u_p(r))$ for all $t < \tau$, then

Theorem 2 *Any causal and continuous forward model* $J \times C \to \mathbb{R}$ *can be arbitrarily closely approximated by BSs.*

This result has been independently obtained by Yatsenko. In the discrete-time case, the former result is not valid, as can be seen with the classic counter example $y(t) = u^2(t-1)$. However, a similar result can be shown replacing BSs by state-affiance systems:

$$\dot{x}(t+1) = \left[A_0 + \sum_{i=1}^{p} f_i(u_i(t), \ldots, u_m(t))A_i\right] x(t),$$
$$y(t) = \lambda x(t), \tag{17}$$

where $x \in \mathbb{R}^n$, $y(t)$ is scalar and $f_i(u_1(t), \ldots, u_m(t))$ are monomials in the inputs $u_1(t), \ldots, u_m(t)$. An input-output map is continuous if and only if for every t the output depends continuously on the inputs $u_1(j), \ldots, u_m(j)$, with $0 < j < t-1$.

Theorem 3 *Any forward model can be arbitrarily closely approximated on a finite time interval and with bounded inputs by state-affine systems.*

At the identification stage it seems convenient, for computational reasons, to use discrete models, as presented by Yatsenko. He used a least-squares method on a single-input-single-output SBE, verifying the following conditions:

i) the input-output map can be linearized at several operating points,
ii) an operating point is described by certain measurable parameters $\theta \in \mathbb{R}^{m-k}$ which, in the state-affine approximation, play the same role as control inputs where $u_{k-1}, \ldots, u_m$ are the parameters θ. In applications, the difficulty lies in the determination of the number of parameters, their physical meaning and their relation with the dynamic behavior of the system.

In spite of what has been said previously, continuous-time analysis should not be discarded, since it is common in practice when processes and systems described by differential equations are found. This is the motivation for our work on the proposed identification method based on an approximating model given by equation (16). The technique used is of the Ritz–Galerkin kind (Mosevich, 1977). The hypotheses i) and ii) mentioned above are not needed, although it is required that the states of the SBE must be observable.

Identification procedure. Let $z(t) \in \mathbb{R}^n$ be the output variables of an SBE when the inputs are $u_1(t), \ldots, u_p(t)$. Both $z(t)$ and $u(t)$ are assumed to be measured. We need to determine coefficients A_i, $i = 1, \ldots, p$, to ensure that the system in equation (16) produces states $x(t)$ close, in some sense, to $z(t)$, given the inputs $u_i(t)$ on the time interval $[0, T]$. To achieve this goal, we follow the ideas presented in (Mosevich, 1977). Let $\{\mu_k(t)\}_{k=0}^{\infty}$ be a basis of $L^2[0, T]$ and define $u_0(t) \equiv 1$. The system in equation (16) can be written as:

$$\dot{x}(t) = \left[\sum_{j=0}^{p} u_j(t) A_j\right] x(t). \tag{18}$$

Observe that the apparently more general

$$\dot{y}(t) = \left[\sum_{j=0}^{p} u_j(t) M_j\right] y(t) + \sum_{j=1}^{p} u_j(t) B_j \tag{19}$$

can be written in the form of equation (17) by defining

$$x = \begin{bmatrix} y \\ 1 \end{bmatrix}. \quad A_j = [M_j \vdots B_j] \tag{20}$$

Integrating equation (18), multiplying by γ_k and integrating from 0 to T, results in

$$b_k^t \triangleq \int_0^{\tau} \gamma_k [x(t) - x(0)] dt = \int_0^T \gamma_k(t) dt \int_0^t \sum_{j=0}^{p} u_j(s) A_j x(s) ds. \tag{21}$$

Let Γ_k denote the integral of γ_k, then

$$b_k^t = \sum_{j=0}^{p} A_j \int_0^T u_j(s) x(s) [\Gamma_k(T) - \Gamma_k(s)] ds. \tag{22}$$

The integral on the right-hand side is denoted by $(q^{jk})^t$. Equation (21) can be repeated in k, until the same number of equations as unknowns is obtained. It can be shown that the number of functions γ_k required is $L = (p+1)n - 1$. Proceeding in this manner, we obtain the system of equations

$$Qa = b, \tag{23}$$

where

$$Q = \begin{bmatrix} q^{00} & q^{10} & \dots & q^{L0} \\ q^{01} & q^{11} & \dots & q^{L1} \\ \cdot & \cdot & \dots & \cdot \\ \cdot & \cdot & \dots & \cdot \\ \cdot & \cdot & \dots & \cdot \\ q^{0L} & q^{1L} & \dots & q^{LL} \end{bmatrix}, \tag{24}$$

$$b = \begin{bmatrix} b_0 \\ b_1 \\ \vdots \\ b_L \end{bmatrix} (L+1) \times n, \quad a = \begin{bmatrix} A_0' \\ A_1' \\ \vdots \\ A_p^t \end{bmatrix} (L+1) \times n. \tag{25}$$

Solving the linear system of equations (22), we calculate the coefficients of the system in equation (18). This solution is always possible because of the linear independence of the γ_k's. Furthermore, it must be noted that we require $u_i(t) \neq u_j(t)$, $0 \leq i, j \leq p$. It is possible to built a rectangular system of the form of equation (22) by adding equations corresponding to γ_k's with $k > L$; and then solving by the least squares method.

5. CONCLUSIONS AND FURTHER WORK

In this chapter, we proposed a framework for solving the inverse EEG problem. The forward modeling via Boundary element method (BEM) for solving Maxwell's equations was adopted. Moreover, we used MRI segmentation for modeling of the inverse problem. At inversion stage, we proposed an alternative bilinear formulation of the inverse problem. The advantages of the bilinear formulation over the classical one are that it allows more refined modeling in terms of current dipoles, and it has a strong inverse solution. The next goal for our research is to apply our model in simulated and real data for localization of epileptic spikes and evoked potentials.

REFERENCES

1. R. D. Pascual-Marqui, C. M. Michel, and D. Lehmann, *International Journal of Psychophysiology* **18**, 49–65 (1994).
2. M.S. Hamalainen, and R. J. Ilmoniemi, *Med Biol Eng Comput* **32**, 35 (1994).
3. I. F. Gorodnitsky , J. S. George , and B. D. Rao, *Electroencephalography and clinical Neurophysiology* **95**, 231–251 (1995).
4. J.C. Mosher, S. Baillet, and R.M. Leahy, *Journal of Clinical Neurophysiology* **16**, 225–238 (1999).
5. R. G. D. Menendez, S. G. Andino, G. Lantz, C. M. Michel, and T. Landis, *Brain Topography* **14**, 131–137 (2001).
6. M. Fuchs, J. Kastner, M. Wagner, S. Hawes, and J. S. Ebersole, *Clin Neurophysiol.* **113**, 702–712 (2002).
7. BrainWeb: Simulated Brain Database, `http://www.bic.mni.mcgill.ca/brainweb/`.
8. C.A. Cocosco, V. Kollokian, R.K.-S. Kwan, and A.C. Evans, *NeuroImage* **5**, S425 (1997). .
9. R.K.-S. Kwan, A.C. Evans, and G.B. Pike, *IEEE Transactions on Medical Imaging* **18**, 1085–1097 (1999).
10. R.K.-S. Kwan, A.C. Evans, and G.B. Pike "An Extensible MRI Simulator for Post-Processing Evaluation" in *Lecture Notes in Computer Science, Visualization in Biomedical Computing (VBC'96)*, **1131**, Springer-Verlag, 1996, pp. 135–140.
11. D.L. Collins, A.P. Zijdenbos, V. Kollokian, J.G. Sled, N.J. Kabani, C.J. Holmes, and A.C. Evans, *IEEE Transactions on Medical Imaging* **17**, 463–468 (1998).

Automated image processing and analysis of cartilage MRI: enabling technology for data mining applied to osteoarthritis

Hussain Z. Tameem[*,¥] and Usha S. Sinha[¥,†]

[*] *Department of Biomedical Engineering, University of California, Los Angeles, 7523 Boelter Hall, Los Angeles, CA 90024 USA.*
[¥] *Medical Imaging Informatics, University of California, Los Angeles, 924 Westwood Boulevard, Suite 420, Los Angeles, CA 90024 USA.*
[†] *Department of Physics, San Diego State University, 5500 Campanile Drive, San Diego, CA 92182 USA.*

Abstract. Osteoarthritis (OA) is a heterogeneous and multi-factorial disease characterized by the progressive loss of articular cartilage. Magnetic Resonance Imaging has been established as an accurate technique to assess cartilage damage through both cartilage morphology (volume and thickness) and cartilage water mobility (Spin-lattice relaxation, T2). The Osteoarthritis Initiative, OAI, is a large scale serial assessment of subjects at different stages of OA including those with pre-clinical symptoms. The electronic availability of the comprehensive data collected as part of the initiative provides an unprecedented opportunity to discover new relationships in complex diseases such as OA. However, imaging data, which provides the most accurate non-invasive assessment of OA, is not directly amenable for data mining. Changes in morphometry and relaxivity with OA disease are both complex and subtle, making manual methods extremely difficult. This chapter focuses on the image analysis techniques to automatically localize the differences in morphometry and relaxivity changes in different population sub-groups (normal and OA subjects segregated by age, gender, and race). The image analysis infrastructure will enable automatic extraction of cartilage features at the voxel level; the ultimate goal is to integrate this infrastructure to discover relationships between the image findings and other clinical features.

Keywords: Atlas, Cartilage, Data Mining, Osteoarthritis
PACS: 87.57.-s

1. INTRODUCTION

Osteoarthritis (OA) is a type of arthritis that is caused by the breakdown and eventual loss of cartilage for one or more joints. OA is the most common form of arthritis and the major cause of activity limitation and physical disability in older people [1]. It is characterized clinically by pain, enlargement and deformity of the joints, and limitation of motion. The prevalence of OA in a population is difficult to assess because the degree of radiological change in symptomatic individuals varies greatly. Many individuals with radiographic evidence of OA have no symptoms [2]. Among the most commonly affected sites in OA, the knee is the major source of

CP953, *Data Mining, Systems Analysis, and Optimization in Biomedicine*
edited by O. Seref, O. E. Kundakcioglu, P. M. Pardalos

reported disability and loss of function [3, 4]. At present there is no medical treatment available for curing or halting the progress of osteoarthritis. An obstacle to assessing cartilage loss has been the absence of validated, non-invasive methods of quantifying articular cartilage degradation. Although substantial advances have been made through radiographic techniques the ability to assess cartilage loss has been fundamentally limited due to the fact that they are an insensitive measure of OA pathology and reflect mainly on a more advanced disease. Moreover, radiography is a projection technique, which collapses three-dimensional (3D) anatomy into 2D images [5].

Significant advances in MRI and digital image analysis technology have revolutionized non-invasive assessment of cartilage macro-morphology such as volume and thickness. These techniques have enabled systematic assessment of the impact of various factors on normal knee cartilage morphology, including sex, maturity and age, body weight and height, body mass index (BMI), knee bone size, bone mineral density and level of physical exercise [6, 7]. Monitoring by serial MR imaging enables evaluation of potential effects of medical or surgical treatment on disease progression, in clinical research and eventually clinical practice. To derive quantitative morphological data from sequential, contiguous images, the articular cartilage must first be segmented from neighboring structures. Due to relatively low contrast in some regions of the joint's surface, fully automated segmentation of the joint cartilage from MR images has yet to be developed and requires different degrees of human interaction [8-11].

Data mining in medicine can be defined as a way to discover new knowledge from the existing data, and then applying the knowledge to aid in diagnosis and prognosis of new patient or data [12, 13]. There are many clinical trails in progress for evaluation of subjects with OA. For example, the Osteoarthritis Initiative (OAI), described in detail below, contains comprehensive information including extensive serial imaging data. Image data mining from such a comprehensive collection will certainly yield previously unknown and potentially useful information from the data. However, imaging data is not directly amenable to analysis and requires processing to extract characteristic image features that will serve as the signature for that image. Imaging features can then be integrated with other types of information stored in the database to perform comprehensive data mining.

The Osteoarthritis Initiative (OAI) is a multi-center, longitudinal, prospective observational study of knee osteoarthritis (OA). The overall aim of the OAI is to develop a public domain research resource to facilitate the scientific evaluation of biomarkers for osteoarthritis as potential surrogate endpoints for disease onset and progression. The OAI will establish and maintain a natural history database for osteoarthritis that will include clinical evaluation data, radiological (x-ray and magnetic resonance) images, and a bio-specimen repository from 4796 men and women ages 45-79. This seven-year project will recruit participants who have, and those who are at high risk for developing, symptomatic knee osteoarthritis. When complete, the OAI should provide an unparalleled state-of-the-art database documenting both the natural progression of the disease, information on imaging, biochemical biomarkers and, outcome measures. Valuable information can be obtained by data mining the OAI and our focus is on MR imaging data. In order to extract imaging features for data mining, automated robust image processing techniques are critical; these methods will be the focus of this chapter. Ultimately, image data mining will help establish one of the primary objectives of the OAI initiative: to help find biochemical, genetic and imaging biomarkers for development and progression of OA.

2. RELEVANT MR IMAGING FEATURES IN OSTEOARTHRITIS

2.1. Overview

Measurement of the dimensions of articular cartilage is believed to be important for assessing the progression of articular joint diseases; in particular, ability to quantify accurately small changes in cartilage thickness and volume may be useful not only in evaluation of disease progression but also for monitoring the efficacy of drug treatments or to monitor cartilage repair [14].

Magnetic resonance imaging (MRI), owing to its excellent soft tissue contrast, can visualize cartilage directly in three dimensions and distinguish it from the bone, other neighboring soft tissues, and synovial fluid. Comparisons of MR evaluation of the morphology of articular cartilage with data from histology have suggested good correlation across those two modalities. Cartilage thinning detected by MRI has been associated with arthritis in human subjects and, importantly, have been found to manifest earlier than that detected by conventional x-ray examinations. A number of studies have used MRI to quantify cartilage volume and thickness in human articular joints, particularly the knee.

In addition to morphological features, MR also provides additional information that has been shown to have potential in disease characterization of OA. These MR specific parameters include the spin-spin relaxation time, T2; Magnetization Transfer contrast, MTC; and Spin-lattice relaxation in the spin locked state, T1rho. Of these, the T2 relaxation time has been established more than the other parameters as relevant in OAI and is an image available in the OAI collection. Earlier studies have also demonstrated spatial dependency of cartilage T2 in both isolated cartilage explants and in vivo human patella. Studies have correlated changes in T2 with early symptomatic degeneration in patellar cartilage. T2, a time constant also known as the spin-spin relaxation time, characterizes the rate at which the M_{xy} component of the magnetization vector decays in the transverse magnetic field. T2 decay occurs 5 to 10 times more rapidly than T1 recovery, and different tissues have different T2s. For example, fluids have the longest T2s (700-1200 msec), and water based tissues are in the 40-200 msec range, while fat based tissues are in the 10-100 msec range. Immobilization of water protons in cartilage by the collagen-proteoglycan matrix promotes T2 decay, while mobile water protons in synovial fluid retain their high signal.

The cartilage is a very thin structure, and the variations in diseased state, especially at the early onset of disease, are small. It is important to be able to detect these subtle morphological and relaxivity changes. The ability to detect these imaging based changes automatically will enable researchers to determine correlation between imaging features with clinical outcomes in large data collections such as the OAI. Previous methods developed for measuring cartilage morphology and relaxivity measures were manual or semi-automatic and have primarily been validated on smaller datasets. Such methods would not be efficient in data mining tasks where a large corpus of images have to be examined.

Clearly, there is now an increasing need to establish fully automated and reproducible methods to quantify articular cartilage changes in morphology and relaxivity for diagnostic and prognostic purposes. In this chapter we emphasize the development of a fully automated image analysis technique which is designed to serve as an infra-structure for mining image information from a large database such as OAI.

2.2. Cartilage morphology

2.2.1. Volume

After segmentation, the cartilage volume can be readily derived as the number of voxels enclosed within the segmented structure. A number of studies have shown direct relation of volume reduction with cartilage degeneration in OA. However, the distribution of cartilage damage in articular joints tends to be uneven and can have higher statistical prevalence on load bearing sites than in the peripheral regions. Hence, the total cartilage volume may be less sensitive in identifying focal defects because small changes in some focal regions may be lost statistically against a background of the far larger volume of cartilage that remains unchanged across a series of measurements. To overcome this variability, the femoral and tibial cartilages of the knee are usually divided in their lateral and medial compartments and variations of thickness and volumes are studied for each compartment individually [15].

2.2.2. Cartilage thickness

The cartilage thickness is defined as the distance between the cartilage-bone and cartilage-synovium interface (articular surface of cartilage) and is determined from the 3D reconstruction of the cartilage by a 3D Euclidean distance algorithm. This algorithm calculates the cartilage thickness as the minimal distance between the bone-cartilage interface and the cartilage surface, taking into account deviations of the minimal distance vector out of the previous original section plane.

With the edges of the cartilage-bone and cartilage-synovium identified, the normal at each point on the bone surface is determined by a local surface-fitting technique. Each normal is extended to meet the cartilage-synovium interface, and the cartilage thickness then is calculated as the distance between the points of intersection of the normal to the two interfaces. The cartilage volume is calculated simply by voxel count for each femoral and tibial compartment. The values for mean and maximal thickness and volume for the cartilage are assessed separately for the medial and lateral tibial and femoral plateaus [11].

2.2.3. *Cartilage relaxometry*

As mentioned briefly above, MR provides information in addition to morphology. Cartilage transverse relaxation time (T2) mapping exploits the sensitivity of MRI to biophysical properties of tissue. These techniques have the potential to identify the earliest stages of matrix degeneration that precede visible cartilage damage. Cartilage T2 mapping uses intrinsic cartilage water as a probe to study the structural integrity of the extracellular matrix. The major component of cartilage is water, which is non-uniform in distribution, increasing from 67% near sub-chondral bone to 74% near the articular surface. The remaining 25 to 35% of the wet-weight of cartilage is solid matrix, primarily type II collagen and large aggregating proteoglycans [16]. The sequence of events leading to visible changes has been postulated in the following order. Collagen fatigue and breakdown are the earliest events in a destructive chain of events; the loss of the collagen framework integrity allows swelling of the proteoglycans, resulting in an increase in cartilage water, and a concurrent increase in cartilage permeability. Compressibility and permeability of cartilage are highly correlated with water content. As water content increases, the matrix of the tissue becomes more compressible, and a greater portion of the load is carried by the solid components of the extracellular matrix. This subsequently leads to increased stress, structural fatigue, fragmentation of cartilage, and ultimately, visible changes such as cartilage fibrillation and ulceration.

The spin lattice relaxation, T2, is a sensitive marker of free water content and mobility. T2 mapping methods have confirmed the non-uniform distribution of water in normal cartilage. E.g., T2 changes from the deep radial zone (32±1 ms) to the outer transitional zone of cartilage (67±2 ms) [17]. Further, T2 values also reflect the changes seen in OA: increased water mobility translates to increased T2 value. In osteoarthritis specimens, water content is increased by 9% compared to normal cartilage, suggesting an increase in T2 of 16 ms.

It is clear from the discussion on morphometry and relaxometry that complicated patterns are seen in normal cartilage and these patterns change with the disease state. The challenge is to extract these features (and their spatial variation) automatically and then reduce them to a compact set for future data mining applications. Image segmentation of the cartilage is a pre-requisite for feature extraction. In the next sections, we first review the status quo in cartilage segmentation from MR images. The review is then followed by a summary of our novel approach to cartilage segmentation and localized feature extraction that will address some of the limitations in the current approaches.

3. IMAGE PROCESSING

3.1. Segmentation

To derive quantitative data from sequential, contiguous images, the anatomical structure on interest (articular cartilage) must first be isolated from neighboring structures (segmented out). Due to relatively low contrast in some regions of the joints surface (e.g. articular contact areas, areas adjacent to the synovial folds, muscle, tendons, ligaments, Hoffa's fat pad, and repair tissue), fully automated segmentation of the joint cartilage from MR images has yet not been developed and the current algorithm require different degrees of human interaction. Verification (and some degree of correction) by an experienced user is generally required on a section by section basis.

There are number of algorithms which have been validated for articular cartilage segmentation and a brief overview follows: 1) Region Growing Method by Peterfly et.al. [18], Piplani et.al. [8], and Eckstein et. al. [19] In this method a connected component classification was performed, after which the object was manually edited to isolate segmented objects by inserting a line of null pixels to separate articular cartilage surfaces from neighboring surfaces. 2) Edge Detection by Kshirsagar et. al. [11]. Here, the binary image for the contour of the femoral bone-cartilage interface was derived from a 3-D edge-map obtained from each 3-D MRI data set using custom-written 3-D edge-detection. Additional manual refinements are made to eliminate any edges that are obviously artifactual. 3) Fitting B-Spline by Cohen et. al. [10]. In this approach, an initial set of points was digitized manually along each articular contour curve, with a coarse spacing of 1–2 cm. An interpolated cubic B-spline curve was fitted to the initial set of points. The B-spline curve, which followed the contour of the desired surface, was then sampled at 0.5 mm intervals. 4) Watershed Algorithm by Ghosh et. al. [9]. This segmentation procedure is initiated by sorting the pixels into a simple array according to intensity, and assigning a threshold pixel value. The cartilage contour is detected automatically by selecting contours with maximum perimeter and minimum cluster label. 5) Active Contours by Lynch et. al. and Kauffmann et. al. [20, 21]. The basic idea of active contour models or snakes is to evolve a curve which is subjected to constraints from the given image and hence is able to detect or segment the region of interest from the same image.

Most of the algorithms mentioned above are either semi-automated requiring some degree of human interaction or are not very accurate in segmenting the cartilage from the neighboring structures. Active contour models [20, 21] based on a 2D model and integrating edge information to stop the evolution of the curve work well but are limited by image noise, edge strength, and presence of iso-intense adjacent tissues, especially at the end of the cartilage volume. We address the limitations of 2D segmentation as well as the presence of fuzzy edges by implementing a 3D algorithm that does not depend on edge gradients as a stopping function for the curve evolution.

3.2. T and Z scores for cartilage abnormality detection

T and Z scores have been used in the diagnosis of osteoporosis providing a measure of deviation of bone density in patients versus normal subjects. Studies have shown these scores are applicable in quantifying the cartilage loss in osteoarthritis as well. T score is an indicator of the difference between patients and young healthy subjects whereas Z score is an indicator of difference between patients and age-matched healthy subjects. A relatively large inter-subject variability of normal cartilage volume has been reported; this variability is reduced by normalizing cartilage volume to joint surface area. In statistics, Z score also called standard score or normal score is a dimensionless quantity derived by subtracting the population mean from an individual raw score and then dividing the difference by the population standard deviation. This conversion process is called standardizing or normalizing. The standard score indicates how many standard deviations an observation is above or below the mean. It allows comparison of observations from different normal distributions.

The patient data are compared with those from young healthy subjects and, if appropriate, also with those from age-matched (elderly) healthy subjects of the same sex. The absolute difference is then divided by the standard deviation of the mean value in the normal volunteers, to compute the T and Z score for each patient. Studies show T and Z scores calculated for imaging features such as cartilage mean and maximum thickness, and cartilage volume for better characterization of OA. Dunn et. al. [22] has calculated T and Z score in T2 mapping to improve the discriminatory power. Since significant differences in quantitative cartilage parameters have been reported between men and women, T score systems must be derived from sex-specific reference data; otherwise, cartilage loss is underestimated in men and overestimated in women. Studies show that physiologic changes of cartilage morphology occur with aging, even in the absence of joint disease. Therefore, T scores should be supplemented by Z scores, to express the difference between the patient and an age-matched (elderly) volunteer, if appropriate [23].

The spatial T and Z maps of relaxometry measurements are generated in the following way. Per-pixel and per-compartmental z scores are generated by using the mean and standard deviation of T2 values from healthy subjects in each compartment as given in Eq. 1:

$$z = \frac{(voxel_i - mean_{healthy})}{sd_{healthy}} \tag{1}$$

where $voxel_i$ is the T2 value in the voxel of interest, $mean_{healthy}$ is the mean T2 value for all voxels of healthy knee in the same compartment, and $sd_{healthy}$ is the standard deviation of the same T2 distribution. The quantitative measure such as cartilage volume and thickness are global measures of the cartilage. The method we propose is a voxel based method which can characterize the cartilage changes taking place in the subject more comprehensively and enable one to identify even the most subtle changes.

4. NOVEL IMAGE ANALYSIS APPROCAHES FOR CARTILAGE IMAGE DATA MINING

We build upon the extensive image analysis methods developed by other groups with focus on deriving features for data mining the OAI database. Our specific aims toward achieving this goal are: (1) to automate the image segmentation algorithm so that it works robustly for a wide variety of image quality including blurred cartilage edges and overlapping intensities of adjacent structures; (2) the creation of morphological and relaxivity atlases of the cartilage using normal subjects segregated by age, sex, and gender. These atlases capture the variation of shape/relaxivity patterns in normal subjects; (3) to identify regions of abnormality in cartilages of subjects with OA by comparison to the normal atlas; this will be based on the analysis of the 3D deformation field required to move voxels to their corresponding locations in the atlas. Deformations beyond ±2SD of normal variations constitute regions with large morphological changes; (4) generating active shape models from the normal subject data and using the shape coefficients to classify cartilage morphology using a compact set of features; these features will be the basis for image data mining.

4.1. Segmentation

The first important task is to develop a fully automated segmentation technique to extract the cartilage from the surrounding structures in the MR images. Our segmentation model is an extension of the active contour without borders proposed by Chan and Vese [24]. The basic idea of active contour models or snakes is to evolve a curve which is subjected to constraints from the given image to segment a region of interest. In classical active contours and snakes models an edge detector is used which basically depends on gradient of the image to the stop the evolving curve. In practice, the discrete gradients are bounded and then the stopping function is never zero on the edges, and the evolving curve may pass through the boundary. The Chan-Vese model does not incorporate a stopping edge-function, and is ideal for segmenting cartilage from MR knee images (which often present with blurry edges).

The new model is described as follows: Let us define the evolving curve C in Ω, as the boundary of an open subset ω of Ω (i.e. $\omega \subset \Omega$, and $C = \partial\omega$). In what follows, inside (C) denotes the region ω, and outside (C) denotes the region $\Omega \setminus \bar{\omega}$. The method is minimization of energy-based segmentation. Assume that an image u_0 is formed by two regions of approximately piece-wise constant intensities, of distinct values u_0^i and u_0^o. We now assume that the region u_0^i is the region of interest to be segmented. Then we have $u_0 \approx u_0^i$ inside the object [or *inside* (C_o)], and $u_0 \approx u_0^o$ outside the object [or *outside* (C_o)]. Now we define the following "fitting" term:

$$F_1(C)+F_2(C)= \int_{inside(C)} |u_0(x,y)-c_1|^2\,dxdy+ \int_{outside(C)} |u_0(x,y)-c_2|^2\,dxdy \tag{2}$$

where C is any other variable curve, and the constraint c_1, c_2, depending on C, are the averages of μ_0 inside C and respectively outside C. In this simple case it is obvious that C_0, the boundary of the object will minimize the fitting term;

$$\inf_{c}\{F_1(C)+F_2(C)\} \approx 0 \approx F_1(C_0)+F_2(C_0) \tag{3}$$

Consider the following cases: 1) if the curve C is outside the object, then $F_1(C) > 0$ and $F_2(C) \approx 0$; 2) If the curve C is inside the object, then $F_2(C) > 0$ but $F_1(C) \approx 0$; 3) If the curve C is both inside and outside the object, then $F_1(C) > 0$ and $F_2(C) > 0$. Finally, the fitting energy is minimized if $C = C_o$, i.e., if the curve C is on the boundary of the object. The new active contour model minimizes the above mentioned fitting term and some new regularizing terms are added for example the length of the curve C, and (or) the area of the region inside C. Thus a new energy function is formed which is defined by Eq. 4

$$F(c_1,c_2,C)=\mu\cdot Length(C)+\nu\cdot Area(inside(C))+ \lambda_1\int_{inside(C)}|u_0(x,y)-c_1|^2\,dxdy+\lambda_2\int_{outside(C)}|u_0(x,y)-c_2|^2\,dxdy \tag{4}$$

where $\mu\geq 0$, $\nu\geq 0$, $\lambda_1,\lambda_2>0$ are fixed parameters. In all numerical calculations the values for $\lambda_1=\lambda_2=1$ and $\nu=0$. The minimization problem is solved by level sets. The final implementation integrates the active shape model to limit the evolution to known anatomical variability. The preliminary segmentation results based on the 2D algorithm is shown in Fig. 1.

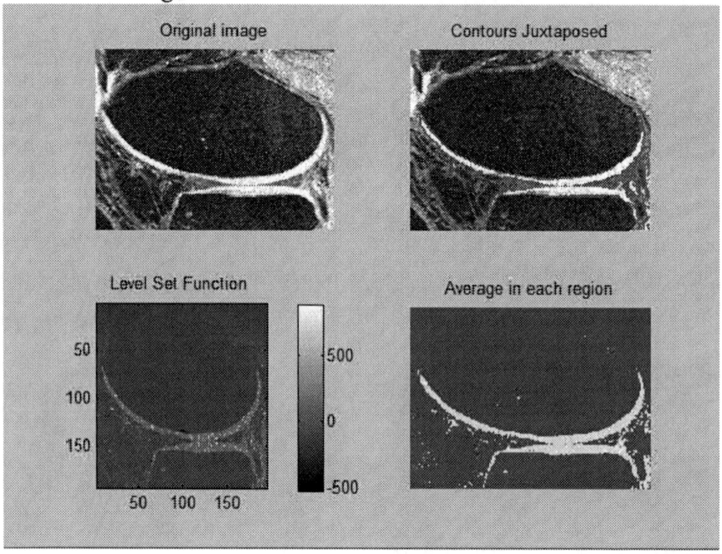

FIGURE 1. The bottom right image in the figure shows the final segmentation of the cartilage obtained on a test knee image shown in the top left corner.

4.2. Creation of morphological atlases

The methodology used to create the atlases includes both affine and non-linear transformations. The segmented cartilage obtained from the previous step is then interpolated to a pixel resolution of $0.365\times0.365\times0.365$ mm^3. The atlas was created using the following steps:

Step I: One subject was randomly chosen from the pool of 20 to serve as a reference. Mutual information based affine transformation corrects for subject positioning and global size differences (with respect to reference).

Step II: An elastic registration based on demons algorithm [25] was employed to locally map all the images in the group of subjects to the reference image using the affine transformation parameters as an initial estimation. This provides 3D deformation fields that can map the spatial locations on an individual in the group to the coordinate system of the reference. The registration algorithm computes the transformation iteratively using Eq. 5.

$$v_{n+1} = G_\sigma \otimes \left(v_n + G_\sigma \otimes \frac{1}{2} \left[\frac{C(T-S\|\nabla S\|\|\nabla T\|)}{\left(\|\nabla T\|^2 + \|\nabla S\|^2\left(\|\nabla S\|^2 + \|\nabla T\|^2 + 2(T-S)^2\right)\right)} \nabla S \right] \right) \quad (5)$$

where v_{n+1} is the correction vector field at iteration n+1, G_σ is the Gaussian filter with variance σ_2, $\otimes$ denotes the convolution, C is the scaling factor and T and S are the target and transformed images respectively. The algorithm estimates the displacement which maps a voxel at location (x, y and z) in T to the corresponding anatomical location in S. The algorithm is implemented hierarchically and to preserve the morphology, deformation vector fields were computed utilizing both the forward and backward transformation.

Step III: A mean intensity image with the shape of the reference image is created by averaging the globally and locally transformed images of the group.

Step IV: A mean deformation field that encodes the shape variation between the reference image and average shape of the elements in the subject group is created by averaging over 3D deformation vector fields of the individual subjects of the group.

Step V: Average deformation field is applied to the average intensity image to generate an average intensity and deformation image template for the group under study.

Step VI: Steps 1-5 are iterated until no significant change in the deformation field is observed relative to the previous computation. At the end of iteration the original reference image is replaced by the average template constructed at Step V generating both average shape (morphometric) and intensity atlases that represent the centroid of the population data set. Fig. 2 visualizes the atlas generated and also visually validates the accuracy of the non-linear registration algorithm.

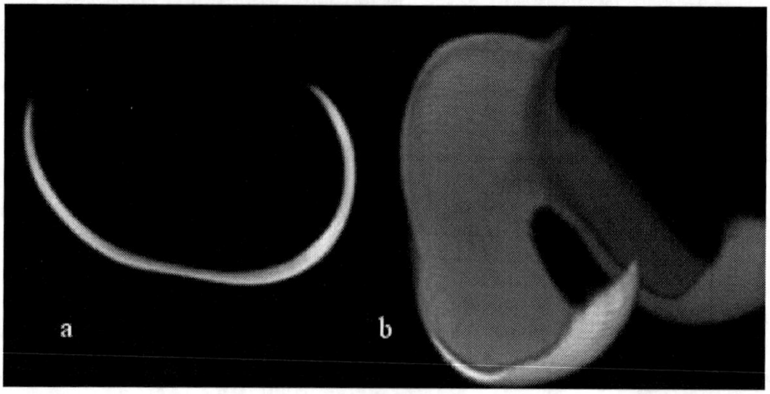

FIGURE 2. Image a is a 2D image showing sharp edges on the atlas created from 20 subjects. Image b is a 3D volume reconstruction of the cartilage from the atlas.

The goal is to create atlases segregated by age, race and gender. The differences in average atlases from each sub-group can be used for data mining: e.g. to determine correlations in shape differences between sub-groups to body hiatus and/or life-style factors. The atlases will be further employed in detection of morphological abnormalities (based on the deformation maps) in patient with OA.

4.3. Deformation maps

Image registration aims to determine a correspondence between each pixel $\boldsymbol{q}$ in the reference image I_R to a pixel $\boldsymbol{p}$ in the test image I_T. When a reference image is warped to match a subject image, some regions may get enlarged and some may be reduced. It is possible to determine the amount of scaling applied to an infinitesimal area around each point of the reference image, by computing the Jacobian determinant of the spatial transformation. In the case of three-dimensional images the displacement vector field from the reference image to the subject i, $\boldsymbol{u}_i$, can be decomposed into its components u_i, v_i and w_i. Similarly $\boldsymbol{q}$ can be expressed in terms of its co-ordinates (x, y, *and* z). The Jacobians n determinant $Ji(\boldsymbol{q})$ is defined as the determinant of the gradient of the mapping function $\boldsymbol{q} + \boldsymbol{u}_i(\boldsymbol{q})$ given in Eq. 6. The value of J is positive as the registration produces a continuous one to one mapping. As part of our preliminary work, deformation fields generated from warping a new subject to the 'normal' atlas were visually evaluated using a color encoding for the magnitude of the deformation.

$$J_i(\boldsymbol{q}) = \left|\nabla(\boldsymbol{q}+\boldsymbol{u}_i(\boldsymbol{q})\right| = \left|\begin{pmatrix} \frac{\partial u_i(\boldsymbol{q})}{\partial x} & \frac{\partial u_i(\boldsymbol{q})}{\partial y} & \frac{\partial u_i(\boldsymbol{q})}{\partial z} \\ \frac{\partial v_i(\boldsymbol{q})}{\partial x} & \frac{\partial v_i(\boldsymbol{q})}{\partial y} & \frac{\partial v_i(\boldsymbol{q})}{\partial z} \\ \frac{\partial w_i(\boldsymbol{q})}{\partial x} & \frac{\partial w_i(\boldsymbol{q})}{\partial y} & \frac{\partial w_i(\boldsymbol{q})}{\partial z} \end{pmatrix}\right| \tag{6}$$

Deformation fields were evaluated for subjects used in the atlas creation, normal subjects from 2 different age groups not included in the atlas, and subjects in three different age groups from the progression cohort. The deformation fields are shown in Fig. 3. The top row shows deformation fields of 3 subjects from the incidence group and different ages. Similarly, the bottom row shows subject from the progression group of three different ages as shown. The region of maximum deformations is indicated by the color red. The image clearly shows that the oldest subject in the progression group with the maximum deformation indicating maximum difference in cartilage morphology with the 'normal subject' atlas. The first subject (denoted as (a) in Fig. 3) was one of the subjects used to create the atlas and as anticipated, shows the least deformations. Normal subjects who belonged to different age groups from those used in the atlas presumably show some morphological changes from 'normal' aging.

The 'progression' group subject in each abnormal image will be labeled by the location, extent and degree of morphological abnormality captured in the deformation maps. This feature set will be used to index the cartilage for data mining the OAI database.

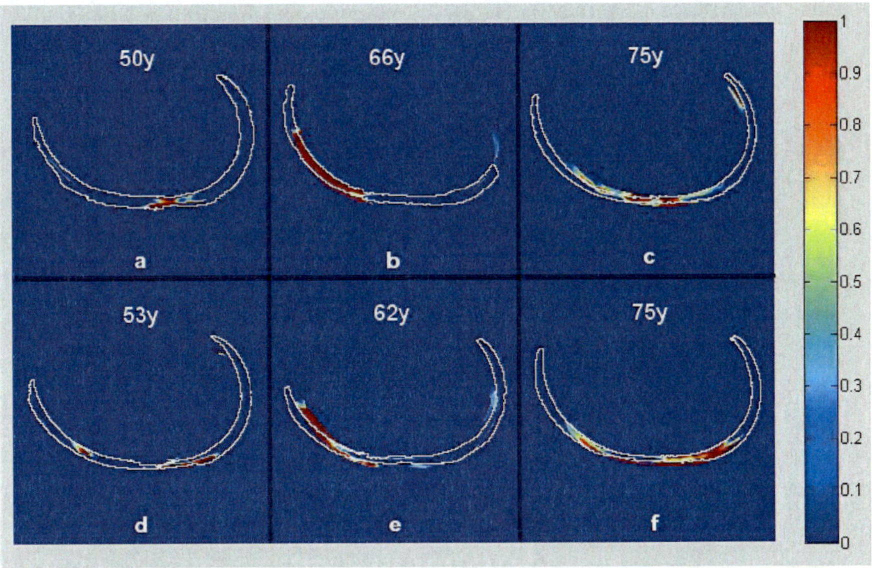

FIGURE 3. Deformation fields of subjects in incidence and progression group. Maximum deformations are seen in subjects from the progression cohort.

4.4. T2 mapping

In addition to morphological atlases, relaxivity atlases capturing the variation in T2 in normal subjects are generated and then used to detect changes in subjects with OA. Cartilage T2 maps are generated from the multi-echo images obtained from the OAI database, by fitting the signal intensity (SI) for the ith, j^{th} pixel as a function of time, t, as shown in the equation.

$$SI_{i,j}(t) = S0_{i,j} * \exp\left(\frac{-t}{T2_{i,j}}\right)^{1} \quad (7)$$

Here $S0_{i,j}$ is the pixel intensity at t=0, and $T2_{i,j}$ is the T2 relaxation time constant of pixel i and j. A T2 map is generated from the $T2_{i,j}$ data. Once the T2 maps are generated a color coded look-up table is generated to display the T2 maps as shown in Fig. 4.

Although it is rather simple to calculate a bulk cartilage T2 value, this method of analysis is insensitive to localized areas of cartilage damage and results in large standard deviations due to the inherent spatial variation in T2 of normal cartilage. It is possible to account for the normal spatial variation by generating normalized cartilage T2 profiles, which describe the mean T2 value as a function of normalized distance from bone. With this method of analysis, it is possible to identify small regional differences in cartilage T2 between groups, for example, as a function of age.

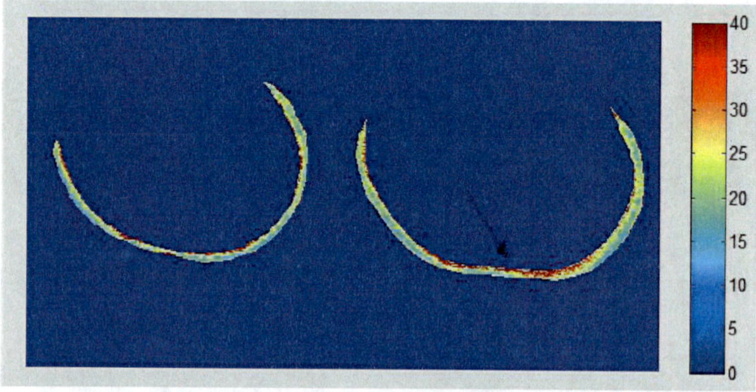

FIGURE 4. Color map showing increased T2 values on the subject from the progression group (right) as compared to the subject chosen from the incidence group (left). Unit in msec.

5. SUMMARY

The novel segmentation strategy is robust to image noise and is less sensitive to fuzzy edges. The atlas based method allows one to model normal physiological variations and detection at the voxel level providing a detailed characterization not possible with current tools. The morphological and relaxivity features provide a compact signature for the images; these features can then be used for data mining the OAI database to uncover new correlations. The same methodology can be applied to the longitudinal studies in the database to infer knowledge on disease progression. In addition, the normal atlases will, in themselves provide a wealth of information on cartilage morphometry and relaxivity which can be used to construct age, gender, and ethnicity-specific models.

ACKNOWLEDGMENTS

"The OAI is a public-private partnership comprised of five contracts (N01-AR-2-2258; N01-AR-2-2259; N01-AR-2-2260; N01-AR-2-2261; N01-AR-2-2262) funded by the National Institutes of Health, a branch of the Department of Health and Human Services, and conducted by the OAI Study Investigators. Private funding partners include Merck Research Laboratories; Novartis Pharmaceuticals Corporation, GlaxoSmithKline; and Pfizer, Inc. Private sector funding for the OAI is managed by the Foundation for the National Institutes of Health. This manuscript was prepared using an OAI public use data set and does not necessarily reflect the opinions or views of the OAI investigators, the NIH, or the private funding partners."

REFERENCES

1. D. Dunlop, L. M. Manheim, E. H. Yelin, J. Song and R. W. Chang, "The cost of arthritis," in Arthritis & Rheumatism, 2003, pp. 101-113.
2. E. Bagge, A. Bjelle, H. A. Valkenburg and A. Svanborg, "Prevelance of radiographic osteoarthritis in two elderly European populations," in Rheumatology International, 1992, pp. 33-38.
3. M. Urwin, D. P. M. Symmons, T. Allison, T. Brammah, H. Busby, M. Roxby, A. Simmons and G. Williams, "Estimating the burden of musculoskeletal disorders in the community: the comparative prevalence of symptoms at different anatomical sites and the relation to social deprivation," in Annals of Rheumatic Diseases, 1998, pp. 649-655.
4. G. Peat, R. McCarney and P. Croft, "Knee pain and osteoarthritis in the older adults: a review of community burden and current use of primary health care," in Annals of Rheumatic Diseases, 2001, pp. 91-97.
5. F. Eckstein, M. Reiser, K. H. Englmeier and R. Putz, "In vivo morphometry and functional analysis of human articular cartilage with quantitative magnetic resonance imaging – from image to data, from data to theory," in Anatomy & Embryology, 2001, pp. 147-173.
6. F. Cicuttini, A. Forbes, K. Morris, S. Darling, M. Bailey and S. Stuckey, "Gender differences in knee cartilage volume as measured by magnetic resonance imaging," in Osteoarthritis and Cartilage, 1999, pp. 265-271.
7. G. Jones, M. Glisson, K. Hynes and F. Cicuttini, "Sex and site difference in cartilage development: a possible explanation for variations in knee osteoarthritis in later life," in Arthritis and Rheumatism, 2000, pp. 2543-9.
8. M. Piplani, D. Disler, T. R. McCauley, T. J. Holmes and J. P. Cousins, "Articular cartilage volume in the knee: semi-automated determination from the three-dimensional reformations of MR images," in Radiology, 1996, pp. 855-9.
9. S. Ghosh, M. Ries, N. Lane, C. Ghajar and S. Majumdar, "Segmentation of the high resolution articular cartilage MR images," in Transaction of Orthopedic Research Society, 2000, pp. 246.
10. Z. Cohen, D. McCarthy, S. Kwak, P. Legrand, F. Fogarasi, E. Ciaccio and G. Ateshian, "Knee cartilage topography, thickness, and contact areas from MRIL in-vitro calibration and in-vivo measurements," in Osteoarthritis and Cartilage, 1999, pp. 95-109.
11. A. Kshirsagar, P.Watson, J. Tyler and L. Hall, "Measurement of localized cartilage volume and thickness of human knee joints by computer analysis of three-dimensional magnetic resonance images", in Investigative Radiology, 1998, pp. 289-99.
12. N. Lavrac, "Machine Learning for Data Mining in Medicine," in *Artificial Intelligence in Medicine*, edited by W. Horn, Berlin/Heidelberg: Springer, 2004, pp. 47.
13. K. J. Cios and G. W. Moore, "Uniqueness of Medical Data Mining", in Artificial Intelligence in Medicine, 2002, pp. 1-24.

14. F. Eckstein, F. Cicuttini, J. P Raynauld, J. C. Waterton and C. Peterfly, "Magnetic Resonance Imaging of articular cartilage in knee osteoarthritis (OA): morphological assessment", in Osteoarthritis and Cartilage, 2006, pp. A46-A75.
15. R. Burgkart, C. Glaser, A. Hyhlik-Dürr, K. Englmeier, M. Reiser and F. Eckstein, "Magnetic Resonance Imaging-Based Assessment of Cartilage Loss in Sever Osteoarthritis," in Arthritis and Rheumatism, 2001, pp. 2072-2077.
16. T. Mosher and B. Dardzinski, "Cartilage MRI T2 Relaxation Time Mapping: Overview and Applications," in Seminars in musculoskeletal radiology, 2004, pp. 355-368.
17. B. Dardzinski, T, Mosher, S. Li, M. Van Slyke and M. Smith, "Spatial variation of T2 in human articular cartilage," in Radiology, 1999, pp. 546-550.
18. C. Peterfly, C. Van Dijke, D. Janzen, C. Gluer, R. Namba, S. Majumdar, P. Lang and H. Genant, "Quantification of articular cartilage in the knee with pulsed saturation transfer subtraction and fat-suppressed MR imaging: optimization and validation," in Radiology, 1994, pp. 485-491.
19. F. Eckstein, M. Schneir, M. Haubner, J. Priebsch, C. Glaser, K. Englmeier and R. Maximilan, "Accuracy of Cartilage Volume and Thickness Measurements With Magnetic Resonance Imaging," in Clinical Orthopedics and Related Research, 1998, pp.137-148.
20. J. Lynch, S. Zaim, J. Zhao, A. Stork, C. Peterfly and H. Genant, "Cartilage segmentation of 3D MRI scans of the osteoarthritic knee combining user knowledge and active contours," in Proceedings of SPIE. International Society for Optical Engineering, 2000, pp. 925-935.
21. C. Kauffman, P. Gravel, B. Godbout, A. Gravel, G. Beaudoin, J. Raynauld, J. Martel-Pelletier, J. Pelleteir and J. de Guise, "Computer aided method for quantification of cartilage thickness and volume changes using MRI: validation study using synthetic model," in IEEE transactions on Biomedical Engineering, 2003, pp. 978-988.
22. T. Dunn, Y. Lu, H. Jin, M. Ries and S. Majumdar, "T2 relaxation time of the cartilage at MR imaging: Comparison with severity of knee osteoarthritis," in Musculoskeletal Imaging, 2004, pp. 592-598.
23. R. Burgkart, C. Glaser, S. Hinterwimmer, M. Hudelmaier, K. Englmeir, M. Reiser and F. Eckstein, "Feasibility of T and Z score from magnetic resonance imaging data for quantification of cartilage loss on osteoarthritis," in Arthritis and Rheumatism, 2003, pp. 2829-2835.
24. T. Chan and L. Vese, "Active contours without edges," in IEEE Trans on Image Processing, 2001, pp. 266-278
25. J. Thirion, "Image matching as a diffusion process: an analogy with Maxwell's demons," in Medical Image Analysis, 1998, pp. 243-260.

Phase-rotated MR spectroscopy using dual-PRESS: theory and application in human brain

Saadallah Ramadan[*,¥], M Albert Thomas[†] and Carolyn E Mountford [*,¥]

[*] *Department of Radiology, Brigham & Women's Hospital, Harvard Medical School, 75 Francis St Boston MA 02215, USA*
[†] *Radiology and Psychiatry, David Geffen School of Medicine at UCLA, Los Angeles, CA 90095, USA.*
[¥] *Previous Address: Institute for Magnetic Resonance Research University of Sydney NSW 2065 Australia*

Abstract. Phase-rotation spectroscopic acquisition is inherently different from the popular signal-averaging method. Phase-rotation will be described theoretically and experimentally in this article. Traditionally, a single echo is acquired in a PRESS or STEAM sequence at a particular TE. If a long-TE spectrum is desired, then another echo is usually acquired at a longer echo time. We here propose a method by which a pair of echoes, at short-TE and a long-TE, are acquired in one experiment, thus saving 50% of total acquisition time without significant sacrifice of spectral quality. The phase-rotation approach has been implemented with the proposed method. An additional benefit of the proposed Dual-PRESS method, is that it gives an insight into the transverse relaxation time constant, T2, for the various metabolites. The Dual-PRESS method is applied in phantom and in-vivo.

Keywords: Magnetic Resonance, Spectroscopy, phase-rotation, Dual-PRESS, T2 relaxation
PACS: 87.80.Lg, 87.85.-d, 87.55.kh

1. INTRODUCTION

Traditionally, one-dimensional (1D) magnetic resonance (MR) spectra are acquired using phase-cycling [1-3], where desired signals add coherently and unwanted signal add destructively. This selective constructive/destructive addition is implemented by carefully cycling the phases of the radio-frequency (RF) pulses and receiver (Rec) to particular values. Phase-rotation is an alternative approach by which signals from various coherences can be sorted out in a two-dimensional (2D) Fourier transformed (FT) map based on their unique phase characteristics. To date, the application of phase-rotation has been implemented in stimulated echo spectroscopy (STEAM) [4], point resolved spectroscopy (PRESS) [5] and double quantum localised spectroscopy [6]. In this article, a phase-rotation based acquisition will be described, both theoretically and experimentally, and its application in acquiring two echoes (Dual PRESS) in a single experiment will be introduced.

Hennig [5] was the first to propose a new method to acquire localised spectroscopic data, where by various scans are stored as rows in a two-dimensional (2D) matrix, which is then doubly Fourier transformed (FT) and a specific row is extracted to represent the 1D spectrum. A prime motivation of this new data acquisition/processing was to record MR spectra using a short echo time (TE) [4, 7-12]. Short TE-based sequences minimize signal loss due to T_2

CP953, *Data Mining, Systems Analysis, and Optimization in Biomedicine*
edited by O. Seref, O. E. Kundakcioglu, P. M. Pardalos

relaxation and reduce signal cancellation due to *J*-modulation, especially for short-T_2^* metabolites. However, TE in localised MR spectroscopy pulse sequences using phase-cycling schemes has to be long enough to allow time for the 'spoiler' or 'crusher' gradients to be applied, since the signals will be co-added in the end. In the phase-rotation method, however, spoilers are still needed but to a lesser extent since the undesired signal can be separated from the resonances of interest. Additionally, the demand on spoiler gradients to reduce unwanted signal increases in inhomogeneous magnetic or RF field [13]. Importantly, long TE-based acquisitions amplify motion artefacts and spoiling gradients also increase sensitivity to motion [14].

The minimum number of RF pulses in a localised pulse sequence is three, due to the requirement for localisation by three orthogonal slices where the intersection constitutes the region of interest or volume of interest (ROI/VOI). The number of single quantum signals occurring after the last RF pulse in a pulse sequence containing n RF pulses is $(3^{n-1} - 1)/2$. A list of existing signals with their coherence transfer pathways (CTPs) [15] is shown in Table 1 gives the reader an insight into what signals need to be eliminated to obtain an uncontaminated spectrum.

TABLE 1. Types of existing signals in a three-pulse localisation sequence and their corresponding phase equations. Echo 12, 13 and 23 are obtained from two RF pulses only. STEAM N and STEAM P are stimulated echo and anti-echo signals [36], respectively.

Signal	Phase equation	CTP
FID 1	$\phi1$	−1,−1,−1
FID 2	$\phi2$	0,−1,−1
FID 3	$\phi3$	0,0,−1
Echo 12	$2\,\phi2 - \phi1$	+1,−1,−1
Echo 13	$2\,\phi3 - \phi1$	+1,+1,−1
Echo 23	$2\,\phi3 - \phi2$	0,+1,−1
STEAM N	$\phi3 + \phi2 - \phi1$	+1,0,−1
STEAM P	$\phi3 - \phi2 + \phi1$	−1,0,−1
PRESS	$2\,\phi3 - 2\,\phi2 + \phi1$	−1,+1,−1
DQF 1	$3\phi2 - \phi1 - \phi3$	+1,−2,−1
DQF 2	$3\phi3 - \phi1 - \phi2$	+1,+2,−1
DQF 3	$\phi1 + \phi2 - \phi3$	−1,−2,−1
DQF 4	$\phi1 + 3\,\phi3 - 3\,\phi2$	−1,+2,−1

In phase-rotation, each transient during, say a 128-average experiment, is acquired with a slightly different RF phase and saved as a row in a two-dimensional (2D) matrix, and the resulting time-phase matrix is then Fourier Transformed (FT) in both dimensions, then each signal in Table 1 will be shifted in the phase dimension by an amount proportional to the size of phase increment during the 128-average experiment and the phase equation in TABLE 1. Fourier Shift Theorem [16, 17] is the basis of signal separation along the phase dimension in phase-rotation technique. When a phase shift, Exp($i\ \omega_o$ t), is applied to the time-domain signal, Exp($i\ \omega$ t), then the frequency domain signal F(ω) can be shifted to F($\omega-\omega_o$) upon Fourier transformation.

It is worth noting that as the phase increment, $\Delta\phi$, decreases, signals from different CTPs become less dispersed in the phase dimension, which makes the process of signal separation more difficult and increases the chance of signal contamination. However, if $\Delta\phi$ is too large then dispersion becomes too high so that peaks from some CTPs will fall off the top and bottom limits of phase dimension, without folding back into the 2D matrix. It is important to observe signals from all CTPs in a 2D phase-rotation map to be able to separate desired signal from undesired ones and to tailor the design of the pulse sequence to minimise undesired signal. Minimising undesired signal is useful in phase-cycling based acquisition protocols, since if undesired signal is weak, then imperfect signal subtraction artefacts are generally weak. An optimum value for $\Delta\phi$ can be selected by calculating the maximum phase change (mpc) from the phase equations of existing signals as a function of $\Delta\phi$ (Table 1) and making sure that 'nrow $\Delta\phi$ /360' evaluates to a whole number, where 'nrow' is the number of increments in the phase dimension. If 'nrow $\Delta\phi$ /360' is not a whole number, then signal will be partitioned between different rows, instead of one row for each type of signal. Thus, $\Delta\phi$ can have any value described by:

$\Delta\phi = 360$ wnum/nrow, where wnum = 1,2,3,4, IntegerPart $(\frac{\text{nrow}}{2\,\text{mpc}})$ where IntegerPart gives the integer part of $\frac{\text{nrow}}{2\,\text{mpc}}$, with an upper limit of $\Delta\phi = \frac{180}{\text{mpc}}$.

The nature of a phase-rotation experiment is cyclic, that is, the phases of the three RF pulses in say, a PRESS sequence phase-rotation experiment, are simply repeated every 16 increments in a 22.5°-increment-size experiment. Thus, the spectrum obtained with phases set to 22.5° is equivalent to the spectrum obtained with the phases set to (360° + 22.5°) and (720° + 22.5°) and so on. This is not the case for 2D spectroscopic correlation experiments [18, 19] were a time delay is incremented continuously and no two spectra acquired at different time delays are the same. One might envisage, then, to acquire only the first 16 increments in a phase-rotation experiments and fill up the remaining 112 increments, in a 128-increment experiment, with the first 16 increments. This approach will save time, but is flawed due to the fact that noise averaging and signal averaging will both be coherent during the 2DFT and as a result, there will be no overall improvement in the signal to noise ratio (SNR) of the final desired row.

2. EXPERIMENTAL IMPLEMENTATION OF PHASE-ROTATION SCHEMES ON MR SCANNERS

At the time of writing, phase-rotation schemes are not implemented by manufacturers of MR scanners, since the software was not written to accommodate 2D spectroscopic acquisition and processing. The interested reader will need to manually implement aspects of the protocol to get the phase-rotation experiment working properly.

First, the 1D pulse sequence must be modified so that the phase of particular RF pulses is incremented during the experiment. Free induction decays (FID) must be written to disk individually without averaging as is the case with phase-cycling based methods. At the end of the experiment, the total number of raw FIDs must be concatenated to form one 2D raw data matrix. The concatenation can either be undertaken by modifying the processing software on the scanner, or offline. Bruker software Xwinnmr [20] and MestReC [21] both have built in protocols to do this concatenation. The raw 2D data is then processed (phase-sensitive mode), and the row of interest is extracted, inverse Fourier-transformed, zero-padded (filled) to double its size, and then Fourier transformed with the appropriate processing parameters. The user must avoid setting $\Delta\phi$ to a value where the desired row is located on row number nrow/2, since signal that is independent of phases of RF pulses will fall on this row, and this might reduce SNR.

3. PHASE-ROTATION VERSUS PHASE-CYCLING

The advantage of the phase-rotation method is that the imperfections due to incomplete signal subtraction are avoided and pure signal can be located at a particular row in the two-dimensional experiment. Artefacts arising from sudden subject movement, a serious problem in in-vivo studies using phase-cycling schemes, can be easily eliminated in phase-rotation schemes due to the unique phase value of the motion-related signal. All effects of signal variation (sudden, periodic, or slow types) are eliminated in phase-rotation technique due to the phase dependence of these signal variations [5].

Phantom studies show that peaks are more symmetrical in phase-rotation than in phase-cycling methods. Resolution is also improved in phase-rotation methods. Phase-rotation based single quantum (STEAM and PRESS) spectra have comparable SNR to phase-cycling ones, but DQF phase-rotation SNR was found to be higher than phase-cycling SNR by about 10-20 %.

4. EXPERIMENTAL

Data were collected on a standard Magnetom 3T Trio system (Siemens AG, Erlangen, Germany), using a circularly polarized (CP) head coil (Siemens AG, Erlangen, Germany) to transmit and receive. A spherical GE brain MRS phantom MRS-HD (GE Medical Systems, Milwaukee, WI) containing the standard brain metabolites (N-acetyl-L-aspartic acid (NAA, 12.5 mM), creatine hydrate (Cr, 10 mM), choline chloride (Ch, 3 mM), myo-inositol (mI, 7.5 mM), L-glutamic acid (Glu, 12.5 mM), DL-lactic acid (Lac, 5 mM), sodium azide (0.1%), potassium phosphate monobasic (KH2PO4, 50 mM), and sodium hydroxide (NaOH, 56 mM) containing 1 ml/l Gd-DPTA (Magnevist)) [22] was used. Water suppression was performed before the pulse sequence by using a WET scheme [23]. Processing was done in Matlab [24] and MestReC [21] programs.

The method was tested on apparently healthy volunteers. This study was approved by the local institutional ethics committee and written informed consent was obtained from participants.

5. STEAM AND PRESS SPECTROSCOPY USING PHASE-ROTATION

STEAM spectroscopy [13, 25] and PRESS spectroscopy [26] sequences were previously implemented in phase-rotation mode by Knight-Scott et al. [4, 27] and Hennig et al. [5], respectively. In PRESS phase-rotation experiment, Hennig et al. [5] was able to reduce TE to 20 ms. This was considered to be quit short in 1992, and was achieved by reducing the duration of spoilers to 2 ms on either side of the 180° RF refocussing pulses with the same amplitude as the slice-refocussing gradient. Phase increments values were $\Delta\phi1 = -\Delta\phi2 = \Delta\phi3 = 22.5^o$. Thus, in a 128-increment experiment, PRESS signal was located on row 24, after 2DFT was applied.

A PRESS phase-rotation spectrum of the brain MRS phantom, acquired without spoilers, is shown in Fig. 1. Five types of signals were obtained and are assigned in Fig. 1. Note that only one of them is considered a "proper" signal. Notice that there was no need to eliminate or cancel undesired signal, but instead it was possible to keep them away from the proper signal on row 24. Phase-cycling will never work if all spoilers are omitted, and the final spectrum in this case will be equivalent to the co-addition of all signals (a-e) in Fig. 1.

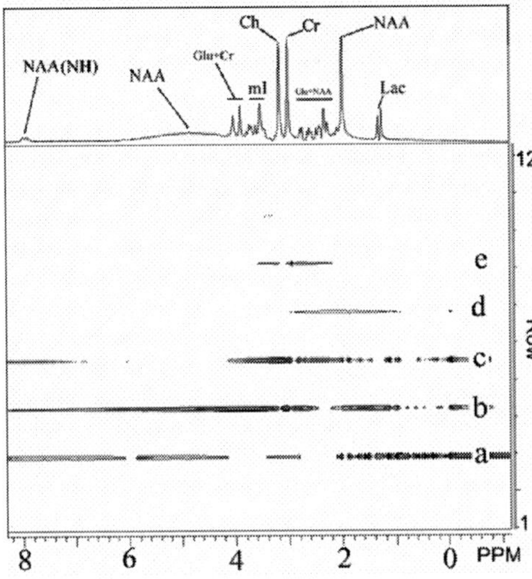

FIGURE 1. PRESS (90° SP1 180° SP1 SP2 180° SP2 Acquire, SP1 and SP2 are spoilers) phase-rotation 2D map of a brain phantom acquired with $\Delta\phi1 = -\Delta\phi2 = \Delta\phi3 = 22.5^o$, all spoilers were set to zero, number of averages per increment =1, Number of increments = 128, TR = 2s, bandwidth = 2000 Hz, total experimental time= 4 minutes, water suppression [23] was inserted before the PRESS sequence. Assignment of rows is as follows: a: PRESS signal, b: Echo 23, c: FID 1 + FID 3 + Echo 13, d: FID 2, and e: Echo 12. 1D spectrum on top is a PRESS spectrum (row 24, a). Turning SP1 and SP2 on will eliminate all signals except signal a.

Phase-rotation based STEAM spectroscopy [4, 27] was only implemented some thirteen years after Hennig's original paper [5]. The aim here again was to reduce TE further. A TE of 6 ms was achieved by implementing phase-rotation based acquisition and by shifting all spoilers and slice-refocusing gradients into the 2nd TE/2 delay. This helped also in the spoiling of the relatively large and undesired signal of FID 3. An illustration of STEAM phase-rotation experiment on a phantom is shown in Fig. 2.

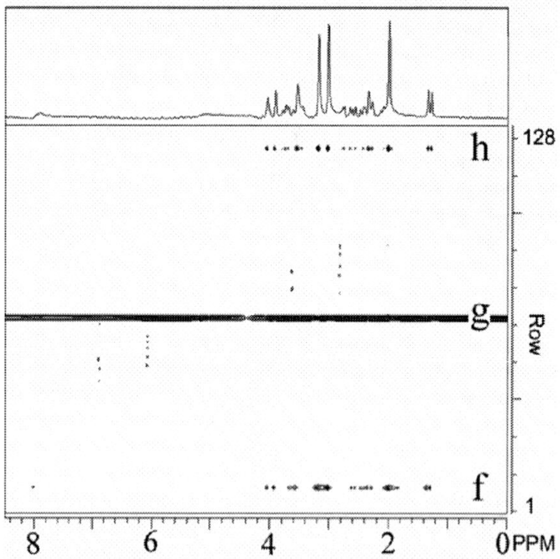

FIGURE 2. STEAM (90° SP1 90° SP2 90° SP1 Acquire, SP1 and SP2 are spoilers) phase-rotation 2D map of phantom acquired with $\Delta\phi1 = 157.5°$, $\Delta\phi2 = \Delta\phi3 = 0°$, all spoilers were set to zero, number of averages per increment =1, Number of increments = 128, TR = 2s, bandwidth = 2000 Hz, total experimental time= 4 minutes, water suppression [23] was inserted before the STEAM sequence. Assignment of rows is as follows: f: FID 1 + STEAM P +PRESS + DQF 3 +DQF 4, g: FID 3 +FID 2 + Echo 23, h: STEAM N + Echo 12 + Echo 13 + DQF 1 + DQF 2. 1D spectrum on top is STEAM spectrum (row 120, h). Turning SP1 and SP2 on will eliminate all signals except signal h. Spectral assignment is the same as in Fig. 1.

The use of a 157.5° phase-step for the 1st RF pulse in STEAM sequence while leaving the phase of all other pulses constant [4] does not seem to be the best design of a STEAM phase-rotation sequence. Whilst such parameters will separate FID 3 from the desired STEAM N signal, but the phase change experienced by STEAM N will be the same as that experienced by Echo12, Echo13, DQF1 and DQF2, which partly defies the purpose of the phase-rotation technique. Even if these unwanted signals do not contain large amounts of signal, they will add noise to the desired row, which will decrease the overall S/N ratio. A better phase-rotation setup is one with $\Delta\phi1 = 22.5°$, $\Delta\phi2 = 67.5°$, $\Delta\phi3 = 45°$, where STEAM N signal can be detected solely on row 32 in a 128 increment phase-rotation experiment. Thus, STEAM N signal can be detected without any signal or noise contamination.

6. DUAL-PRESS

An additional implementation of phase-rotation acquisition is using Dual-PRESS whereby two spectra, instead of one, are acquired in the same experiment. Usually, short and long echo times (TE) localised spectra are acquired from a tissues or organs to show rapidly and slowly relaxing metabolites. In single echo based methods, the echo is acquired until it decays completely. In Dual-PRESS, the short TE is acquired for a shorter time, and then the remaining signal is refocussed using a refocussing radio frequency (RF) pulse, and another echo-with a longer TE- is collected. Thus, Dual-PRESS effectively saves 50% of the acquisition time, by collecting the two echoes in a single experiment. An additional advantage of the proposed sequence is that T2 can be reasonably estimated for the observed metabolites. Ideally, a multiple point fitting curve at many values of TE is preferred for an optimal T_2 measurement.

7. THEORY

In this experiment two echoes are acquired in the phase-rotation mode. The proposed pulse sequence is shown in Fig. 3.

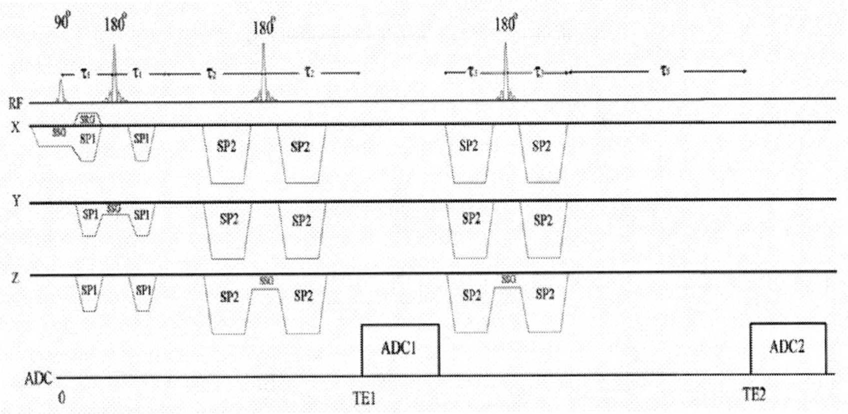

FIGURE 3. A dual-PRESS pulse sequence showing TE1=2*τ1+2*τ2, TE2=TE1+2*τ3+2*ADC1, spoiler gradient pulses SP1 and SP2, slice selection gradients (SSG), slice refocussing gradients (SRG), ADC1 and ADC2. SRG and SP1 are applied as one gradient, and are shown separately for clarity. Note that τ5 must be equal to ADC1 if a phase-sensitive echo2 is desired, otherwise τ5 can be set to zero to collect maximum signal.

Phase-rotation approach was used to collect data using the Dual-PRESS protocol. The data was not averaged in time domain, but was arrayed into a 2D matrix. 2D-FT was implemented and the desired signal detected on a specific row in the 2D map. The position of the row depends on phase-increments size, and number of increments [6].

Note that each 180° RF pulse is surrounded by a pair of spoiler gradients pulses along the three logical axes. The three spoiler pulse pairs have different amplitudes and durations to eliminate any stimulated echo signal. The spoiler gradients of last 180° RF pulse must be made strong enough to spoil any free induction decay (FID) which is usually produced by a tissue that does not experience a perfect refocusing pulse. This FID can be detected as a weak signal on row 48, if the present phase-increments are used and 128 phase-increments are acquired and if spoilers of last 180° RF are not strong enough. Replacing the last refocusing pulse by a hard 180° RF pulse will produce a second echo, but will also produce a large unwanted FID from inside and outside the voxel/slice. The terminal slice-selective 180° RF was applied in the transverse plane to reduce any spurious signal associated with sagittal or coronal orientations [28].

The coherence transfer pathway (CTP) [15] for Echo1 and Echo2 are (0, −1,+1,−1) and (0, +1,−1,+1,−1), respectively, assuming quadrature detection. From the CTP of each coherence, the phase equations of Echo1 and Echo2 as a function of RF phases can be calculated:

Echo1: $2\,\phi 3 - 2\,\phi 2 + \phi 1$

Echo2: $2\,\phi 4 - 2\,\phi 3 + 2\,\phi 2 - \phi 1$

where $\phi 1, \phi 2, \phi 3, \phi 4$ are the phases of the first, second, third and fourth RF pulses, respectively.

The phases of receivers during acquisition of Echo1 and Echo2 signals were kept constant whereas phases of first, second, third and fourth RF pulses where incremented by 11.25°, 33.75°, 78.75° and 45.0°, respectively from one row to another. Thus, in a 128 increment experiment, Echo1 will be detected on the row 28 after applying two-dimensional (2D) Fourier transform (FT) for the 128 concatenated Echo1 raw data files. Similarly, Echo2 will be observed on row 68 after 2DFT is applied to the concatenated 128 Echo2 raw data files. Post-processing details are as described previously [6]. The proposed combination of phase-increments described above is not unique, but was chosen because it keeps the desired signal on a separate row and distinct from any unwanted signal in the final 2D map.

There are many different coherences that can be produced by the proposed combination of four RF pulses [29]. The suggested values of phase-increments ensure that no two coherences occur on the same row after 2D Fourier transform. This ensures the purity of Echo1 and Echo2 signals.

A phase-cycling analogue of the proposed pulse sequence was also implemented, but the quality of the phase-rotated spectrum was superior. Due to imperfect addition/subtraction of unwanted signal, especially in-vivo, some residual spectral feature were observed in the phase-cycled spectrum [4-6, 27].

It is worth noting that Echo1 is fully refocussed due to the balance of time delays and gradient momentums between RF pulses, which means that it can be processed in a phase-sensitive mode. To maximize the amplitude of the heavily relaxed second echo, data acquisition was started immediately after the last spoiler gradient pulse after the terminal 180° RF pulse, i.e. a full echo is acquired. This explains why the second echo was processed in magnitude mode. Acquiring the second echo in phase-sensitive mode can be done if a delay equal to the duration of the first echo is inserted before the second echo. This, however, will reduce the signal amplitude of the second echo due to heavy T_2* weighting. In case of long T_2 species, Echo2 can be phased during post-processing even if full echo signal was acquired by shifting the time-domain data to the left by a duration equal to Echo1.

8. PHANTOM STUDIES

Echo1 and Echo2 were acquired from the brain phantom, using the proposed pulse sequence in phase-rotation mode. 128 phase-increments were acquired, concatenated and doubly Fourier Transformed. Rows corresponding to Echo1 and Echo2 were extracted, inverse-Fourier transformed, zero-filled to 4096 points, and re-Fourier transformed. Phantom Results are shown in Fig. 4. Echo2 was acquired in phase-sensitive mode due to the relatively long T2 values of the metabolites present in the phantom.

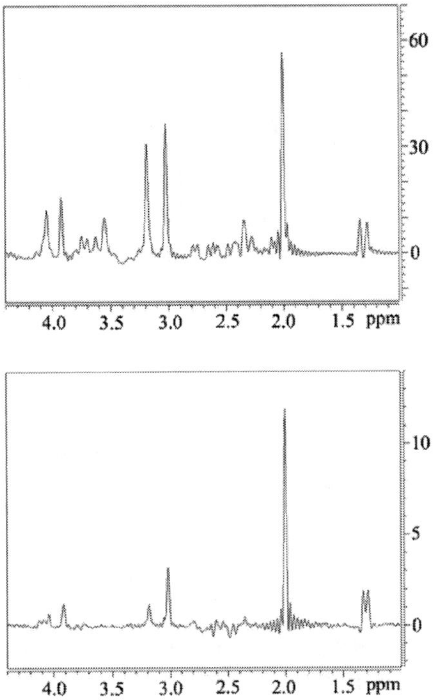

FIGURE 4. 3T MR spectra from a phantom using Echo1 and Echo2 of Dual-PRESS localised single voxel spectrum processed in Mestrec. Acquisition parameters for Echo1 (top) are: TE=30 ms, acquired vector size=512 points, zero-filling to 2048 points, acquisition duration=256 ms, spectral width= 2000 Hz, number of phase-increments=128, voxel size=30x30x30 mm^3. Echo1 was Fourier transformed in phase-sensitive mode after an exponential (line broadening 0.3 Hz) window functions was applied. Acquisition parameters for Echo2 (bottom) are: TE=550 ms, acquired vector size=512 points, zero-filling to 2048 points, acquisition duration=256 ms, spectral width= 2000 Hz, number of phase-increments=128. Echo2 was Fourier transformed in the phase-sensitive mode after an exponential (line broadening 0.3 Hz) window function was applied. WET water suppression scheme [23] was applied prior to the Dual-PRESS sequence. Further removal of water was achieved by convolution method during post-processing. The two spectra are drawn to scale. See text for other details. Spectral assignment is the same as in Fig. 1.

9. HEALTHY VOLUNTEER

An in-vivo Dual-PRESS experiment was undertaken on the brain of an apparently healthy 23 year old and resultant data shown in Fig. 5 and Fig. 6. For comparison, a single voxel, single echo PRESS (traditional) spectrum was acquired from the same region of interest as shown in Fig. 7.

Also, in-vivo T_2 values for Ch, Cr and NAA were measured from a dual-echo experiment, and were found to be 217, 147 and 214 ms, respectively.

A) B)

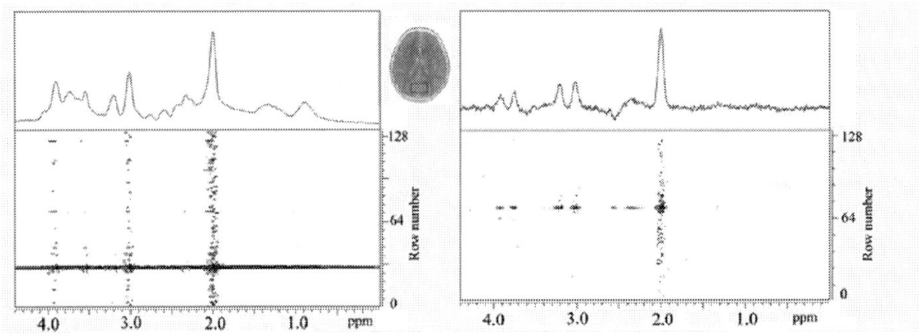

FIGURE 5. 3T in-vivo spectra of Echo1 and Echo2 recorded using the Dual-PRESS in a healthy volunteer and processed in Mestrec. Acquisition parameters for Echo1 (**A**) are: TE=30ms, acquired vector size=256 points, zero-filling to 4096 points, acquisition duration=128 ms, spectral width= 2000 Hz, number of phase-increments=128, voxel size=40x20x30 mm^3. Echo1 was Fourier transformed in the phase-sensitive mode after an exponential (line broadening 0.3 Hz) window functions was applied. The 1D shown on top of 2D is row 28. Acquisition parameters for Echo2 (**B**) are: TE=294ms, acquired vector size=1024 points, zero-filling to 4096 points, acquisition duration=930 ms, spectral width= 1100 Hz, number of phase-increments=128. Echo2 was Fourier transformed in the phase-sensitive mode after an exponential (line broadening of 0.3 Hz) window functions was applied. The 1D shown on top of 2D is row 68. WET water suppression scheme [23] was applied before the Dual-PRESS sequence. Further removal of water was achieved by the convolution method during post-processing. Spectral assignment is the similar to Fig. 1.

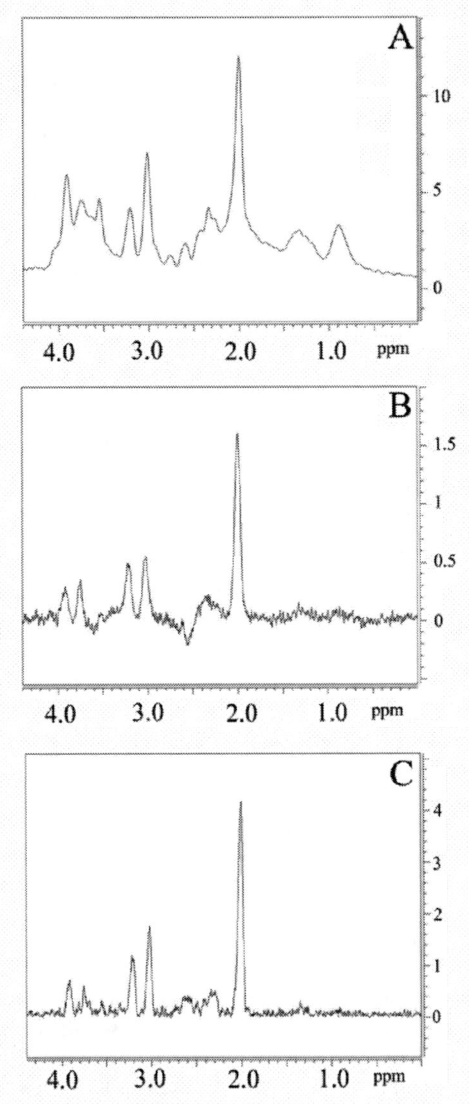

FIGURE 6. 3T in-vivo spectra of Echo1 and Echo2 of Dual-PRESS localised single voxel spectra acquired from a healthy volunteer and processed in Mestrec. Spectra **A** and **B** are the same spectra shown in Fig. 5. Acquisition parameters for spectrum **C**, which is the "full" second echo of a phase-rotation experiment, are: TE=166 ms, acquired vector size=1024 points, zero-filling to 4096 points, acquisition duration=930 ms, spectral width= 1100 Hz, number of phase-increments=128. Echo2 was Fourier transformed in magnitude mode. WET water suppression scheme [23] was applied before Dual-PRESS sequence. Further removal of water was achieved by convolution method during post-processing. The spectra are drawn to scale. Spectral assignment is the similar to Fig. 1.

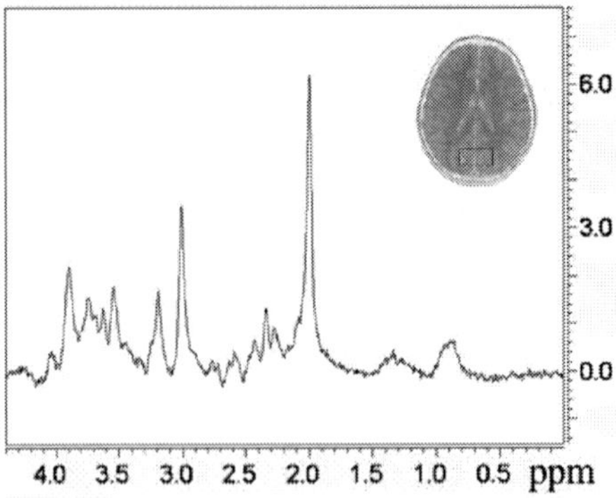

FIGURE 7. 3T in-vivo PRESS localised and water suppressed MR spectrum acquired from a healthy volunteer and processed in Mestrec. Acquisition parameters are: TE=30ms, acquired vector size=1024 points, acquisition time=512 ms, spectral width= 2000 Hz, number of averages=32, voxel size=40x20x30 mm^3. WET water suppression scheme [23] was applied before PRESS sequence. Further removal of water was achieved by convolution method during post-processing. Raw data was zero filled to 2048 points and multiplied by an exponential window function (line broadening factor = 0.3 Hz) before Fourier transform. Spectral assignment is the similar to Fig. 1.

10. DISCUSSION

The short and long TE spectra shown in Fig. 6 have different types of spectral information, but they were both acquired using the same experimental protocol. While the short TE spectrum shows both long and short T2 species, macromolecules and lipids, the long TE spectrum show resonances of Ch, Cr and NAA only which are expected to have long T_2. Strongly J-modulated species (e.g. Glu, glutamine (Gln), γ-aminobutyric acid (GABA) and mI) are heavily attenuated in the long TE spectrum, and are too weak to be detected in the second echo.

The validity of metabolite quantitation by peak fitting in a truncated time-domain signal, was proven by comparing the quantitation results of a truncated time-domain signal with those obtained from a fully acquired one. Differences between the two quantitations were within negligible experimental error.

The implementation of phase-rotation for both echoes ensures that detected signal actually arise from the interrogated voxel. Extraneous signal has a different phase and if survived spoilers, would show up on different row numbers in the final 2D map.

The application of Pade transform [30, 31] to Echo1 due to its relatively short acquisition duration, might bring about some spectral improvement, especially in resolution. However, the traditional Fourier transform results were comparable to those obtained from fully acquired spectral data, and no attempt was done to implement Pade transform.

In addition to the short and long TE spectra obtained in a single experiment, the proposed pulse sequence provides a quick and efficient method to measure T_2 relaxation times of spectral singlets. Singlet peaks of Ch, Cr and NAA methyl groups at 3.18, 3.03 and 2.00 ppm [32] are most suited to illustrate the T_2 evaluation feature due to minimal spectral interference and absence of J coupling. Measuring T_2 from a series of spectra acquired at different TE is well documented [16, 29, 33]. Thus, T_2 can be obtained from the slope of the straight line obtained when TE is plotted against $-\ln(S)$, where S is the peak area [16, 29].

Using in-vivo data from fully refocussed first and second echoes (A and B in Fig. 6), the T_2 values obtained for Ch, Cr and NAA were 217, 147 and 214 ms, respectively. These results match very well with T_2 values reported in the literature [34, 35].

Note that when the second echo is non-refocussed, it cannot be used for T_2 measurements, due to the effect of magnetic inhomogeneities present during the first half of the second echo, and the absence of chemical shift refocussing. However, a more intense, non-refocussed second echo can still be used for non-quantitative purposes.

11. CONCLUSIONS

The phase-rotated Dual-PRESS method allows two localized single voxel short and long TE spectra to be acquired in a single experiment. This method saves 50% of the experimental time and provides T_2 values for different metabolites in water suppressed 1H MR spectroscopy.

ACKNOWLEDGMENTS

This work was supported by a grant from the Australian Research Council (DP0663987).

REFERENCES

1. Kingsley PB. Product Operators, Coherence Pathways, and Cycling Part I: Product Operators, Spin-Spin Coupling, and Coherence Pathways. Concepts in Magnetic Resonance 1995;7:29-47.
2. Kingsley PB. Product Operators, Coherence Pathways, and Phase Cycling. Part III: Phase Cycling. Concepts in Magnetic Resonance 1995;7:167-192.
3. Levitt MH. Spin Dynamics: Basics of Nuclear Magnetic Resonance. Chichester: Wiley & Sons; 2001.
4. Knight-Scott J, Shanbhag D, Dunham S. A phase rotation scheme for achieving very short echo times with localized stimulated echo spectroscopy. Magn Reson Imaging 2005;23:871-876.
5. Hennig J. The application of phase rotation for localized in vivo proton spectroscopy with short echo times. J Magn Reson 1992;96:40-49.
6. Ramadan S. Phase-Rotation in In-vivo Localised Spectroscopy. Concepts in Magnetic Resonance 2007;30:147-153.
7. Frahm J, Michaelis T, Merboldt K, Bruhn H, Gyngell M, Hanicke W. Improvments in localised Proton NMR spectroascopy of Human Brain. Water suppression, short echo times, and 1 ml resolution. J Magn Reson 1990;90:464-473.
8. Mlynarik V, Gruber S, Starcuk Z, Starcuk Z, Jr., Moser E. Very Short Echo Time Proton MR Spectroscopy of Human Brain With a Standard Transmit/Receive Surface Coil. Magn Reson Med 2000;44:964-967.

9. Seeger U, Klose U, Seitz D, Nagele T, Lutz O, Grodd W. Proton spectroscopy of human brain with very short echo time using gradient amplitudes. Magn Reson Imaging 1998;16:55-62.
10. Tkac I, Andersen P, Adriany G, Merkle H, Ugurbil K, Gruetter R. In Vivo 1H NMR Spectroscopy of the Human Brain at 7 T. Magn Reson Med 2001;46:451-456.
11. Tkac I, Starcuk Z, Choi IY, Gruetter R. In Vivo 1H NMR Spectroscopy of Rat Brain at 1 ms Echo Time. Magn Reson Med 1999;41:649-656.
12. Zhong K, Ernst T. Localized In Vivo Human 1H MRS at Very Short Echo Times. Magn Reson Med 2004;52:898-901.
13. Moonen CT, von Kienlin M, Van Zijl CPM, et al. Comparison of Single-Shot Localisation Methods (STEAM and PRESS) for In Vivo Proton NMR Spectroscopy. NMR Biomed 1989;2:201-208.
14. Diehl P, Fluck E, Gunther H, Kosfeld R, Seeling J, editors. In Vivo Magnetic Resonance Spectroscopy II : Localization and Spectral Editing. Volume 27. Berlin, Heidelberg: Springer Verlag; 1992.
15. Ernst RR, Bodenhausen G, Wokaun A. Principles of nuclear magnetic resonance in one and two dimensions. New York : Oxford University Press: Clarendon Press; 1987.
16. Bernstein MA, King KF, Zhou XJ. Handbook of MRI Pulse Sequences. Amsterdam; Boston: Elsevier Academic Press; 2004.
17. De Graaf RA, Rothman DL. In Vivo Detection and Quantification of Scalar Coupled 1H NMR Resonances. Conc Magn Reson 2001;13:32-76.
18. Claridge TDW. High-resolution NMR techniques in organic chemistry: Amsterdam ; Oxford : Pergamon; 1999.
19. Derome AE. Modern NMR techniques for chemistry research. Oxford ; New York: Pergamon Press; 1987.
20. XWINNMR. Bruker BioSpin. Version 3.5. Karlsruhe, Germany; 2006.
21. MestReC. www.mestrec.com. V 4.9.9.1; 1996-2006.
22. Schirmer T, Auer DP. On the reliability of quantitative clinical magnetic resonance spectroscopy of the human brain. NMR Biomed 2000;13:28-36.
23. Ogg RJ, Kingsley PB, Taylor JS. WET, a T1- and B1- insensitive Water-Suppression Method for in vivo Localized 1H NMR Spectroscopy. J Magn Reson Ser B 1994;104:1-10.
24. MathWorks. MATLAB. 7.0.1.24704 (R14); 1984-2004.
25. Frahm J, Merboldt K, Hanicke W. Localized Proton Spectroscopy Using Stimulated Echoes. J Magn Reson 1987;72:502-508.
26. Bottomley PA. Spatial localisation in NMR spectroscopy in vivo. Ann N Y Acad Sci 1987;508:333-348.
27. Knight-Scott J, Dunham SA, Shanbhag DD. Phase Rotation with Asymmetric RF Pulses in Localized Stimulated Echo Spectroscopy (PRAISES) : A 2.5-ms TE Sequence for Clinical Spectroscopy. Proceedings of the Internation Society of Magnetic Resonance in Medicine 2006;14:3052.
28. Ernst T, Chang L. Elimination of Artifacts in Short Echo Time 'H MR Spectroscopy of the Frontal Lobe. Magnetic Resonance in Medicine 1996;36:462-468.
29. Haacke EM, Brown RW, Thompson MR, Venkatesan R. Magnetic resonance imaging: physical principles and sequence design. New York; Chichester: J. Wiley & Sons; 1999. 494 p.

30. Belkic D, Belkic K. In vivo magnetic resonance spectroscopy by the fast Pade transform. Physics in Medicne and Biology 2006;51:1049-1075.
31. Williamson DC, Hawesa H, Thacker NA, Williams SR. Robust Quantification of Short Echo Time 1H Magnetic Resonance Spectra using the Pade′ Approximant. Magnetic Resonance in Medicine 2006;55:762-771.
32. Govindaraju V, Young K, Maudsley AA. Proton NMR chemical shifts and coupling constants for brain metabolites. NMR Biomed 2000;13:129-153.
33. Kreis R, Ernst T, Ross BD. Absolute Quantititation of Water and Metabolites in the Human Brain. II Metabolite Concentration. J Magn Reson Ser B 1993;102:9-19.
34. Barker PB, Hearshen DO, Boska MD. Single-Voxel Proton MRS of the Human Brain at 1.5T and 3.0T. Magnetic Resonance in Medicine 2001;45:765-769.
35. Mlynarik V, Gruber S, Moser E. Proton T1 and T2 relaxation times of human brain metabolites at 3 Tesla. NMR in Biomedicine 2001;14:325-331.
36. Zhu JM, Smith ICP. Stimulated Anti-Echo Selection in Spatially Localized NMR Spectroscopy. Journal of Magnetic Resonance 1999;136:1-5.

Early seizure detection in an animal model of temporal lobe epilepsy

Sachin S. Talathi*, Dong-Uk Hwang*, William Ditto* and Paul R Carney*,†

*J Crayton Pruitt Family Department of Biomedical Engineering, University of Florida, FL 32611
†Department of Pediatrics, Division of Neurology, Department of Neurosceince, University of Florida, FL 32611

Abstract. The performance of five seizure detection schemes, i.e., Nonlinear embedding delay, Hurst scaling, Wavelet Scale, autocorrelation and gradient of accumulated energy, in their ability to detect EEG seizures close to the seizure onset time were evaluated to determine the feasibility of their application in the development of a real time closed loop seizure intervention program (RCLSIP). The criteria chosen for the performance evaluation were, high statistical robustness as determined through the predictability index, the sensitivity and the specificity of a given measure to detect an EEG seizure, the lag in seizure detection with respect to the EEG seizure onset time, as determined through visual inspection and the computational efficiency for each detection measure. An optimality function was designed to evaluate the overall performance of each measure dependent on the criteria chosen. While each of the above measures analyzed for seizure detection performed very well in terms of the statistical parameters, the nonlinear embedding delay measure was found to have the highest optimality index due to its ability to detect seizure very close to the EEG seizure onset time, thereby making it the most suitable dynamical measure in the development of RCLSIP in rat model with chronic limbic epilepsy.

Keywords: Epilepsy, Rat model, EEG, Seizure detection
PACS: 87.80.Tq, 87.80.Xa

1. INTRODUCTION

Epilepsy is a syndrome of episodic brain dysfunction characterized by recurrent seemingly unpredictable spontaneous seizures [1]. The occurrence of the seizure in patients without any forewarning is the most debilitating aspect of the disease. A lot of scientific research has therefore focused on developing methods for predicting, anticipating, forecasting or detecting the seizure early enough [2] such that it could then facilitate timely therapeutic intervention and thereby improve the life of an epileptic patient. The final goal of any seizure prediction or detection scheme is in its application in development of system that not only forewarns the occurrence of seizure but also takes measures to prevent the impending seizure from happening.

A survey of the current seizure prediction tools by [2] suggest that many of these algorithms have not been subject to rigorous statistical evaluation in terms of their predictive power (defined as the ration of true positive to the sum of true positive and false positives), sensitivity (defined as ratio of true positive to the sum of true positive and false negative), and specificity (defined as the ratio of true negative to sum of true negative and false positive) to warrent a closed-loop seizure prevention study. The authors suggest that in order to advance towards a clinical application, future studies on seizure forecasting or prediction should place a strong emphasis on a sound methodology

CP953, *Data Mining, Systems Analysis, and Optimization in Biomedicine*
edited by O. Seref, O. E. Kundakcioglu, P. M. Pardalos

and include rigorous statistical validation.

The absence of sound methodology in the development of a robust seizure prediction algorithm is intrinsically linked to the inherent variability in the EEG data recorded from epileptic patients. Possible reasons for such high variability are the confounding effects of epilepsy types and severity in patients, the pathology of the disease, the effects of anticonvulsant drug treatments, the recording sites and the over all environment surrounding patient care. Animal models of epilepsy that emulate human epileptic condition provide an opportunity to generate experimental EEG data under controlled experimental conditions, thereby minimizing the effect of confounding variables that affect the performance of seizure prediction algorithms. The EEG data generated from these animal models can then form a test bed to evaluate the performance of any given seizure prediction algorithm.

While seizure prediction algorithms aim to detect pre-ictal state minutes to hours in advance of an impending seizure, early seizure detection algorithms aim to identify the seizure onset zone in the EEG data, which may occur few seconds in advance of the clinical manifestation of the seizures. Statistical robustness of early seizure detection schemes [3, 4, 5] make them more attractive for use in the development of closed loop intervention system [6]. However there is still a lack of any comparative study for seizure detection methods in terms of their performance for use in real-time closed loop intervention system.

In this chapter we present a comparative study on a number of different univariate measures derived from the EEG data for seizure detection. We used EEG data from rat model of chronic limbic epilepsy (CLE) in order to compare different dynamical measures for seizure detection which can be utilized in the development of a closed loop seizure intervention system. The following parameters are compared to determine the most efficient measure for seizure detection : (a) Sensitivity (b) Specificity (c) Predictability (d) Computational efficiency and (e) The mean delay in the seizure detection by the algorithm with respect to electrographic seizure onset as determined by visual inspection.

The chapter is organized as follows. In Section 2, we first describe the experimental procedure for preparing the CLE rat model. We then present the mathematical details on the number of measures studied in this chapter. We then describe the statistical approach used to determine the sensitivity and specificity for each measure used for seizure detection and present an empirical optimality function that is used to compare various measures for seizure detection. In Section 3, we begin with the demonstration of various EEG seizure morphology and the common features shared by each EEG seizure. We then present the results derived from each of the measure used for seizure detection. Table 2, summarizes the final results.

2. MATERIAL AND METHODS

2.1. Experimental procedure for developing rat model for chronic limbic epilepsy

2.1.1. Animals

Experiments were performed on 2-month old male Sprague Dawley rats (n= 9) weighing 210-265 g. Protocols and procedures were approved by the Institutional Animal Care and Use Committee of the University of Florida.

2.1.2. Surgery and electrode implantation

The top of the head was shaved and chemically sterilized with iodine. The skull was exposed by a midsagittal incision that began between the eyes and extended caudally to the level of the ears to expose the bregma and lambdoidal suture. A peroxide wash was applied to the skull to remove excess soft tissue. Four 0.8 mm stainless steel screws (Small Parts, Miami Lakes, FL) were placed in the skull to anchor the acrylic headset: 1) two screws were AP 2 mm and bilaterally 2 mm; 2) one screw was AP -3 mm and left 2 mm and served as a ground electrode; and 3) one screw was AP -2 mm to the lambdoidal suture and right 2 mm and served as a reference electrode. Holes were drilled to permit insertion of 2 stainless steel bipolar twist electrodes (1 mm tip separation) into the left and right ventral hippocampi for electrical stimulation. In 3 of the 9 CLE rats, in addition holes were drilled to insert 2 recording stainless steel electrodes (AP -5.3 mm, bilateral 4.9 mm, vertical -5 mm below the dura) in the left and the right ventral hippocampi. 2 stainless steel monopolar recording electrodes were also placed in the bilateral frontal cortices (AP 3.2 mm, bilateral 1 mm, vertical -2.5 mm below the dura). Electrodes were labeled according to their relative positions on the ratÕs skull as LF/RF (left/right frontal cortex) and LH/RH (left/right hippocampus). Electrode pins were positioned in a plastic strip connector and the entire headset was glued into place using cranioplast cement (Plastics One, Inc., Roanoke, VA). In 6 of the 9 rats with CLE, recordings were done with sixteen microwire recording electrodes (50 μm polyimide insulated tungsten microwires) were implanted to the left of the midline suture in a diagonal fashion in the CA1, CA2, CA3 and the dentate gyrus region of the hippocampus. The orientation of the microelectrodes was such that the furthest right microwire was positioned at 4mm caudal to bregma, 1.7mm to the left of midline suture and at a depth of 4mm from the midline suture. The microelectrodes were configured in four bundles (four microelectrodes in each bundle) arranged in rectangular pattern to conform with the morphology of the hippocampus. On the long axis of the bundle each microelectrode was seperated by 200μm and on the short axis, the separation was 400μm. A second array of sixteen microelectordes was place to the right of the midline in the same orientation. The two seizing rats also contained one bipolar twisted Teflon sheathed stainless steel electrode of 330μm diameter, implanted in the contralateral posterior ventral hippocampus (5.3 mm caudal to bregma, 4.9mm to the right of midline suture and at a depth of 5mm from the cortical surface) for stimulating the rat into self sustaining status epilepticus. All the electrodes were chronically secured with dental cement. Following surgery, animals were allowed to recover for 1 week prior to additional procedures.

2.1.3. Induction of status epilepticus

All animals were stimulated electrically 1 week after surgery to induce status epilepticus. During electrical stimulation and iEEG acquisition, animals were housed in specially-made chambers (Bertram et al., 1997). Stimulus trains were delivered for 50-70 min with a duty cycle of 10 sec on/ 2 sec off, consisting of biphasic square wave pulses at a frequency of 50 Hz, a pulse duration of 1 ms, and intensities of 300Ð400 mA. During the stimulus, a normal behavioral response was for the animal to display Òwet dog shakesÓ and increased exploratory activity. After approximately 20-30 min of stimulation, convulsive seizures (up to 1 min in duration) were usually observed about every 10 min. At the end of the stimulus period, continuous iEEG recordings were observed for evidence of slow waves in all recorded channels. If slow waves were not demonstrated, then the stimulus was re-applied for 10 min intervals 1Ð3 times until continuous slow waves appeared following termination of the stimulus. Lack of response to this stimulation protocol was infrequent (<10% of animals) and was attributed, in part, to inaccurate placement of the stimulating electrode in the ventral hippocampus. Upon termination of continuous hippocampal stimulation, the iEEG continued to demonstrate activity below 5 Hz for 12Ð24 h and intermittent spontaneous 30-60 sec electrographic seizures for 2-4 h. Animals were observed for seizure activity and adequate food and water intake for 12Ð24 h after stimulation. Following behavioral stabilization, animals were returned to the vivarium for 6 to 8 weeks during which time spontaneous seizures developed.

2.1.4. iEEG/Video acquisition

Following the induction of limbic status epilepticus, the rats were moved to a specially designed cages that allow full mobility of animals, good visualization for video monitoring and stable recording environment. Each rat cage is a 10-inch diameter cast acrylic tube which is 12 inch high and has a plastic mesh floor. The rat cages are housed in a controlled environment (21 1 oC; humidity 60 %, and standard 12 hour light-dark cycle). Continuous multielectrode activity through all 32 channels in the 6 rats with micro electrodes implanted was recorded over a period of 6 to 8 weeks post stimulation using Tucker Davis (Alachua Florida) Pentusa recording unit sampling at 12KHz. In house software was used to store the data acquired through the Tucker Davis system, in a 16bit binary format for post processing. During this period, the rats that were stimulated seized multiple times at varying time intervals. iEEG and video data were transferred to a 1.4 TB RAID server for subsequent off-line review and analysis. The 3 CLE rats with macro electrode implantation, were connected by a 6-channel commutator and shielded cable to the recording system, which consisted of an analog amplifier (Grass Telefactor-Model

10), a 12 bit A/D converter (National Instruments, Inc), and recording software (HARMONIE 5.2, Stellate Inc., Montreal). The system was synchronized to a video unit for time-locked monitoring of behavioral changes. Each channel was sampled at a uniform rate of 200 Hz and filtered using analog high and low pass filters at cutoff frequencies of 0.1 Hz and 70 Hz, respectively. The recording system used a 4-channel referential montage and was set to a continuous mode so that data sets containing ictal as well as interictal data could be collected for analysis. iEEG and video data were transferred to a 1.4 TB RAID server for subsequent off-line review and analysis

2.1.5. Data characteristics

Long-term continuous iEEG recordings of 9 epileptic animals yielded a total of 45 seizures. Seizures were identified by review of technician logs and visual scanning of all recordings. Seizures were confirmed and classified by a board-certified electroencephalographer who made an independent determination of the time and anatomical location of electrographic seizure onsets. Thirty of the identified seizures showed behavioral characteristics with Racine grading of 1-5. These seizures were then used to analyze the performance of the five different dynamical measures studied in this work in terms of their ability to detect seizures at the onset.

2.2. Dynamical measures for seizure detection

Each measure for seizure detection discussed below is computed on single channel of EEG which depicted the best visual characteristics for seizure as identified by trained epileptologist with a moving window of 10 seconds in duration (2000 data points), with an overlap of 5 second between the neighbouring time windows. The EEG data is preprocessed by band limiting the signal between 1 and 70 Hz.

2.2.1. Nonlinear embedding delay : τ

We define the nonlinear embedding delay function $\tau = min(\tau_1, \tau_2)$. τ_1 is obtained as the first minimum of average mutual information, I define by [7],

$$I(T) = \sum_{x(t),x(t+T)} P(x(t),x(t+T)) \log_2 \left[\frac{P(x(t),x(t+T))}{P(x(t)).P(x(t+T))} \right]$$

where $P(x(t),x(t+T))$ is the joint probability density for EEG measurement data taking values $x(t)$ and $x(t+T)$ respectively, while $P(x(t))$ and $P(x(t+T))$ are individual probability densities for EEG data to take values $x(t)$ or $x(t+T)$ respectively. τ_2 is obtained as first zero crossing of the autocorrelation function C defined by,

$$C(T) = \sum_t [x(t) - \bar{x}][x(t+T) - \bar{x}]$$

with $\bar{x} = \frac{1}{T}\sum_{t=0}^{T} x(t)$.

2.2.2. *Hurst scaling: α*

We use the trend corrected rescaled range analysis [8] to determine the Hurst scaling function α as follows: Let $\bar{x_\tau} = \frac{1}{\tau}\sum_{t=0}^{\tau} x(t)$. For given τ we define the deviation sample mean at any given time t as $\bar{e_t} = x(t) - \bar{x_\tau}$. The cumulative sum of the deviations is then given by $X(t_0,\tau) = \sum_{t=1}^{\tau} \bar{e}_{t_0+t}$. The trend corrected range of the data is then given by, $R(t_0,\tau) = max(X(t_0,\tau)) - min(X(t_0,\tau))$ for the range $1 < t_0 \leq \tau$. If $S^2(t_0,\tau)$ represents the sample variance of the data set $\{\bar{e}_{t_0+i}\}_{i=1}^{\tau}$, the rescaled range is then given by $R(\tau) = \left\langle \frac{R(t_0,\tau)}{S(t_0,\tau)} \right\rangle_{t_0}$. Assuming a power law scaling for the rescaled range function $R(\tau)$, we obtain the Hurst scaling function α as, $R(\tau) = \alpha.\tau^{\beta}$.

2.2.3. *Wavelet scaling: ζ*

The wavelet scaling function ζ is computed as follows: We use a sombrerro mother wavelet [9], $\psi(t) = \frac{1}{\sqrt{2\pi}}(1-t^2)e^{\frac{-t^2}{2}}$ to compute the continuous wavelet transform of time series $x(t)$, as

$$W(\tau,s) = \int x(t)\frac{\psi(\frac{t-\tau}{s})}{\sqrt{s}}dt$$

The nomalized wavelet energy is then obtained as, $\tilde{W}(\tau,s) = \sqrt{\frac{W^2(\tau,s)}{\sigma^2}}$, where σ^2 is the sample variance of the data set $\{x(t)\}$. The wavelet scaling function ζ is then given by $\zeta = max_s\left(\sum_\tau \tilde{W}(\tau,s)\right)$

2.2.4. *White autocorrelation: ω*

We define an autocorrelation measure, which we call White autocorrelation: ω, based on algorithm for seizure detection reported in [10] as follows: The EEG data x(t) in the given moving window of 10 second duration, is divided into two subbands each of 5 second in duration (1000 data points). For data in each subband we then compute maximum and minimum over a group of 10 points S_i, denoted by max(S_i) and min(S_i). Following [10], we then compute relative maximum (*RU*) and minimum (RL) for each of the 100, 10 point groups S_i as,

$$RU_i = min[max(S_i), max(max(S_{i+1}), max(S_{i+2}))]$$

and

$$RL_i = max[min(S_i), min(min(S_{i+1}), min(S_{i+2}))]$$

For each subband we then define a metric for autocorrelation given by

$$M = \sum_{i=1}^{100} (RU_i - RL_i)$$

We finally compute the White autocorrelation measure ω, as the average of the autocorrelation metric M, computed over each subband.

2.2.5. *Gradient of accumulated energy:* ε

Accumulated energy at t-th time window is is obtained by cumulatively summing the variance in the EEG data over all past time windows,

$$AE(t) = \sum_{i=1}^{t} \sigma_i^2$$

where σ_i^2 is the variance of the EEG data in the i-th time window.

The gradient of the accumulated energy at t-th time window is obtaines as, $\varepsilon(t) = \frac{1}{10}(AE(t) - AE(t-1))$

2.3. EEG data set and statistical test

We analyzed data from a total of 9 CLE rats. The analyzed data is comprised of atleast one hour of EEG segment containing a minimum of one seizure and two hours of EEG segment from inter-ictal period selected randomly. In total we analyzed 30 EEG data sets with seizures and 60 EEG control data sets. All the analyzed data is made publicly available at http://ebdl.bme.ufl.edu/~stalathi/EEGData. In order to test the statistical robustness of each measure in its ability to detect seizure, we compute the following there statistical quantities:

- **Predictability** *P*, defined as the likelihood of a detected event being a seizure event and is given by the ratio of number of true positive events detected to the sum total of all the events detected.
- **Sensitivity** *S*, defined as the likelihood of detecting a seizure if it is present and is given by the ratio of the number of true positive events detected to the sum total of the number of seizures present in the data set.
- **Specificity** *K*, defined as the likelihood of not triggering a false event detection and is given by the ratio of the number of true negative events, to the sum total true negative events and the number of false positives.

It is desirable for any seizure detection scheme to be used in real time closed loop intervention program to have a very high sensitivity and specificity.

3. RESULTS

3.1. Raw EEG recordings

EEG signal from long term continuous in-vivo recordings is comprised of several components: seizure activity, interictal spiking activity, artifacts, state dependent EEG modulation and noise [10]. Artifacts are the component of EEG signal not emanating from the brain, but primarily contributed to through movements of the jaw muscles, during chewing, scratching of the headstage mounted on the head of the rat, electrical 60 Hz line-noise and clipping of the EEG signal during data acquisition. State dependent changes are evident in the EEG recordings, through the presence of variable low frequency (1-10 Hz) component in the signal corresponding to sleep v/s awake state of the animal. In addition the high resolution in-vivo recordings also pick up interictal spikes. Spike morphology is dependent on the location of the spike origin, the placement of recording electrodes and the field created by the neuronal circuitry producing the spikes. In addition detereortion of the recording electrodes in long term recordings, also contribute to varying amplitude of interictal spikes.

In Fig. 1 we show EEG segment of about 1.5 hours containing an EEG seizure, and demonstrate some of the other contributions to the EEG signal through all the sources mentioned above.

In Fig. 2 we demonstrate three examples of EEG seizures from CLE rat model, as described by Bertram [11] with the corresponding time-frequency plot. In the first example, the EEG seizure begins with a very high amplitude spike (pop) followed by suppression of the background activity followed by gradual increase in amplitude of the EEG signal, leading to a sustained high amplitude rhythmic activity during the entire seizure epoch. In example two, the EEG seizure begins, sudden increase in amplitude of EEG signal followed by sustained high amplitude rhythmic activity during the seizure epoch. Finally in example three the EEG seizure begins with suppression of background activity which is then followed by a gradual sustained increase in EEG amplitude during the entire seizure epoch. It has been our observation that the clinical manifestation of the seizure becomes evident during the sustained high amplitude EEG activity in the seizure epoch. Since the common feature in all the EEG seizure types that we have observed is the gradual increase in EEG amlitude that is sustained for the duration of the seizure, we mark, the EEG seizure onset time as the first time point in the EEG seizure epoch when the amplitude of EEG signal begins to rise. This EEG seizure onset time (ESOT) was agreed upon by there expert epileptologist. The time at which each dynamical measure considered in this work, crosses the threshold for triggering a warning, is referred to as the algorithmic seizure onset time (ASOT). One of the key parameters in accessing the performance of different dynamical measures in their ability to detect seizure, which is important for the development of a real time closed loop system, is the time difference between ESOT and ASOT. This is one of the key parameters in the optimality function that we use to access the performance of the seizure detection measures presented in this chapter.

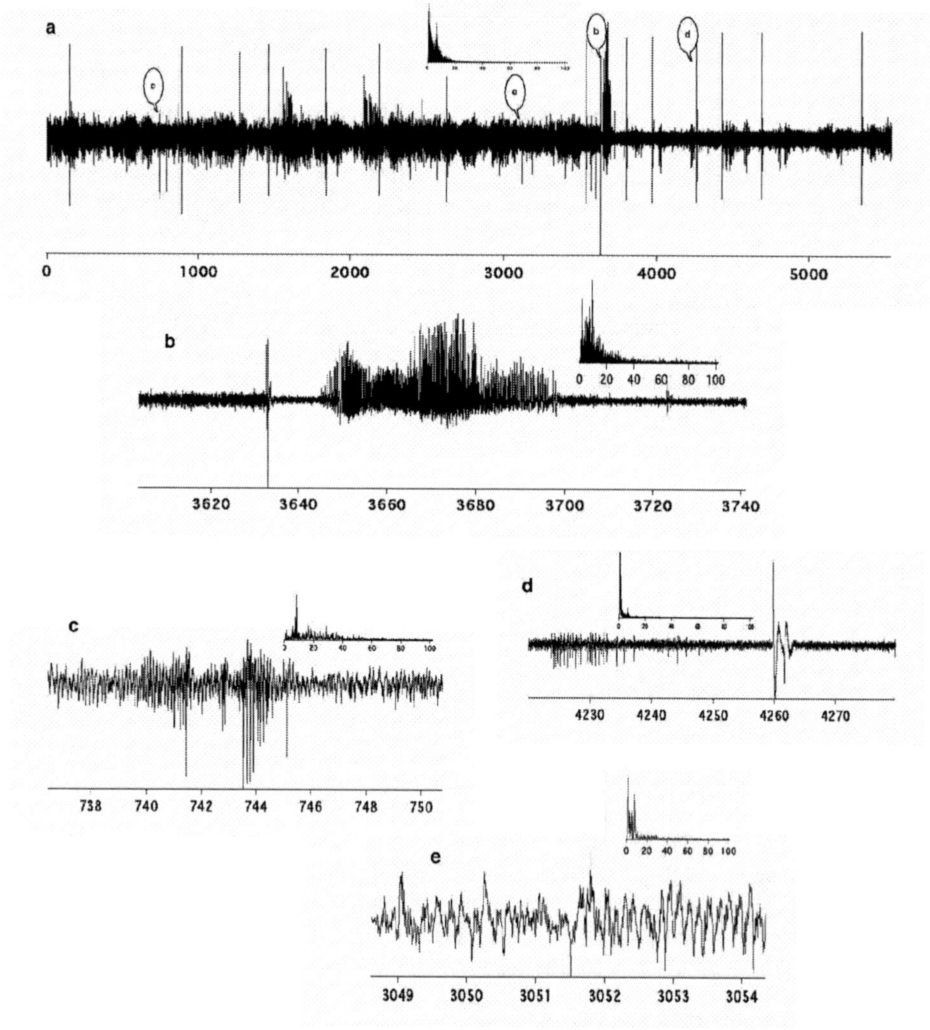

FIGURE 1. Raw EEG data of about 1.5 hour demonstrating different components present in EEG data (a) The entire 1.5 hour epoch of EEG data, and its power spectrrum (shown in inset). Various time points marked on the data correspond to different features that can be identified in the EEG. (b) A typical EEG seizure starting with a large amplitude spike, followed by suppression of the background activity and the subsequent buildup of rhythmic spiking activity. (c) An example of typical chewing artifact present in the EEG data. (d) An example of interictal spike in the EEG data. (e) The background activity is shown. Peaks in the delta and theta frequency band are evident, corresponding to the sleep/ preconscious wake state of the rat. In all the figures, above we show the corresponding power spectrum is given in the inset

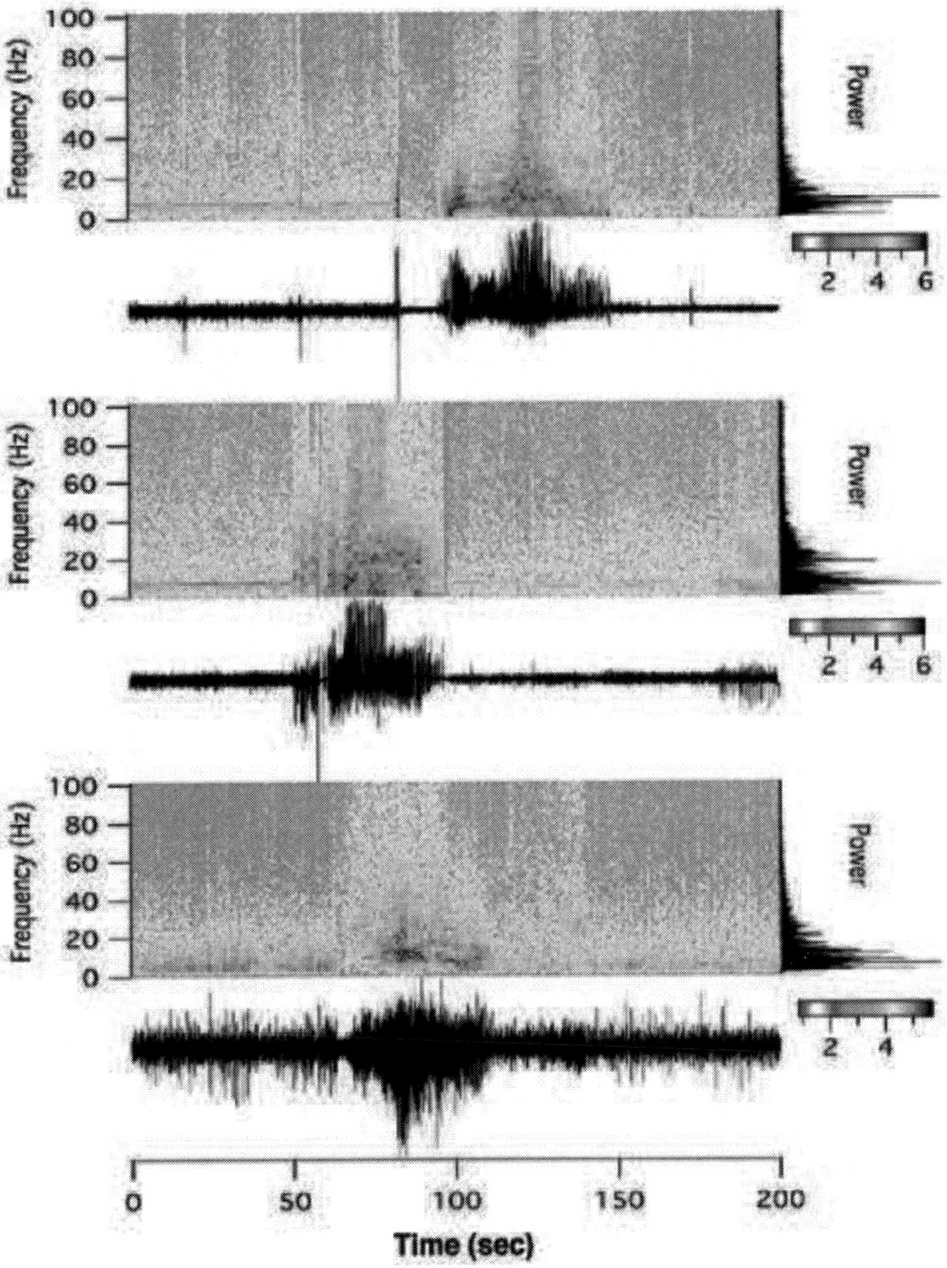

FIGURE 2. Examples of EEG seizure from Rat model of chronic limbic epilepsy. We see that there is sustained increase in energy in the frequency band 0-50 Hz for entire duration of the EEG seizure.

3.2. Threshold choice for seizure detection

The choice of threshold for a given measure i.e., $\tau, \alpha, \zeta, \omega$, and ε, which triggers an algorithmic seizure detection time, was optimized to maximize $P+S+K$ which measures the performance of a given dynamical measure to detect seizure, in terms of its statistical robustness. In Table 1 we summarize the threshold values used for each of the dynamical measures studied in this work.

TABLE 1. Threshold values of the dynamical measures, to trigger an event. The trigger direction determines the direction in which the value of a given measure has to cross the threshold to trigger an event

Algorithm index	Dynamical measure	Threshold	Trigger Direction
1	τ	6	↓
2	α	2.6	↓
3	ζ	10.5	↑
4	ω	2.75×10^{5}	↑
5	ε	2.5×10^{-3}	↑

3.3. Performance analysis

In Fig. 3 we present a sample 5 minute EEG epoch containing an EEG seizure. The time evolution of each of the 5 dynamical measures, for seizure detection are presented below the sample trace in order to demonstrate the ability of each of the dynamical measures, to successfully pick out an EEG seizure.

In order to compare the the performnace of the dynamical measures, in their ability to detect EEG seizure, closest to the ESOT, in Fig. 4 we plot the time difference $\Delta T = ASOT - ESOT$ as function of the seizure number N.

In Fig. 5, we plot the mean delay in the seizure detection time from ESOT, as function of the seizure detection measure used.

In order to compare the overall performance of the different detection algorithms, with the aim of its utility in the development of a closed loop intervention system, we define an optimality index for a given detection algorithm i as follows,

$$O_i = \frac{(P_i + K_i + S_i)}{300} - \frac{(\Delta T_i + C)}{\overline{D}}$$

where O is the dimensionless optimality function, P, S and K are the statistical estimates of predictability, sensitivity and specificity respectively, ΔT is the difference in time of the seizure detection by the given algorithm (ASOT) and the EEG seizure onset time

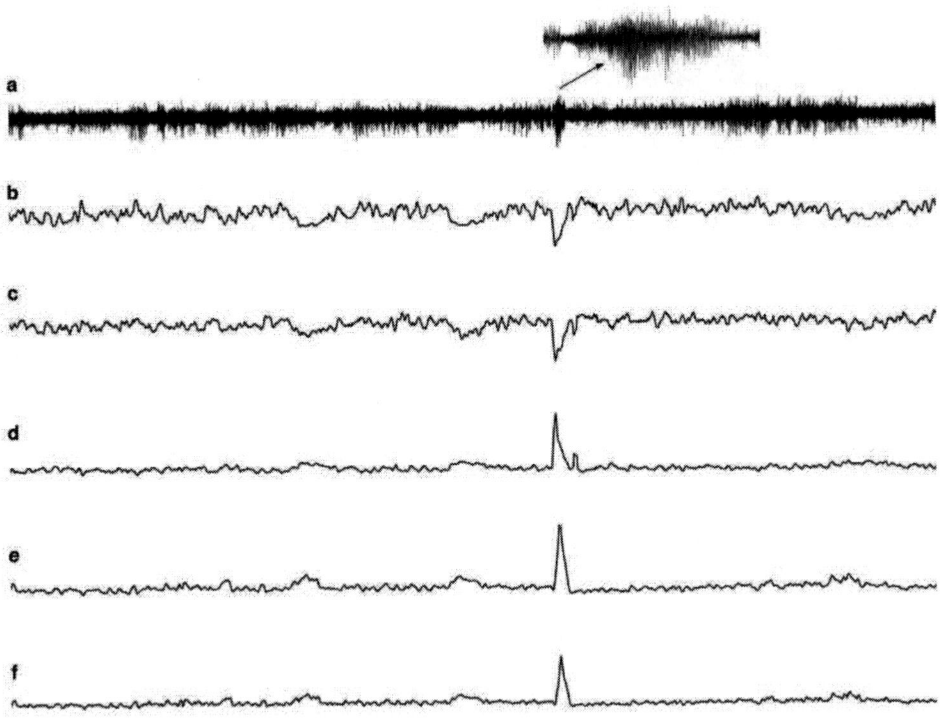

FIGURE 3. (a) 5 minute EEG epoch containing a seizure. (b) Time evolution of the nonlinear embedding delay measure (c) Time evolution of the Hurst scaling measure (d) Time evolution of the wavelet scaling measure (e) Time evolution of the White autocorrelation measure (f) Time evolution of the gradient of accumulated energy measure

(ESOT), $\overline{D}$ is the mean duration of the EEG seizures, from the data set analyzed and C is the computational efficiency of seizure detection algorithm, measured in terms of time in seconds required to compute the measure for EEG data set of 10 seconds duration, which is the time window size considered in all the analysis. All the algorithms were computed using matlab script on a 2GHz intel core duo iMac. In Table 2 we summarize the results for all the detection algorithms presented in this work and their corresponding optimality values as defined above.

We see that although the nonlinear embedding delay computation is most computationally intensive, it has the highest optimality index value, in part due to its ability to detect seizure very close to the EEG seizure onset zone, making it more suitable as a tool in development of a closed loop control device.

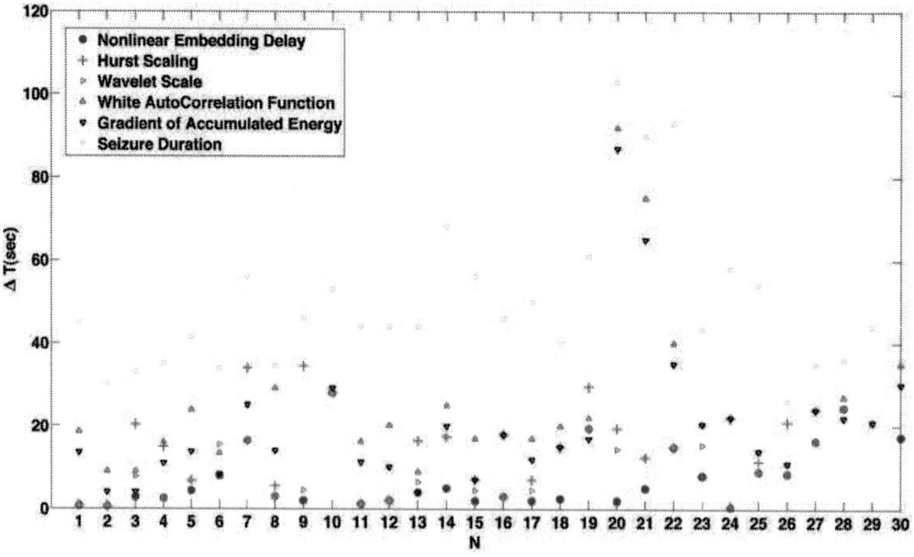

FIGURE 4.

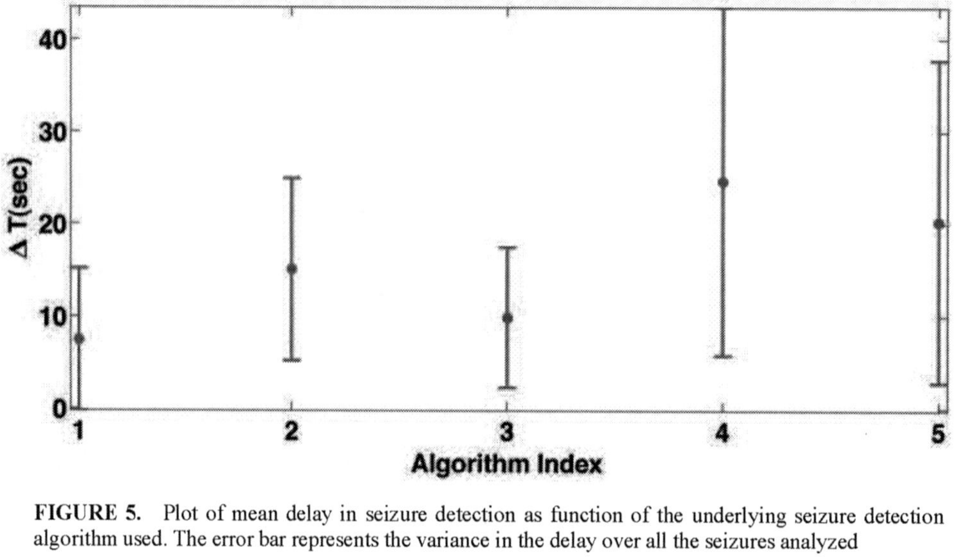

FIGURE 5. Plot of mean delay in seizure detection as function of the underlying seizure detection algorithm used. The error bar represents the variance in the delay over all the seizures analyzed

TABLE 2.

Algorithm Index	Dynamical Measure	Predictability	Sensitivity	Specificity	Mean Delay in detection	Computation Time	Optimality index
1	τ	90.6%	96.6%	95.4%	7.7 sec	1.23 sec	0.76
2	α	100%	86.6%	100%	15.1 sec	.52 sec	0.64
3	ζ	90.3%	93.3%	95.4%	9.9 sec	.53 sec	0.72
4	ω	100%	86.6%	100%	24.7 sec	0.027 sec	0.45
5	ε	90.6%	96.6%	95.4%	20.3 sec	0.015 sec	0.53

4. DISCUSSION

In this chapter we reviewed a number of dynamical measures for early seizure detection in rat model of temporal lobe epilepsy. Performance analysis of each seizure detection algorithm was based on the suitability of the algorithm in development of a real-time closed loop seizure intervention program. The main findings of this study were: 1) Early detection of spontaneous seizures in a rat model of temporal lobe epilepsy (TLE) model could be identified based on quantitative measures of the brain dynamics with very high statistical robustness 2) Although the performance of each seizure detection algorithms was comparable in terms of statistics for sensitivity and specificity, the nonlinear embedding delay, has the highest optimality index value, in part due to its ability to detect seizures very close to the EEG seizure onset time, which in turn, makes it most suitable as a method for closed-loop seizure detection.

Previous animal studies have incorporated seizure prediction in models of provoked seizures. In *in vivo* studies, evidence of preictal transitions was reported for 3-mercaptopropionic acid-induced seizures using stochastic methods [12], as well as for pentylenetetrazol-induced seizures using wavelet-based residual entropy [13]. In-vitro studies using brain slices in low $[Mg^{2+}]$ solution have also reported prediction of seizure-like event onsets in hippocampal networks using wavelet artificial neural networks [14, 15]. The use of a standard *invivo* model of unprovoked seizures in the present study allowed a more direct comparison to the continuous data sets recorded from patients with epilepsy undergoing pre-surgical evaluation. Compared to the spontaneous seizures recorded in the rat TLE model, typical patient data sets include fewer seizures, with higher variability and too few interictal dataset for reliable estimation of false warnings. Use of a standard animal model of epilepsy under controlled experimental conditions can therefore provide numerous insights into the interpretation of human data that can be complicated due to differences in etiology, lesion type and location, and effects of anti-epileptic drug use.

Since the aim of the this study was to identify the simplest computationally efficient and statistically robust early seizure detection algorithm, none of the dynamical measures studied exploited the spatial information in terms of EEG recordings from different brain regions available for study. A prior knowledge of the location of seizure focus based on visual inspection of EEG data from all channels was used for the study. We hypothesize that the analysis of data from all available regions in the brain where electrodes are implanted in addition to utilization of multivariate analysis techniques can result in improved performance for seizure detection time by the algorithms thereby, facilitating the development of an early seizure warning scheme.

In summary, the performance of several seizure detection methods presented here, were acceptable to warrant further testing and refinement of the methods. In particular the nonlinear embedding delay computation was the most efficient in terms of its ability to trigger a warning at the EEG seizure onset. Continued use of the rat TLE model can provide controlled experimental conditions to enable further testing and systematic improvement in seizure prediction software. Ultimately, these studies are intended to translate into the development of implantable closed-loop control systems capable of activating physiological or pharmacological interventions to prevent impending seizures in patients with intractable epilepsy.

ACKNOWLEDGMENTS

This research was supported by the National Institutes of Biomedical Imaging and Bioengineering (NIBIB) through Collaborative Research in Computational Neuroscience (CRCNS) Grant Number R01NS050582-01A1, Bioengineering Research Partnership Grant Number R01EB002089, the Wilder Center of Excellence for Epilepsy Research, and the ChildrenÕs Miracle Network. Stephen Myers, Jason Winters and Wendy Norman provided their kind assistance in scanning the EEG/video data to identify the seizures and creating the control data sets for the statistical analysis.

REFERENCES

1. C. Shin, and J. O. McNamara, *Ann Rev Med* **45**, 379–89 (1994).
2. F. Mormann, R. G. Andrzejak, C. E. Elger, and K. Lehnertz, *Brain* **130**, 314–333 (2007).
3. I. Osorio, M. Frei, and S. Wilkinson, *Epilepsia* **39**, 615–27 (1998).
4. H. Qu, and J. Gotman, *Clin Neurophysiol* **86**, 79–87 (1993).
5. M. Saab, and J. Gotman, *Clin Neurophysiol* **116**, 427–42 (2005).
6. I. Osorio, M. Frei, S. Sunderam, J. Giftakis, N. C. Bhavaraju, S. F. Schaffner, and S. Wilkinson, *Ann Neurol* **57**, 258–68 (2005).
7. H. D. I. Abarbanel, *Analysis of observed chaotic data*, Springer-Verlag New York, Inc, 1996.
8. J. B. Bassingthwaighte, L. S. Liebovitch, and B. J. West, *Fractal Physiology*, Oxford University Press, New York, 1994.
9. M. Latka, Z. Was, A. Kozik, and B. West, *Physical Rev E* **667**, 052902 (2003).
10. A. White, P. Williams, D. Ferraro, S. Clark, S. Kadam, F. Dudek, and K. Staley, *Journal of Neurosci Methods* **152**, 255–266 (2006).
11. E. H. Bertram, *Epilepsia* **38**, 95–105 (1997).
12. S. Sunderam, I. Osorio, M. Frei, and J. Watkins, *J Clinical Neurophysiol* **118**, 275–282 (2001).
13. J. Paul, C. Patel, H. Al-Nashash, N. Zang, W. Ziai, M. Mirski, and D. Sherman, *IEEE Trans Biomed Engg* **50**, 640–648 (2003).
14. A. Chiu, S. Daniel, H. Khosravani, P. Carlen, and B. Bardakjian, *Ann Biomed Eng* **33**, 798–810 (2005).
15. A. Chiu, E. Kang, M. Derchansky, P. Carlen, and B. Bardakjian, *Ann Biomed Eng* **34**, 282–294 (2006).

Optimization of epilepsy treatment with vagus nerve stimulation

Basim Uthman*,†, Michael Bewernitz**, Chang-Chia Liu** and Georges Ghacibeh‡

*Neurology Services, North Florida/South Georgia Veterans Health System, Gainesville, FL, 32608, USA.
†Department of Neurology, the University of Florida, Gainesville, FL 32608, USA.
**Department of Biomedical Engineering, University of Florida, Gainesville, FL, 32611, USA.
‡Northeast Regional Epilepsy Group 20 Prospect Ave, Suite 800 Hackensack, NJ, 07601, USA.

Abstract. Epilepsy is one of the most common chronic neurological disorders that affects close to 50 million people worldwide. Antiepilepsy drugs (AEDs), the main stay of epilepsy treatment, control seizures in two thirds of patients only. Other therapies include the ketogenic diet, ablative surgery, hormonal treatments and neurostimulation. While other approaches to stimulation of the brain are currently in the experimental phase vagus nerve stimulation (VNS) has been approved by the FDA since July 1997 for the adjunctive treatment of intractable partial onset epilepsy with and without secondary generalization in patients twelve years of age or older. The safety and efficacy of VNS have been proven and duplicated in two subsequent double-blinded controlled studies after two pilot studies demonstrated the feasibility of VNS in man. Long term observational studies confirmed the safety of VNS and that its effectiveness is sustained over time. While AEDs influence seizure thresholds via blockade or modulation of ionic channels, inhibit excitatory neurotransmitters or enhance inhibitory neurotransmitters the exact mechanism of action of VNS is not known. Neuroimaging studies revealed that VNS increases blood flow in certain regions of the brain such as the thalamus. Chemical lesions in the rat brains showed that norepinephrine is an important link in the anticonvulsant effect of VNS. Analysis of cerebrospinal fluid obtained from patients before and after treatment with VNS showed modest decreases in excitatory neurotransmitters. Although Hammond et al reported no effect of VNS on scalp EEG by visual analysis and Salinsky et al found no effect of VNS on scalp EEG by spectral analysis, Kuba et al suggested that VNS reduces interictal epileptiform activity. Further, nonlinear dynamical analysis of the electroencephalogram in the rat and man have reportedly shown predictable changes (decrease in the short term Lyapunov exponent STLmax and T-index) more than an hour prior to the clinical or electroencephalographic seizure onset. It is possible that intermittent VNS maintains chaoticity of brain activity in patients with epilepsy that respond to this therapy. The most optimal stimulation parameters of VNS are not known and further study of nonlinear dynamics of brain activity may shed some light on more effective interception or prevention of seizures. Online real time analysis may allow on-demand stimulation rather than hit-or-miss approach.

Keywords: Vagus Nerve Stimulation, Optimization, Data Mining, Quantitative EEG Analysis, Epilepsy
PACS: 89.20.Kk, 87.80.-y

1. BACKGROUND OF INTRACTABLE EPILEPSY

Epilepsy is one of the most common chronic neurological disorders affecting the lives of over 50 million patients and their families world wide [1]. Epilepsy is a serious disorder that can cause injuries, social problems, and death. Patients and their health

CP953, *Data Mining, Systems Analysis, and Optimization in Biomedicine*
edited by O. Seref, O. E. Kundakcioglu, P. M. Pardalos

providers may at times assume the attitude that occasional seizures are acceptable, not realizing that complete seizure freedom is associated with significantly improved safety and quality of life. Baker et al [2], clearly showed reductions in head injuries, burns, dental and other injuries in patients who have less than one seizure per month compared to those whom have one or more seizures per month. Some of our patients have lost fingers while using sharp objects, sustained third degree burns while cooking, black eyes as they fall during a seizure, or drown in a bathtub.

From a large seizure surgery series, it seems that successful surgery resulting in 100% seizure freedom reduces the risk of sudden unexplained death and other epilepsy related deaths [3]. In addition, psychosocial risks of recurrent seizures include unemployment, underemployment, anxiety, depression, stigma, loss of driving privileges, and loss of independence. Thus, it is no wonder that the rates of depressed mood may approach 40% in epilepsy patients. It is for these reasons that hundreds of millions of dollars are spent in search of new anti-epileptic drugs (AEDs) or other treatment modalities.

In spite of multiple new drugs, seizures remain uncontrolled in one out of three patients. The most common drugs available for the treatment of partial onset epilepsy in the United States include carbamazepine, phenotoin, valproate, phenobarbital, primodone, gabapentin, lamotrigine, levitiracetam, oxcarbazepine, pregabiline, tiagabine, topiramate, zonisamide, and felbamate. Various modalities of neurostimulation are currently being investigated for adjunctive treatment of intractable epilepsy. Vagus nerve stimulation (VNS) has been approved for the adjunctive treatment of partial onset epilepsy with or without secondary generalization in patients twelve years of age or older. In this manuscript, we'll describe the historical background and preclinical data leading to clinical trials, surgical procedure, the neurocybernetic prosthesis (NCP) system operation, clinical trials, and the putative mechanisms of action.

2. VAGUS NERVE STIMULATION: HISTORICAL BACKGROUND AND PRECLINICAL DATA

The first animal studies demonstrating the anticonvulsant effect of VNS [4] based on previous observations that VNS blocked interictal EEG spiking [5] or desynchronized EEG in the thalamus or cortex [6, 7] in cats. Zabara extended this idea to demonstrate anticonvulsant activity of VNS in dogs [4]. Later on, McLachlan demonstrated suppression of inter ictal spikes and seizures by VNS in rats [8]. Furthermore, chronic intermittent VNS has been shown to prevent recurrence of epileptic seizures in monkeys [9].

These observations lead to clinical trials to investigate the feasibility, safety, and efficacy of VNS in human patients. The first human implant was performed in 1988. The first randomized active control study was performed in 1992. Encouraging results from the first two pilot studies lead to randomized double blind clinical trials that resulted in FDA approval of VNS as adjunctive therapy for intractable epilepsy in July, 1997. Since then, VNS has shown antidepressant effects and has been approved for adjunctive treatment of refractory depression. So far, more than 30,000 VNS implantations have been performed world wide.

3. SURGICAL PROCEDURE

Implantation of the NCP system (VNS pulse generator and electrical leads) is normally performed under general anesthesia as an outpatient basis. The procedure usually takes two hours and requires two skin incisions; one in the left upper chest for housing the pulse generator in a pocket under the skin and the other is in left neck area above the collar bone to gain access to the vagus nerve and place two semi-circular electrodes and a neutral tether around the left vagus nerve (see Fig. 1).

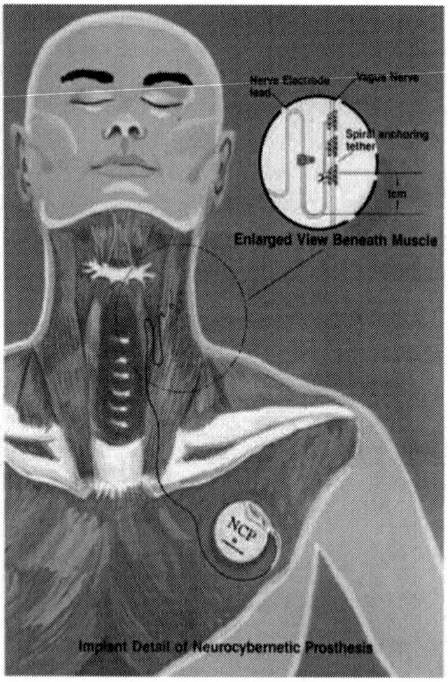

FIGURE 1. VNS pulse generator, nerve electrode lead, and spiral anchoring tether [10].

Connected to the electrodes, wires are tunneled under the skin caudally towards the left upper chest where they are plugged into the pulse generator and secured in place with screws. The semi-circular electrodes are embedded in a spiral silastic (see Fig. 2) which is wound around the left vagus nerve 2.5 times.

Proper wiring and function of the pulse generator are validated intraoperatively using a wand interfacing program and a hand held diagnostic device. Once proper functioning of the device is demonstrated, skin is sutured and the patient is allowed a few hours for recovery and sent home. Adverse events related to anesthesia or complications of surgery are rare

FIGURE 2. Electrode embedded in a silastic spiral anchoring tether.

4. NCP SYSTEM OPERATION

Between 10-14 days postoperatively, patients return to the medical facility for follow-up by the surgeon. At that time, the pulse generator is programmed to start stimulation. Other providers have started stimulation the day of the operation. Using a wand and hand held diagnostic device, the VNS system is interrogated and the stimulation parameters are varied as patient tolerability allows.

Initially, patients are seen every two to four weeks to ramp up the output current for the first three visits. Initial stimulation parameters are: output current, 0.25 mA; frequency, 30 Hz; pulse width, 500 microseconds; on time, 30 seconds; off time, 5 minutes; magnet output current, 0.25 mA above the regular output current; magnet on time, 60 seconds. Output current is ramped up by increments of 0.25 mA. Duty cycle (100 * on time / [on time + off time]) may be modified in future visits according to patient tolerance and seizure control outcome. Patients are provided with a small magnet that can illicit an epoch of VNS on demand for when they experience an aura (a warning of an impending seizure). VNS can be temporarily deactivated by continuously keeping the magnet positioned over the pulse generator.

The interrogating wand allows diagnostics to be performed on the NCP system to assure integrity of all connections and appropriate delivery of stimulation to the nerve. Furthermore, interrogation of the device provides alerts about the status of the battery and when it is approaching the end of service.

5. CLINICAL TRIALS

Two single-blind blind studies demonstrated preliminary safety and efficacy of VNS as adjunctive therapy in the treatment of medically refractory partial onset epilepsy [11] and [12]. Subsequent to the encouraging results of the pilot studies two large double-blind, prospective, active-control studies [13, 14] confirmed the safety and efficacy of VNS in similar populations with intractable partial onset epilepsy. Long term follow-up studies showed that prevention of recurrent seizures was maintained and adverse events decreased significantly over time [15, 16].

The vagus nerve is long and supplies branches to many organs such as the heart, gastrointestinal tract, lungs, larynx and others. It is then not surprising that adverse events include hoarseness, throat pain or tingling, dyspnea, and coughing. These adverse events are mild and occur during the stimulation epoch. They are dose related and can be managed by decreasing the output current or pulse width or frequency. As VNS may cause a change in voice or shortness of breath during exertion, patients may wish to adhere (e.g. with adhesive tape) the magnet over the device while singing or exercising. It is possible that patients become tolerant to the adverse events; however, the intensity of hoarseness clearly decreases over time to almost undetectable levels.

When VNS was interrupted for up to a few weeks we observed a lasting anticonvulsant effect of VNS as seizure frequency did not return to baseline until 3-4 weeks after discontinuation of VNS therapy. A similar lasting anticonvulsant effect of VNS was demonstrated in animal seizure models when animals were pretreated with VNS [17].

Electrocardiogram recordings, Holter monitors, pulmonary function tests, and gastric secretion analysis revealed no clinically significant abnormalities. The lack of AED-induced acute and chronic adverse events with add-on VNS treatment is noteworthy. Serious adverse events like allergic reactions (rash or hepatic failure) or bone marrow depression have not been seen with VNS treatment.

6. PUTATIVE MECHANISM OF ACTION

The vagus nerve is a mixed nerve with 80% of its fibers being afferent and connecting with the nucleus of the solitary tract [18]. From there it projects to multiple areas in the brain including the thalamus, hippocampus, insular cortex, amygdale, hypothalamus, dorsal raphe and the nodose ganglion. It is possible that VNS may interrupt the spread of epileptiform activity if delivered at a theoretically critical time. Thus it could hit or miss. Positron emission tomography and functional MRI studies showed that VNS activates or increases blood flow in certain areas of the brain such as the thalamus [19, 20]. Cerebrospinal fluid (CSF) was analyzed in 16 subjects before, 3 months after, and 9 months after VNS. GABA (total and free) increased in low or high stimulation groups, aspartate marginally decreased and ethanolamine increased in the high stimulation group suggesting an increased inhibitory effect [21]. Krahl et al [22] suggested that seizure suppression induced by VNS may depend on the release of norepinephrine. They observed that acute or chronic lesions of the locus coeruleus attenuated VNS-induced seizure suppression.

Some of our patients report that they can "fight off" their impending seizures and at times abort them by engaging in intense thinking, hyperventilation or rubbing a forearm where a sensory may herald a secondarily generalized seizure. It is possible that what they are doing is stimulating the brain and interrupting the spread of focal epileptiform firing. In fact Rajna and Lona [23] have elegantly demonstrated a halting of absence seizures with auditory stimuli. Finally, Iasemidis et al [24] reported that higher degrees of entrainment occur several minutes prior to the clinical onset of a partial seizure. It is possible that VNS causes small changes in brain dynamics that result in larger effects that inhibit the brain from becoming dynamically entrained, thus interrupting any progression towards a clinical seizure.

7. QUANTITATIVE EEG ANALYSIS IN EPILEPSY

There is a great interest in the application of quantitative EEG analysis for the diagnosis and treatment of epilepsy. The main applications of quantitative EEG analysis are localization of epileptic foci [25, 26] and seizure prediction [27, 28], [29, 30, 31]. Of particular interest is the application of real-time or rapid quantitative EEG analysis for emergency room, bedside, or implantable therapeutic applications [28, 32], [33]. The appealing characteristic of such devices is the ability to monitor some physiologic measure and provide therapeutic intervention such as electrical stimulation or drug dosage on demand. A crucial step in developing an on-demand therapy delivery system is to establish a reliable marker (or signature pattern) in the analyzed physiologic measures indicating when therapeutic intervention is needed.

An example of an electrographic marker is the statistical t-index measure of the short-term Lyapunov exponent, a non-linear EEG measure which can not only predict seizures, but "resets" to a baseline value after the occurrence of a seizure [24]. In addition, studies have demonstrated EEG responses to pharmacological treatment [34, 35], [36] and electrical stimulation [37], [32, 36].

Unfortunately, studies have not reported consistent EEG effects of VNS [38, 39, 13]. However, the world-wide availability of scalp-EEG acquisition systems, state of data mining research, and possibility of real-time analysis application due to modern computer technology make EEG analysis a desirable approach to studying VNS effect.

Thus, identification of an EEG marker that is sensitive to electrical stimulation parameters is a sought after achievement. If such an EEG marker could be identified, and if a correlation to clinical efficacy could be proven, then perhaps the quantitative EEG analysis method could be someday be used to determine optimal VNS stimulation parameters for individual patients.

ACKNOWLEDGMENTS

The Evelyn F. & William L. McKnight Brain Institute of the University of Florida
Neurology Services, North Florida / South Georgia Veterans Health System, Gainesville, FL, 32608, USA.

REFERENCES

1. WHO | Mental Health. Copyright World Health Organization 2007. All rights reserved. Date accessed: June 30th, 2007. http://www.who.int/mental_health/en/
2. Baker GA, Jacoby A, Buck D, Stalgis C, Monnet D. Quality of life of people with epilepsy: a European study. Epilepsia. 1997 Mar; 38(3):353-62.
3. Sperling MR, O'Connor MJ, Saykin AJ, Plummer C. Temporal lobectomy for refractory epilepsy. JAMA. 1996 Aug 14;276(6):470-5.
4. Zabara J. Inhibition of experimental seizures in canines by repetitive vagal stimulation. Epilepsia. 1992 Nov-Dec;33(6):1005-12.
5. Zanchetti A, Wang SC, Moruzzi G: The effect of vagal afferent stimulation on the EEG pattern of the cat. Electroencephalogr Clin Neurophysiol 4:357, 1952.

6. Magnes J, Moruzzi G, Pompeiano O: Synchronization of the EEG produced by low frequency electrical stimulation of the region of the solitary tract. Arch Ital Biol 99:33-67, 1961.
7. Chase MH, Nakamura Y, Clements CD: Afferent vagal stimulation: neurographic correlates of induced EEG synchronization and desynchronization. Brain Res 5:236-249, 1967.
8. McLachlan RS. Suppression of interictal spikes and seizures by stimulation of the vagus nerve. Epilepsia. 1993 Sep-Oct;34(5):918-23.
9. Lockard JS, Congdon WC, DuCharme LL (1990) Feasibility and safety of vagal stimulation in monkey model. Epilepsia 31[Suppl 2]:S20-S26.
10. Cyberonics®, Inc., Houston, Texas. Epilpesy Patient's Manual for Vagus Nerve Stimulation with the VNS Therapy™ System. U.S. Version. Copyright August 2006, Cyberonics® .
11. Penry JK, Dean JC. Prevention of intractable partial seizures by intermittent vagal stimulation in humans: preliminary results. Epilepsia. 1990;31 Suppl 2:S40-3.
12. Uthman BM, Wilder BJ, Hammond EJ, Reid SA. Efficacy and safety of vagus nerve stimulation in patients with complex partial seizures. Epilepsia. 1990;31 Suppl 2:S44-50.
13. Salinsky MC, Burchiel KJ. Vagus nerve stimulation has no effect on awake EEG rhythms in humans. Epilepsia. 1993 Mar-Apr;34(2):299-304.
14. Handforth A, DeGiorgio CM, Schachter SC, Uthman BM, Naritoku DK, Tecoma ES, Henry TR, Collins SD, Vaughn BV, Gilmartin RC, Labar DR, Morris GL 3rd, Salinsky MC, Osorio I, Ristanovic RK, Labiner DM, Jones JC, Murphy JV, Ney GC, Wheless JW. Vagus nerve stimulation therapy for partial-onset seizures: a randomized active-control trial. Neurology. 1998 Jul;51(1):48-55.
15. Uthman BM, Wilder BJ, Penry JK, Dean C, Ramsay RE, Reid SA, Hammond EJ, Tarver WB, Wernicke JF. Treatment of epilepsy by stimulation of the vagus nerve. Neurology. 1993 Jul;43(7):1338-45.
16. Morris GL 3rd, Mueller WM. Long-term treatment with vagus nerve stimulation in patients with refractory epilepsy. The Vagus Nerve Stimulation Study Group E01-E05. Neurology. 1999 Nov 10;53(8):1731-5. Erratum in: Neurology 2000 Apr 25;54(8):1712.
17. Takaya M, Terry WJ, Naritoku DK. Vagus nerve stimulation induces a sustained anticonvulsant effect. Epilepsia. 1996 Nov; 37(11):1111-6.
18. Rutecki P. Anatomical, physiological, and theoretical basis for the antiepileptic effect of vagus nerve stimulation. Epilepsia. 1990; 31 Suppl 2:S1-6. Review.
19. Ko D, Heck C, Grafton S, et al. Vagus nerve stimulation activates central nervous system structure in epileptic patients during PET blood flow imaging Neurosurgery 1996;39:426-431
20. Henry TR, Bakay RA, Votaw JR, Pennell PB, Epstein CM, Faber TL, Grafton ST, Hoffman JM. Brain blood flow alterations induced by therapeutic vagus nerve stimulation in partial epilepsy: I. Acute effects at high and low levels of stimulation. Epilepsia. 1998 Sep;39(9):983-90.
21. Ben-Menachem E, Hamberger A, Hedner T, Hammond EJ, Uthman BM, Slater J, Treig T, Stefan H, Ramsay RE, Wernicke JF, et al. Effects of vagus nerve stimulation on amino acids and other metabolites in the CSF of patients with partial seizures. Epilepsy Res. 1995 Mar; 20(3):221-7.
22. Krahl SE, Clark KB, Smith DC, Browning RA. Locus coeruleus lesions suppress the seizure-attenuating effects of vagus nerve stimulation. Epilepsia. 1998 Jul;39(7):709-14.
23. Rajna P, Lona C. Sensory stimulation for inhibition of epileptic seizures. Epilepsia. 1989 Mar-Apr;30(2):168-74.
24. Iasemidis LD, Shiau DS, Sackellares JC, Pardalos PM, Prasad A. Dynamical Resetting of the Human Brain at Epileptic Seizures: Application of Nonlinear Dynamics and Global Optimization Techniques. IEEE transactions on biomedical engineering, vol. 51, no. 3, March 2004.
25. Ding L, Wilke C, Xu B, Xu X, Van Drongelen W, Kohrman M, He B. EEG Source Imaging: Correlating Source Locations and Extents with Electrocorticography and Surgical Resections in Epilepsy Patients. (J Clin Neurophysiol 2007;24: 130-136).
26. Schevon CA, Cappell J, Emerson R, Isler J, Grieve P, Goodman R, McKhann G Jr, Weiner H, Doyle W, Kuzniecky R, Devinsky O, Gilliam F. Cortical abnormalities in epilepsy revealed by local EEG synchrony. Neuroimage. 2007 Mar;35(1):140-8.
27. Chaovalitwongse WA, Iasemidis LD, Pardalos PM, Carney PR, Shiau DS, Sackellares JC. Performance of a seizure warning algorithm based on the dynamics of intracranial EEG. Epilepsy Res. 2005 May;64(3):93-113.
28. Chaovalitwongse WA, Prokopyev OA, Pardalos PM. Electroencephalogram (EEG) time series classification: Applications in epilepsy. Ann Oper Res (2006) 148: 227-250.

29. Iasemidis LD, Sackellares JC. Chaos theory and epilepsy. Neuroscientist 2 (2): 118-126 MAR 1996.
30. Pardalos PM, Yatsenko V, Sackellares JC, Shiau DS, Chaovalitwongse WA, Iasemidis LD. Analysis of EEG data using optimization, statistics, and dynamical system techniques. Computational Statistics & Data Analysis, Volume 44, Issues 1-2, 28 October 2003, Pages 391-408.
31. Pardalos PM, Sackellares JC, Carney PR, Iasemidis LD. Quantitative Neuroscience, Kluwer Academic Publishers, (2004).
32. Good LB, Sabesan S, Iasemidis LD, Treiman DM, Real-time control of epileptic seizures, Proc. 3rd European Medical and Biomedical Engineering Conf., Prague, Czech Republic 2005a
33. Iasemidis LD, Shiau DS, Pardalos PM, Chaovalitwongse WA, Narayanan K, Prasad A, Tsakalis K, Carney PR, Sackellares JC. Long-term prospective on-line real-time seizure prediction. Clin Neurophysiol. 2005 Mar;116(3):532-44. Epub 2005 Jan 6.
34. Ghacibeh GA, Shiau DS, Dance L, Eisenschenk S, Sackellares JC. Influence of Antiepileptic Medications on EEG Dynamics. Epilepsia, Vol. 26 Suppl 8, 2005.
35. Good LB, Sabesan S, Iasemidis LD, Garvey KJ, Treiman DM. Brain dynamical, disentrainment following successful antiepileptic drug treatments in rat and human status epilepticus. Epilepsia 45: 42-43 Suppl. 7 2004.
36. Good LB, Sabesan S, Marsh ST, Iasemidis LD, Treiman DM. Nonlinear dynamical analysis of deep brain stimulation for control of epileptic seizures in rats. Epilepsia 46: 330-330 Suppl. 8 2005b.
37. Nair SP, Sackellares JC, Shiau DS, Norman WM, Dance LK, Pardalos PM, Principe JC, Carney PR. Effects of Acute Hippocampal Stimulation on EEG Dynamics. Engineering in Medicine and Biology Society, 2006. EMBS '06. 28th Annual International Conference of the IEEE Aug. 2006 Page(s):4382 - 4386.
38. Fisher RS, Handforth A. Reassessment: vagus nerve stimulation for epilepsy: a report of the Therapeutics and Technology Assessment Subcommittee of the American Academy of Neurology. Neurology. 1999 Sep 11;53(4):666-9.
39. Hammond EJ, Uthman BM, Reid SA, Wilder BJ. Epilepsia. Electrophysiological Studies of Cervical Vagus Nerve Stimulation in Humans: I. EEG effects. 1992 Nov-Dec; 33(6):1013-20.

Author Index

B

C

D

E

F

G

H

K

L

M

P

R

S

T

U

W

X

Y

Z